Advanced TRIGONOMETRIC RELATIONS through Nbic FUNCTIONS

Dr. Nisith K. Bairagi
B.C.E., M.E. (Civil), Ph.D.
Retired Professor of Structural Engineering
IIT-Bombay

Publishing Globally

NEW AGE INTERNATIONAL (P) LIMITED, PUBLISHERS
New Delhi • Bangalore • Chennai • Cochin • Guwahati
Hyderabad • Kolkata • Lucknow • Mumbai
Visit us at www.newagepublishers.com

Published by New Age International (P) Ltd., Publishers
First Edition: 2012

BRANCHES

- **Bangalore** 37/10, 8th Cross (Near Hanuman Temple), Azad Nagar, Chamarajpet, Bangalore-560 018
  Tel.: (080) 26756823, Telefax: 26756820, E-mail: bangalore@newagepublishers.com
- **Chennai** 26, Damodaran Street, T. Nagar, Chennai-600 017
  Tel.: (044) 24353401, Telefax: 24351463, E-mail: chennai@newagepublishers.com
- **Cochin** CC-39/1016, Carrier Station Road, Ernakulam South, Cochin-682 016
  Tel.: (0484) 2377004, Telefax: 4051303, E-mail: cochin@newagepublishers.com
- **Guwahati** Hemsen Complex, Mohd. Shah Road, Paltan Bazar, Near Starline Hotel, Guwahati-781 008
  Tel.: (0361) 2513881, Telefax: 2543669, E-mail: guwahati@newagepublishers.com
- **Hyderabad** 105, 1st Floor, Madhiray Kaveri Tower, 3-2-19, Azam Jahi Road, Nimboliadda, Hyderabad-500 027
  Tel.: (040) 24652456, Telefax: 24652457, E-mail: hyderabad@newagepublishers.com
- **Kolkata** RDB Chambers (Formerly Lotus Cinema) 106A, 1st Floor, S.N. Banerjee Road, Kolkata-700 014
  Tel.: (033) 22273773, Telefax: 22275247, E-mail: kolkata@newagepublishers.com
- **Lucknow** 16-A, Jopling Road, Lucknow-226 001
  Tel.: (0522) 2209578, 4045297, Telefax: 2204098, E-mail: lucknow@newagepublishers.com
- **Mumbai** 142C, Victor House, Ground Floor, N.M. Joshi Marg, Lower Parel, Mumbai-400 013
  Tel.: (022) 24927869, Telefax: 24915415, E-mail: mumbai@newagepublishers.com
- **New Delhi** 22, Golden House, Daryaganj, New Delhi-110 002
  Tel.: (011) 23262368, 23262370, Telefax: 43551305, E-mail: sales@newagepublishers.com

---

ISBN: 978-81-224-3023-3

₹ 495.00

C-12-02-6167

Printed in India at Glorious Printers, Delhi.

Typeset at In-house.

PUBLISHING FOR ONE WORLD

NEW AGE INTERNATIONAL (P) LIMITED, PUBLISHERS
4835/24, Ansari Road, Daryaganj, New Delhi-110002
Visit us at **www.newagepublishers.com**

Dedication

Praying humbly
for their blessings
from
the Great Masters of Mathematics,
this work is presented
to
my little beloved granddaughter
Ankana Datta
and
to
my readers to
relish interesting theme attempted

Preface

The Nbic function proposed in this book, is better forwarded as that class of function which belongs to trigonometric-hyperbolic type. The basic Nbic functions, as they are termed by the name, single Nbic functions, are actually the derived form of the product of single complex circular function (Coci function) e^{ix}, in the form: $C(x) = f_1(x) + ig_1(x)$, and single complex hyperbolic function (Cohy function) f^{iy}, in the form: $H(y) = f_2(y) + ig_2(y)$. The product function in complex form is termed as the *Extended DeMoivre's form*, and on separating the real as well as imaginary parts from which, the required Nbic functions can be collected.

In fact finally, the single Nbic functions are the same as obtained from the direct product of the basic trigonometric (circular) functions, such as, $c = c(x) = (\sin x, \cos x)$, and the basic hyperbolic functions, such as, $h = h(y) = (sh\ y, ch\ y)$. For example, the basic trigonometric-hyperbolic unit functions appear like, $c(x).h(y) = \sin x\ ch\ y, \cos x\ ch\ y, \ldots$, etc.

Other higher order Nbic functions, on the other hand, can be generated from the product (which appears in the form of *Extended DeMoivre's form)*, of lower order Nbic function, in the form: $f_3(x, y) + i g_3(x, y)$, and either of the Coci or Cohy function, in the form: $f_4(x) + i g_4(x)$, or, $f_5(y) + i g_5(y)$.

When the Coci function is the multiplier, the corresponding higher order Nbic function is to be designated as E type. Otherwise, it is F type, if Cohy function does the same job, and is identified by the subscript, **e** or **f**, respectively, as the case may be.

In this sense, the single Nbic functions, both with single variable (x, x) [or simply (x)], and with two variables (x, y), is symbolically represented by, $N(x, y)$ or $N(x, x)$ [$= N(x)$], etc. The symbolic representation of higher order Nbic functions appear like: $N_{2e}(x, y)$ as the Double Nbic function (E type), $N_{3f}(x, x)$ as the Triple Nbic function (F type), etc. In fact, as it appears, the class and variety of Nbic functions that can be generated, are limitless.

The use of single Nbic functions $N(x, y)$ like, $(\sin x\ ch\ y, \cos x\ ch\ y)$, or $(\cos x\ ch\ x, \sin x\ ch\ x)$, etc., are not new in the field of solution of differential equations.

The purpose and scope of the present book mainly is to propose the various types of Nbic functions developed, starting from the simple Nbic functions and subsequently proceeding towards the higher order Nbic functions, and to present the various relations of the Nbic functions in the form of formula or identity, derived in the form similar to the relations of the plane trigonometric (circular) or hyperbolic functions, and side by side, to bring out and correlate the respective Nbic function with the standard circular functions and/or hyperbolic functions and/or exponential function, wherever possible, by providing the corresponding relations of the interchangeability rules, or criteria.

The subject matter dealt in this book is with a view to contribute a stepping stone, whatever small it may be, in the field of trigonometry, which existed ***Never before***. This view has rightfully reflected in the name of the proposd title of the book with **Nbic functions** (to be pronounced as: en-bic, similar to, hyperbolic, trigonometric, etc.). These **N**ew **b**eautiful class of functions seems to have no limit, and so may be taken as open ended, which means that, their varieties know **no** **b**ounds.

The operators of the circular functions are: sin, cos, tan, cosec, sec and cot, and those of the hyperbolic functions (identified by the additional letter, **h**) are: s**h** (= sin**h**), c**h** (= cos**h**), th (= tan**h**), csc**h** (= cosec**h**), sc**h** (= sec**h**) and ct**h** (= cot**h**). Keeping with the tradition, the operators of the present Nbic functions (identified by the additional letter **N**, leaving no gap between the letter N and the letters to the left), are maintained throughout as: sin**N**, cos**N**, tan**N**, csc**N** (= cosecN), sec**N** and cot**N**.

The scope of the subject matter on the Nbic functions and the various related aspects have been spread over eight chapters, followed by Concluding Remarks, and four Appendices.

In this book, the beautiful world of Nbic functions and Nbic trigonometric relations derived thereof, have been explored, exposed and proposed.

The author sincerely believes that, this class of functions, namely Single, Double, Triple Nbic functions, and so on.., is a definite addition in the field of mathematics, and though small, but a positive step beyond the existing boundary of mathematics.

The operational tool of mathematics used throughout the entire body of the text of this book is kept simple by setting aside altogether any complexity, so that, a student of secondary/ higher secondary school level, can easily follow the step-by-step development of the various chapters on the subject of this newly proposed Nbic Function and Nbic Trigonometric Relations. Anyone, who has the basic knowledge of plane trigonometry, elementary hyperbolic and exponential functions, simple differentiation and integration of elementary calculus, and basic matrix operations from algebra, is expected to be sufficient prerequisite background to follow the entire text without any difficulty. Students will feel that a Chapter more they will have to go through in their textbook of mathematics.

To a college going student, the subject matter of this book will appear to him/her as an easy task to understand, and will neither become a matter of headache or night mare to him/her, nor he/she will have to run helter shelter to acquire additional information of higher mathematics at all, to follow the logical flow of the presentation. Care has been taken to explain briefly, the reference background help required along with such topics, wherever they are tackled in the text.

A professor of science or mathematics, who is bestowed with the ability to analyse the logical chain of development and reasoning, will taste this new form of functions and the trigonometric relations arising thereof, as challenging, interesting and easy-to-follow new idea offered in mathematics.

A large number of Numerical Exercise problems along with their detailed full length calculation with numerical data (along with cross reference with the text), associated with each chapter, have been provided, in order to understand how the various relations in the form of principles and formulas, works. Additionally to mention that, the provision of these numerical examples is to serve also the purpose of checking and verifying the correctness of the derived relations offered in various chapters of this book.

During the development of this present mathematical work, many new and never-heard-before terminologies have been created, and used.

To mention, few are:

[Nbic, Complex Nbic, Coci, Cohy, Cof, Con, fexponential] – function, [Coci, Cohy, Cof, Con, Bairagi] – number, Bairagi model, [tehy, teN, sweep] – angle, [Hyperbolic, Nbic (single, double, triple...)] – circle, [(Full/Complete/Zero/C_0), (Incomplete/Inferior/Negative/C_H), (Superior/Overcomplete/Positive/C_N)] – circle, etc.

In preparing the text, no reference book or literature is separately cited under the heading of: Selected Reference, since the work utilises only the type of mathematics which is available in any higher secondary school or, at best in college mathematics books, and the number of which is countless in standard book shops and libraries, and in the internet websites. The author freely browsed through the internet innumerable mathematical websites for brushing up the knowledge displayed on the topics of plane and hyperbolic geometry (Non-Eucledian geometry), plane and hyperbolic trigonometry, representation of complex number in matrix form, and like-wise.

To the best belief of the present author, the subject matter brought in focus and dealt thoroughly in this book, will certainly draw attention of the men of letters, thinkers, mathematicians, scientists, educationists, students, people of recreational mathematics, and interested people at large, and will get acknowledgement and appreciation, as the maiden and original mathematical contribution offered.

—Author

Acknowledgement

The idea of writing a book on a new form of trigonometry came in my mind long back, may be around 1972, when being as a new faculty member at the Civil Engineering Department, IIT Bombay, I was handling the Structural Engineering subjects like, Solid Mechanics, Structural Mechanics and Analysis of Plates and Shell Structures, and for obvious reason, I had to take help of extensive library references.

I am indebted to the Central Library of IIT Bombay, which houses enormous collection of text and reference books and journals, both current and back volume, practically covering every field of all branches of knowledge. This library is a gold mine for me, which I have utilized throughout my long academic professional carrier at IIT Bombay for about 33 plus years.

Towards the later part of my professional carrier as Professor at that prestigious Institute, I started collecting and jotting down derivations, small write ups lying scattered from piles of notes and papers, and the very concept for the present book started taking shape since then. In the background of the reference materials thus collected, and the ones visited through countless internet sites for books, papers/journals, etc., the present book is the result of the continuous effort, working single handedly.

The scientific inquisitiveness in me for doing new and innovative things had been planted at my early age, by my pen friend Anne Elizabeth Figg (later Mrs. Jervis) of England, when I was a high school boy, by showing constant inspiration and encouragement, since then up to a long period of time. She expired untimely at the age of 46. I also owe greatly to my elder sister *(mejdi)* Mrs. Diptilata, who looked after me during my school and college days, till graduation, by safeguarding my studies and other academic activities, since my childhood, after I lost my mother at my very early age of seven. But for her motherly care, I would not have been what I am today.

I acknowledge the constant encouragement rendered and sympathetic attitude shown by my wife Mrs. Sipra Bairagi and my daughter Mrs. Malabika Datta, during the various stages of the work. Psychologically, this kind of lovely family touch has eased much of the severe strain of the book writing programme, like this one, taken in hand, at this advanced age of mine. I am grateful to both of them. They deserve my sincere thanks, and I like to offer them special credit for supporting me this way.

I admit that, this book in its present form would not have been a reality but for the helping hand extended by New Age International Publishers, New Delhi for their constant and untiring effort to bring out this book in a presentable form in a minimum possible time.

Utmost care has been taken to eliminate error and printing mistakes. Even then, it is possible that, a few unnoticed may still remain. Readers are requested to point them out, if detected, with constructive criticism and suggestion, if any, for the betterment of this very new subject presented in this book.

Nisith K. Bairagi

Notation

A = function, angle, matrix
a = angle, length, shell parameter, number, area

B = mix type, function, function type, angle
b = angle, length, number, suffix

C = constant, angle
CC = single Nbic function
CT = single Nbic function
CC^* = single Nbic function
CT^* = single Nbic function
C_0 = Eucledian circle
C_H = hyperbolic circle
C_N = Nbic circle
C_1 = single Nbic circle
C_1, C_2 = centres of curvatures, function
C_1^*, C_2^* = single Nbic functions
C_{2e}, C_{2f} = Double Nbic circles, E and F type, respectively
C_{3e}, C_{3f} = Triple Nbic circles, E and F type, respectively
c = angle, length, number, function

D = constant, angle, plate/shell rigidity
d = length
E = category of Nbic function, modulus of elasticity
e = exponent, base of natural lostgarithm

F = category of Nbic function
EF, FE = category of Nbic function
f = function, fexponent
g = function
gA = geometrical angle

H = tehy angle
h = function, height

I = moment of inertia, Integration symbol, identity matrix
i = imaginary number ($\sqrt{-1}$), matrix

K = Gaussian curvature (= $K_1 . K_2$), suffix indicating Nbic function type
K_1, K_2 = curvatures along the two principal directions
k = foundation modulus, shell parameter, index

ln ( ) = natural logarithm = $\log_e$ ( )

M_1 = longitudinal moment
m = exponent, matrix
N = character used for Nbic function [as in sinN(x, y), cosN(x, $-x$), etc.], single Nbic function matrix
N_1 = single Nbic angle
N_{2e}, N_{2f} = double Nbic function, double Nbic angle (E and F type, respectively)
N_{3e}, N_{3f} = triple Nbic function, triple Nbic angle (E and F type, respectively)
n = integer, suffix, exponent, matrix

P = Nbic function cosNx
Q = Nbic function sinNx
Q_1 = radial shear

R = function, radius (= reciprocal of curvature)
R_1, R_2 = radii along the two principal directions

S_1, S_2 = single Nbic functions
S_1^*, S_2^* = single Nbic functions
SC = single Nbic function
s = coordinate, length
sA = sweep angle

T, t = thickness

U = variable
u = number, coordinate
v = number, coordinate
w = displacement

X, Y = coordinates
x, y = numbers, coordinates, lengths
$\bar{x}$, $\bar{y}$ = non-dimensional coordinates

0 $(\overline{x,\ y})$ = consecutive number pair

$(\overline{x,\ y},\ z)$ = Pythagorean triplets

z = coordinate, number, length

α = angle, semi apex angle

β = constant

ϕ = transverse direction, angle

λ = parameter, constant

γ = parameter, constant

π = ratio of the circumference to diameter of a circle

ψ = function along transverse direction

ρ = shell parameter

δ = symbol of partial derivative

∇^2, ∇^4 = Laplace operator, and biharmonic operator

Δ = symbol for triangle

Contents

Introduction

In the area of structural engineering problems, the applicability of the single Nbic functions may be witnessed in plenty. Few typical examples are cited in refeřence to this.

Example 1:

For prismatic beams of flexural rigidity EI, resting on elasic foundation of modulus k, under uniformly distributed loading of constant intensity q, the governing differential equation (Hetenyi, M, 1946) for the deflection $y = y\ (x)$, is:

$$\frac{d^4 y}{dx^4} + \frac{k}{EI}\, y = \frac{q}{EI}$$

The homogeneous solution in terms of four arbitrary constants is written as:

$$y = y\ (x) = e^{\lambda x}\,[\text{A} \cos \lambda x + \text{B} \sin \lambda x] + e^{-\lambda x}\,[\text{C} \cos \lambda x + \text{D} \sin \lambda x]$$

On converting the exponential functions, $e^{\lambda x}$ and $e^{-\lambda x}$, in terms of the hyperbolic functions, using $e^x = (ch\ x + sh\ x)$, and, $e^{-x} = (ch\ x - sh\ x)$, the equation of the displacement function $y = y\ (x)$ when viewed, will appear as rewritten in terms of the proposed basic unit Nbic functions, $c(x)\ .\ h(y)$.

Though Hetenyi's solution is a sort of idealistic one in regard to the idealistic elastic behaviour of the foundation model adopted (Winkler's model), the applicability of the solution of the very fourth order differential equation of the type:

$\dfrac{d^4 y}{dx^4} + 4\lambda^4 y = 0$, has spread on a very wide area under its umbrella cover. Particularly mention worthy is the field of structural engineering, where the bending analysis of thin shell roofs like, spherical and conical domes, short span circular cylindrical shells etc., are encountered, where only the roof structure is considered there in the absence of any foundation. For instance:

Example 2:

Geckler's equation for **spherical domes** having shell rigidity D (in terms of meridional radial shear Q_1, where ϕ = meridional angle)

$$\frac{d^4 Q_1}{d\phi^4} + 4\lambda^4\, Q_1 = 0 \ \left[\text{where, } 4\lambda^4 = \frac{EtR^2}{D}\right]$$

Example 3:

Approximate equation in non-dimensional form of **conical shells** (in terms of a variable U, where, α = semi apex angle, and s = length along the inclined length measured from the apex):

$$\frac{d^4U}{ds^4} + \lambda^4 U = 0 \text{ [where, } \lambda^4 = \frac{Et}{s^2 D}\cot^2\alpha]$$

Example 4:

Hammad's (1956) approximate equation for **short circular cylindrical shells** in terms of the longitudinal moment $M_1 = M_1(x, \phi)$ is:

$$\frac{\delta^4}{\delta x^4}[\nabla^4 M_1 + \frac{12}{t^2 R^2} M_1] = 0$$

which when solved accordingly, gives the form in terms of the geometrical parameter k. Neglecting the term associated with $e^{k\phi}$, the solution for M_1 as:

$$M_1 = e^{-k\phi}[C \cos k\phi + D \sin k\phi] \,.\, \sin\frac{m\pi x}{L}$$

Example 5:

For medium and moderately long, thin prismatic cylindrical shells of length L, and of constant thickness t, having circular directrix of radius R, under uniform loading of constant intensity, the governing differential equation expressed in terms of the various geometrical parameters of the shell, defined under the orthogonal coordinate system (x, ϕ, z), for the displacement $w = w(x, \phi)$, is obtained as the partial differential equation of eighth order (D K J equation) as:

$$\nabla^8 w + 12/(R^2 t^2) = 0$$

where, the operator $\nabla^8 = \nabla^4\nabla^4$

and, the biharmonic operator $\nabla^4(x, \phi) = \nabla^4 = \nabla^2\nabla^2$

where, the Laplace operator $\nabla^2(x, \phi) = \nabla^2 = \frac{\delta^2}{\delta x^2} + \frac{\delta^2}{R^2 \delta\phi^2}$

The homogeneous solution, in terms of the eight arbitrary constants corresponding to the eighth order partial differential equation, then is

$$w = w(x, \phi) = [A_1\psi_1 + A_2\psi_2 + \ldots + A_8\psi_8] \,.\, \sin\frac{m\pi x}{L}$$

In terms of the various parameters of the shell geometry, namely, (ρ, a, b), the functions $\psi = \psi(\phi)$, can be expressed as:

$$\psi_1 = \text{ch}(\rho a\phi)\sin(\rho b\phi),\ \psi_2 = \text{ch}(\rho a\phi)\cos(\rho b\phi),\ \ldots,\ \psi_8 = \text{sh}(\rho a\phi)\sin(\rho b\phi), \text{ etc.}$$

The various ψ-functions can easily can be identified through the proposed basic single unit Nbic functions, $c(p\phi) \,.\, h(q\phi)$.

The functions like, sin x, cos x, tan x etc., are the plane trigonometric operators, and all are the function of the parameter x, which is an angle. These basic building blocks of plane trigonometry, (particularly, $\sin x$, $\cos x$ and $\tan x$), do satisfy the Pythagorean criterion, or alternatively, the Circularity criterion.

In short, the **Pythagorean criterion**, may be stated as:

If in a right triangle OAP, which is defined in the rectangular coordinate system (X,Y), where X is along the side (or base) OA, and Y is along the altitude AP, so that OP is the hypotenuse, then, $OP^2 = OA^2 + AP^2$.

If additionally, the hypotenuse OP [where O is at (0,0), and P is at the point (X,Y)], is of unit length, then, $OA^2 + AP^2 = 1$.

In other words,

$$OA = X - \text{component of } OP = \cos x, \text{ and}$$
$$AP = Y - \text{component of } OP = \sin x$$

For visualising $\tan x$, extend the line OA up to the point C on the X axis, where the tangent PC drawn at P, meets. Remember that, the radius OP = 1, we get

$$AP/OA = \sin x / \cos x = PC/OP. \text{ Hence, } PC = \tan x.$$

So the **Pythagorean criterion** which is satisfied by the plane trigonometric functions is: $\cos^2 x + \sin^2 x = 1$.

Now visualise the Pythagorean criterion of the right triangle OAP bounded in rectangular coordinate system (X,Y) in the light of **Circularity criterion**. It may be presented as:

If a circle is drawn with centre at O at (0, 0) and the radius OP is of unit magnitude then, for any location of the point P at (X,Y), [which is at the other end of the hypotenuse OP], the triangle OAP satisfies the Pythagorean criterion, that is, $OP^2 = OA^2 + AP^2$. Since, the locus of the point P about O, maintaining a constant distance from it, is a circle, it satisfies the equation: $X^2 + Y^2 = 1$. This is what the **Circularity criterion** is.

In other words, Circularity criterion speaks for the coordinates X and Y, which can be expressed through the angular distance x, that is, $X = \cos x$, $Y = \sin x$. That is to say that, X-axis, and Y-axis are represented by, $\cos x$-axis, and $\sin x$-axis, respectively.

Both the Circularity criterion and the Pythagorean criterion actually are the one and the same. Looking from this angle, the variable parameter x, which is nothing but the angle, is called also as **circular angle**, and the trigonometric functions, $\sin x$, $\cos x$, etc., are rightly termed as **Circular functions**.

Assessment-wise, the parameter x in circular functions, can be interpreted as the measure of the angle x, which actually is twice the area of the circular sector bounded between the two boundary radii of unit length, which produces the angle.

On the other hand, in hyperbolic functions, $sh\ x$, $ch\ x$, $th\ x$, etc., the parameter x, is not an angle, but a parameter. Its measure is twice the area of the hyperbolic sector, enclosed between the two boundary rays originating from the origin (0, 0), one is of unit length along X axis, the other is of length equal to $\sqrt{ch2x}$, and the hyperbolic arc between the apex (1, 0) and the point ($ch\ x$, $sh\ x$) on the hyperbola. Analogous to the Euclidean system, it is often termed as hyperbolic angle, though it is not an angle.

The unit hyperbolic functions, viz., (*ch x*, *sh x*) can be visualized as the coordinates on a unit hyperbola having the equation: $x^2 - y^2 = 1$, which certainly is not that of a circle. So as such, the Pythagorean criterion and, so also the Circularity criterion cannot be applied for the equation of the hyperbola. However, on modifying the coordinates, $X = [(ch\ x)/(\sqrt{ch2x})]$, and $Y = [(sh\ x) / (\sqrt{ch2x})]$, the equation of the unit hyperbola can be reconstructed to that of a unit circle (to be known as, **Hyperbolic circle, C_H**) : $X^2 + Y^2 = 1$, and then the applicability of the said two criteria is thus ensured.

In contrast to the design for the definition of the basic circular functions taking help of unit circle, using the parameter x (which is nothing but an angle subtended at the centre by the radius of unit magnitude, with the X-axis), the definition of the basic hyperbolic functions can be traced through a completely different context.

It is imperative that, these two classes of functions, namely, the circular functions and the hyperbolic functions, are completely different in nature, and also in mathematical structure, and hence for obvious reason, each class of function has restriction within its own respective field, for applicability. Surprisingly, there exist a lot of structural similarities between the relations derived for plane trigonometric functions (circular functions) and their corresponding hyperbolic counterparts. The formulas relating, single angle, two angle (addition and subtraction), multiple angle, half angle, powers of functions, sum, difference and product of functions, derivatives, integrals, law of sine and cosine, area formulas, etc., have striking resemblances in pattern and forms.

So far as the interchangeability of one class of function into another class is concerned, needless to say that, it is possible to convert circular function into the hyperbolic counterpart, or, vice versa, with the help of certain formulas and functions proposed from time to time. Among them, Osborn's rule [due to G. Osborn] and Gudermann function [due to Christoph Gudermann] are mention worthy.

Since, the basic hyperbolic functions are the alternative form of exponential functions, i.e., $ch\ x = (e^x + e^{-x})/2$, and $sh\ x = (e^x - e^{-x})/2$, written using the parameter **e**, the well known logarithmic base, a new class of function f^x, identified by the parameter **f**, on the similar lines, is adopted. This proposed function relates the circular functions as: $f^x = (\cos x + \sin x)$, and, $f^{-x} = (\cos x - \sin x)$. That is, in other words, $\cos x = (f^x + f^{-x})/2$, and $\sin x = (f^x - f^{-x})/2$. The parameter **f** has been exploited and extensively adopted for designating the complex hyperbolic function (Cohy function) as f^{ix}. This new class of function, f^x is given a name **fexponential function,** in keeping harmony with the name of the exponential function, e^x.

The fexponential function, its usage, and interrelation with other functions, have been discussed in detail in the text.

Since the basic Nbic functions are the trigonometric-hyperbolic ones, and can be generated as the product of $c(x)$ and $h(y)$ [or, $c(x)$ and $h(x)$, as the case may be], the ground work for the proposal of the Nbic functions as a maiden contribution into the field of trigonometry, has been done on discussing elaborately first, the various aspects of the said two types of functions (viz., circular and hyperbolic) along with their various respective properties and criteria, laying an emphasis on the interpretation on circular and hyperbolic angles, Pythagorean and Circularity criterion, with a view to bring forth similar characteristic properties of the proposed various types of Nbic functions. Attempting this way, and following the modification applied for the equation of the hyperbola, the tan-equivalent Nbic (teN) angle, analogous to the hyperbolic or circular angles, is identified for the Nbic functions.

The search along the line of Circularity criterion for Nbic functions reveals that, similar to the hyperbolic circle (C_H), all types of the Nbic functions, suitably arranged for the modified coordinates (X,Y), can be represented by the equation of the type, $X^2 + Y^2 = 1$. Being the equation of a circle, they all are identified by the name as: **Nbic circles (C_N)**. This interesting observation immediately necessitates for redefining and categorising the geometric figure, circle, for the first time in the entire history of mathematics (so far the knowledge of the present author goes).

Thus, there would be three categories of circles.

The **Euclidean circle C_0**, is governed by the circular angle x, and is termed as complete circle, since the starting point (corresponding to $x = 0$) on the periphery (1, 0), after travelling an angular distance of 2π along the circular periphery of the circle (of radius unity) ends exactly on the same point which corresponds to $x = 2\pi$. That is why, the **full circle** or **complete circle** is the suitable name earmarked for our year old Euclidean circle.

On the other hand, in Hyperbolic circles $\mathbf{C_H}$, the starting point and the end point corresponding to the equivalent radian measure of the hyperbolic angle $x = 0$ and $x = 2\pi$, does not show geometrically fully closed circular figure. It forms an open circular arc, so that Hyperbolic circles C_H, are to be termed as **Incomplete** or **Inferior circle**.

In the case of **Nbic circles C_N**, the starting point (corresponding to $x = 0$) finally reaches the end point (corresponding to $x = 2\pi$), after encircling more than one complete full turn along the circumference of the respective Nbic circle. In this respect, for obvious reason, Nbic circles are placed in the category of **overcomplete** or **superior circle**.

Again, when judged in terms of the **sweeping angle sA**, which is the geometrical angle (in circular measure) in excess of the corresponding teN angle, the complete, incomplete and overcomplete circles, can be shown to yield zero, negative and positive values.

Thus according to this **new categorisation for circles**, the proposed new identification for circles are:

Hyperbolic circle: C_H circle/negative circle/inferior circle/incomplete circle

Euclidean circle: C_0 circle/zero circle/full circle/complete circle

Nbic circles: C_N circle/positive circle/superior circle/overcomplete circle.

How the different categories [viz., Category 1 or E type: with variable (x, y), and Category 2 or F type: with variable (x, x)] of Nbic functions can be generated from the Complex Nbic functions, has been indicated earlier. Each of the E type and F type, again sometimes, is further sub grouped as A: with variable (x, y), or (x, x), and as B: with variable (x, $-y$), or (x, $-x$). Some typical types of Nbic functions even then do not comply with this categorisation (E type or F type). They are separately placed under EF or FE category.

1

Nbic FUNCTIONS AND Nbic TRIGONOMETRIC RELATIONS

1.1 INTRODUCTION

Nbic functions are the rearranged form of the previously known functions of trigonometric-hyperbolic types, often used for solving Beams on Elastic Foundations and Cylindrical Shell Equations involving functions of one or two variables. The pattern and similarity of the present Nbic functions with the corresponding ones of the trigonometric and also hyperbolic functions are remarkable. The various formulas and properties of the proposed Nbic functions are described and discussed in brief here in this book.

Functions developed and proposed here in this book are with a view to contribute a stepping stone, whatever small it may be, in the field of trigonometry, which existed **N**ever **b**efore. This view has reflected meaningfully while naming the title of this book. The name Nbic function (to be pronounced as: en-b<u>ic</u>, similar to, hyperbol<u>ic</u>, trigonometr<u>ic</u>, etc.), given to this **N**ew **b**eautiful class of functions of the trigonometric–hyperbolic type discussed here, seems to be suitable to impart the meaningful identity. This **N**ew **b**eautiful class of functions, as it appears, are having no upper limit, and so may be taken as open ended, which means that, their varieties have no upper bounds.

The operators of the circular functions are: sin, cos, tan, cosec, sec and cot, and those of the hyperbolic functions (identified by the additional letter, h) are: sh (= sin$h$), ch (= cos$h$), th (= tan$h$), csc$h$ (= cosec$h$), sc$h$ (= sec$h$) and ct$h$ (= cot$h$). Keeping with the tradition, the operators of the present basic Nbic functions (identified by the additional letter, N), are maintained throught out as: sinN, cosN, tanN, cscN (= cosecN), secN and cotN.

These operators are of four letter words, and **N** is attached with sin, cos, tan etc., with no gap in between, which is similar to the operators of hyperbolic functions like, sin$h$, cos$h$, tan$h$ etc.

[Pronounce **sinN(x, x)** as: **sine enbic** (or simply, **sine en**) of **(x,x)**, and so on.]

Nbic functions when set to follow circular representation will be termed as Nbic circle. The interesting features of the hyperbolic circle along with the Nbic circle in relation to the

Euclidean circle, have been brought to light. The concept of Complex Circular DeMoivre's form, as the name suggested to the complex form of circular functions, namely, $\cos x + i \sin x = e^{ix}$, has been extended to hyperbolic functions, namely, $chx + i\, shx = f^{ix}$, and finally to the various types of Nbic functions, namely, $N/(x, y) = \cos N(x, y) + i \sin N(x, y) = e^{i\, N/(x, y)}$, $N_2/(x, y) = \cos N_2(x, y) + i \sin N_2(x, y) = e^{i\, N_2/(x, y)}$, and $N_3/(x, y) = \cos N_3(x, y) + i \sin N_3(x, y) = e^{i\, N_3/(x, y)}$, etc., terming them accordingly as Complex Hyperbolic, Complex Single Nbic, Complex Double Nbic, and Complex Triple Nbic DeMoivre's form, etc., respectively. All these actually are the various forms of the trigonometric–hyperbolic functions.

It has been shown that, the various types of Nbic functions described and discussed in this book, can be generated from the respective Complex Nbic functions, as the real and imaginary parts. The types and varieties of Nbic functions can be said to be open ended, only few types have been considered within the scope of this book.

The basic functions like, the plane trigonometric functions ($\sin x$, $\cos x$, $\tan x$, etc.), are the function of the variable x, which is nothing but an angle. The variable x in the basic hyperbolic functions (shx, chx, thx, etc.), is not an angle but a sort of parameter.

However, both the types of variables *viz.*, x in trigonometric functions, and x in hyperbolic functions, have one thing in common. That is, x of hyperbolic functions is called by the name **hyperbolic angle,** which is analogous to x in trigonometric–functions known as **circular angle**. Further that, both the type of $x - s$ can be interpreted in terms of some area of sector, associated to unit circle and unit hyperbola, in the former and later type of functions, respectively. Question arises whether is there any such parameter in Nbic functions whose interpretation may exist similarly, or otherwise.

1.1.1 Circular Angle

Circular functions, that is, plane trigonometric functions like, $\sin x$, $\cos x$ etc., are defined through circular angle x. The hyperbolic angle x (which is not an angle, but a parameter only), defines hyperbolic functions like, shx, chx, etc. Likewise, quite analogously, the proposed Nbic functions, which are the properly arranged trigonometric–hyperbolic functions similarly, can be defined utilising the concept of Nbic angle. Obviously, the Nbic angles, like hyperbolic angle, are not angles, but are certain parameters.

Definition of circular angle

The functions like, $\sin x$, $\cos x$, $\tan x$, etc., are the plane trigonometric operators, and all of them are the functions of the parameter x, which is an angle. These basic building blocks of plane trigonometry, (particularly, $\sin x$, $\cos x$, and $\tan x$), do satisfy the Pythagorean criterion, or alternatively, the Circularity criterion.

Pythagorean criterion

In short, the Pythagorean criterion, may be stated as:

If in a right triangle OAP, which is defined in the rectangular coordinate system (X, Y), where X is along the side (or base) OA, and Y is along the altitude AP, so that OP is the hypotenuse, then,

$$OP^2 = OA^2 + AP^2 \tag{1.1a}$$

If additionally, the hypotenuse OP [where O is at (0, 0), and P is at the point (X, Y)], is of unit length, then

$$OA^2 + AP^2 = 1 \quad (1.1b)$$

In other words

OA = X-component of OP = cos*x*, and
AP = Y-component of OP = sin*x*

For visualizing tan*x*, extend the line OA up to the point C on the X axis, where the tangent PC drawn at P, meets. Remembering that the radius OP = 1, we get

AP / OA = sin*x* / cos*x* = tan*x* = PC / OP.

Hence, PC = tan*x*. (1.1c)

So the **Pythagorean criterion** which is satisfied by the plane trigonometric functions is:

$$\cos^2 x + \sin^2 x = 1. \quad (1.1d)$$

Circularity criterion

Now visualize the Pythagorean criterion of the right triangle OAP bounded in rectangular coordinate system (X, Y) in the light of **Circularity criterion**. It may be presented as:

If a circle is drawn with centre at O at (0, 0) and the radius OP is of unit magnitude then, for any location of the point P at (X, Y), [which is at the other end of the hypotenuse OP], the triangle OAP satisfies the Pythagorean criterion, that is, $OP^2 = OA^2 + AP^2$. Since, the locus of the point P about O, maintaining a constant distance from it, is a circle, it satisfies the equation:

$$X^2 + Y^2 = 1 \quad (1.2)$$

This is what the **Circularity criterion** is.

In other words, Circularity criterion speaks for the coordinates X and Y, which can be expressed through the angular distance *x*, that is, X = cos*x*, Y = sin*x*. That is to say that, X-axis, and Y-axis are represented by, cos*x*-axis, and sin*x*-axis, respectively.

Both the Circularity criterion and the Pythagorean criterion actually are one and the same. Looking from this angle, the variable parameter *x*, which is nothing but the angle, is called also as **circular angle**, and the trigonometric functions, sin*x*, cos*x*, etc., are rightly termed as **Circular functions.**

Interpretation of circular angle

Assessment wise, the parameter *x* in circular functions, can be interpreted as follows **(Figure 1.1)**:

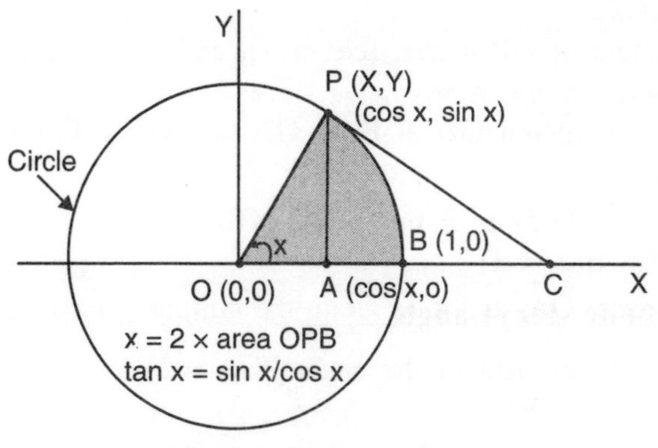

Fig. 1.1: Circular angle, x

Extend the base OA of the triangle OAP up to the point B, such that OB = OP = radius r. Consider that, x is measured in radian, which is produced by the rotational movement of the radius r from the position OB to the position OP about the point O, casuing the point B to travel to the point P, along the circular periphery of the arc length s, same as that of the radius r. In other words, it is the angle whose tangent is the ratio of the Y-component to the X-component of the radial distance measured from the centre O (0, 0) to any point P (X, Y) on the periphery of the circle, $X^2 + Y^2 = 1$, described through the system of coordinates, (X, Y).

We know that, for the full circle, the central angle, circumference and the area are: 2π, $2\pi r$ and πr^2, respectively, and for those corresponding to the circular sector OBP (bounded between two radii OB and OP, and the circular arc BP) are: x, s (= r), and A, respectively. Then

$$\frac{x}{2\pi} = \frac{s(=r)}{2\pi r} = \frac{A}{\pi r^2} \tag{1.3a}$$

Here, for the circle of radius unity, i.e., $r = 1$, we get

$$\frac{x}{2\pi} = \frac{1}{2\pi} = \frac{A}{\pi}, \text{ which suggests that, } x = 2A. \tag{1.3b}$$

Thus, **in plane trigonometry,** the measure of the **angle x, is actually twice the area of the circular sector,** the two boundary radii of unit length of which produces the angle.

1.1.2 Definition of Hyperbolic Angle and Tan-equivalent Hyperbolic (tehy) Angle

Hyperbolic angle

In hyperbolic functions, shx, chx, thx , etc., the parameter x, is not an angle, but analogous to the Euclidean system, it is termed as hyperbolic angle. The hyperbolic angle, as the name suggests, cannot be measured in the same way as done for the circular angles.

Consider the right side branch (i.e., along the positive direction of X axis), of a hyperbola,

$$X^2 - Y^2 = 1, \ [X > 0] \tag{1.4a}$$

which is defined in cartesian coordinates (X, Y). In this branch, the apex point B of the curve is at (0, 1), which is at the minimum distance from the origin O (0, 0), that is OB = 1. Consider a point P at (X, Y) on the first quardrant of the hyperbola. From P draw perpendicular on X-axis to meet at A.

Since P is on the hyperbola

OA = X-component of OP = chx, [$chx > 1$], and

AP = Y-component of OP = shx

For visualizing thx, draw perpendicular at B on OA to meet at C on the ray OP.

Now, remembering that OB = 1

$$AP / OA = shx / chx = thx = BC / OB.$$

Hence, $$BC = thx, \ [thx < 1] \tag{1.4b}$$

Tan-equivalent hyperbolic (tehy) angle

Further, if the circular measurement of the angle POA = H,

we have, tan H = thx

or, $$x = th^{-1}(\tan H) \qquad (1.5)$$

This angle POA of the plane triangle OAP on the hypebola, is identified by plane trigonometric way, i.e., adopting tan-equivalent of the hyperbolic tangent. It is to be clearly understood that, the hyperbolic angle x, which is measured in terms of area, is never same as the angle H, because it is measured by degree, or radian, as the preference may be.

In order to distnguise H from x, a separate name for this geometrically measured angle H is given as, **tehy angle** or, **tan-equivalent hyperbolic angle,** or simply, **geometrical angle, gA.**

Interpretation of hyperbolic angle

The parameter x in hyperbolic functions (shx, chx, thx), unlike the case of circular functions, is not angle, but still it can be interpreted as, twice the area of sector OBP, where the area is enclosed between the boundary hyperbolic arc BP, ray OP [connecting the origin O (0, 0) and the point P (X, Y)], and OB [connecting O (0, 0) and apex B (1, 0)], which is of unit length] (**Figure 1.2a**).

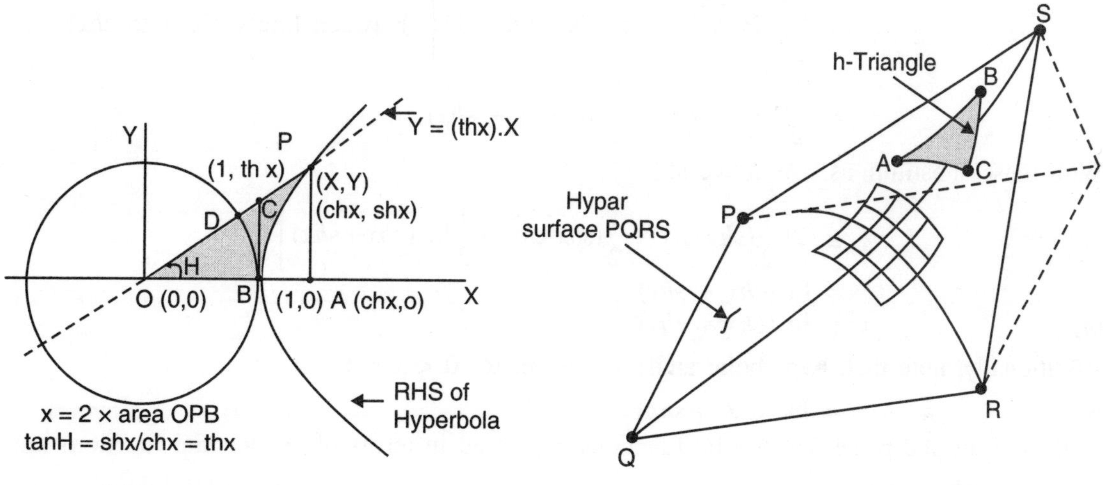

(a) Hyperbolic angle, H

(b) h-triangle in hypar surface

Fig. 1.2

Thus, it can be stated (without proof here) that, in **hyperbolic functions,** the measure of the **parameter x, is twice the area of the hyperbolic sector,** enclosed between the two boundary rays originating from the origin, one is of unit length along X-axis, the other is of length equal to, $\sqrt{ch2x}$, and the hyperbolic arc between the apex (1, 0) and the point (chx, shx).

The slope of the straight line passing through the origin O and the point on the hyperbola P, is tan H = thx. So the equation of the ray through OP is written in the slope intercept form: Y = m X + C, where, m = slope = thx, C = intercept = 0 (since the line passes through the origin).

Then the required equation is: Y = (thx).X

It is interesting to get the very definition of hyperbolic functions in terms of exponential function.

In brief this is done as follows.

We write the equation of the hyperbola, $X^2 - Y^2 = 1$ as, $Y = \sqrt{X^2 - 1}$

Now

$$\text{hyperbolic angle } x = 2 \times (\text{area of hyperbolic sector OBP})$$
$$= 2 \times (\text{area of triangle OAP} - \text{area of sector BAP})$$

$$\text{area of triangle OAP} = \int_0^{chx} (thx) \cdot X \cdot dX = \frac{1}{2} [thx\ ch^2 x] = \frac{1}{2}(shx \cdot chx)$$

[It is directly obtained as: $\frac{1}{2}$ (base = chx) × (altitude = shx)]

$$\text{area of sector BAP} = \int_1^{chx} \sqrt{(X^2 - 1)}\, dX$$

$$= \left[\frac{1}{2} X\sqrt{X^2 - 1} - \ln\,(X + \sqrt{X^2 - 1})\right], \text{ between limits, } (X{=}1 \text{ to } chx)$$

$$= \frac{1}{2} chx\ .\ shx - \frac{1}{2}\ln\,(chx + shx)$$

With these substitutions, we arrive at

$$x = 2\left[\frac{1}{2}(shx\ chx) - \frac{1}{2}(shx\ chx) + \frac{1}{2}\ln\,(chx + shx)\right]$$
$$= \ln\,(chx + shx)$$

or, $\quad x = \ln\,(chx + shx)$

Additionally, note that, hyperbolic angle x is – ve, for $0 < x < 1$.

So, $\quad -x = \ln\,(chx - shx)$

Collecting in one place, hyperbolic functions expressed in terms of exponential functions are

$$chx + shx = e^x$$
$$chx - shx = e^{-x} \qquad (1.6a)$$

Conversely, exponential functions expressed in terms of hyperbolic functions are

$$chx = (e^x + e^{-x}) / 2$$
$$shx = (e^x - e^{-x}) / 2 \qquad (1.6b)$$

Even though the equation of the unit hyperbola, $X^2 - Y^2 = 1$, is not that of a circle, the question arises whether can ever this equation be given the same form of that of a circle, or not. The answer to this quarry is, **yes**.

It is possible, (as discussed in great details later in Chapter 6), if the existing coordinates (X, Y) are suitably modified.

Let us modify the system of the exiting coordinates (X, Y) on replacing X and Y as:

$$X = [(chx) / (\sqrt{ch2x})], \text{ and } Y = [(shx) / (\sqrt{ch2x})], \text{ respectively.}$$

Remembering that, $ch^2x + sh^2x = ch\ 2x$, and with the modified (X, Y), the intended circular representation can be arrived at as, $X^2 + Y^2 = 1$.

This circular representation, henceforth, will be termed as **hyperbolic circle**.

Once the circular form of equation is obtained, representing the unit hyperbola, the Pythagorean criterion, as well as its alternative one, *viz.*, the Circularity criterion, are applicable.

Pythagorean criterion

Now consider that the triangle OAP, drawn inside the hyperbolic circle, exactly in a similar manner as the triangle OAP is drawn for circular functions. But the difference is that, the side and altitude are the modified ones, while the hypotenuse representing the radius becomes = 1.

The sides that are perpendicular to each other are, OA = X-component of OP as the base, AP = Y-component of OP as the altitude, and the third side that is, the radius OP (of length = 1) as the hypotenuse which makes an angle H (= tehy angle = angle POA) with the X-axis.

Thus

$$\text{OA} = \text{X-component of OP} = [(chx) / (\sqrt{ch2x})], \text{ and}$$

$$\text{AP} = \text{Y-component of OP} = [(shx) / (\sqrt{ch2x})] \tag{1.7a}$$

So the **Pythagorean criterion** which is satisfied by hyperbolic functions is:

$$\left[\frac{chx}{\sqrt{ch2x}}\right]^2 + \left[\frac{shx}{\sqrt{ch2x}}\right]^2 = 1 \tag{1.7b}$$

The angle H = angle POA, can be written in terms of the tan-equivalent angle H, as

$$\begin{aligned} \tan H &= \text{Y-component of OP / X-component of OP} = \text{AP / OA} \\ &= [(chx) / (\sqrt{ch2x})] / [(shx) / (\sqrt{ch2x})] \\ &= shx / chx \\ &= thx \end{aligned} \tag{1.7c}$$

Circularity criterion

The circularity criterion which speaks for the coordinates is:

$$X = [(chx) / (\sqrt{ch2x})], \text{ and } Y = [(shx) / (\sqrt{ch2x})]$$

That is to say, X-axis is now, $[(chx) / (\sqrt{ch2x})]$ – axis

and, Y-axis is now, $[(shx) / (\sqrt{ch2x})]$ – axis (1.8)

so that, the location of a point P on the Hyperbolic circle is given by:

$$[(chx) / (\sqrt{ch2x}), (shx) / (\sqrt{ch2x})].$$

1.2 DEFINITION AND INTERPRETATION OF Nbic ANGLE

1.2.1 Nbic Angle and its Interpretation

Speaking in terms of chemistry while categorizing matters, if the basic circular as well as hyperbolic functions are thought to be **elements**, the basic blocks of Nbic functions [like, (sin*x* *chx*), (cos*x* *shx*), etc., which comprise of circular and hyperbolic functions in multiplication], then for the sake of argument, can be taken as, **compounds**. This analogy seems to be valid for all types of Nbic functions, including single and all higher order Nbic functions. Hence it appears that, the task of getting interpretation of the corresponding Nbic angles, similar to the respective process of obtaining circular or hyperbolic angles, is not possible, as it stands today.

Fortunately, all the types of Nbic functions namely, single Nbic function, double Nbic function (both E type and F type), and triple Nbic function (both E type and F type), proposed and discussed in great details along with various other associated aspects, in this book, when set to the corresponding modified coordinates (X, Y), are seen to follow the equation of a unit circle: $X^2 + Y^2 = 1$.

1.2.2 Tan-equivalent Nbic (teN) Angle

Then, extending the same logic, as proposed for the definition of the circular form for the hyperbolic angle, namely the **tehy angle H**, the **Nbic angles N**, are defined through similar tan-equivalent form as, **teN angle**, (**tan-equivalent Nbic angle**), by adopting Nbic functions (which are actually the suitably arranged forms of the trigonometric-hyperbolic functions), expressed through suitably chosen coordinates (X, Y), satisfying the circular form of equation, $X^2 + Y^2 = 1$. This circular representation, henceforth to be termed as **Nbic Circle**, so that, corresponding to the plane geometrical triangular Figure OAP, drawn inside the Nbic circle, the definition of the teN angle N (= geometrical angle gA = angle POA), can be given **(Figure 1.3),** as

$$\tan N = \text{Y-componen t} / \text{X-component} = AP / OA \tag{1.9a}$$

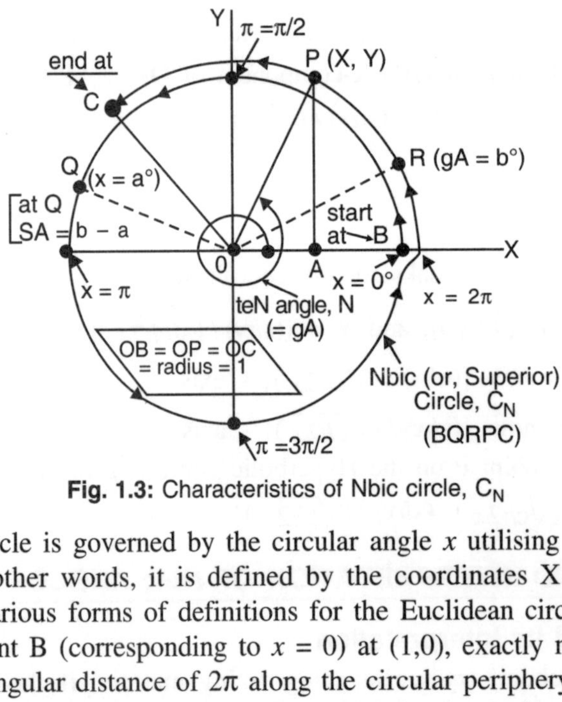

Fig. 1.3: Characteristics of Nbic circle, C_N

The Euclidean circle is governed by the circular angle *x* utilising the definition of radian measure of angle. In other words, it is defined by the coordinates X = cos*x*, Y = sin*x*. The main essence of the various forms of definitions for the Euclidean circle pin points one thing, that is, the starting point B (corresponding to $x = 0$) at (1,0), exactly meets on the same point C, after travelling an angular distance of 2π along the circular periphery of the circle (of radius unity). If the end point C corresponds to $x = 2\pi$, which is also at (1, 0), the Euclidean circle, for obvious reason, is to be termed as **full circle**, and then the point B ($x = 0$) and C ($x = 2\pi$) are the same point (1,0).

On the contrary, the starting point B (corresponding to $x = 0$) at (1, 0), and the end point C (corresponding to $x = 2\pi$) at any point other than (1, 0), for various Nbic circles (hyperbolic circle included), show the interesting characteristics that, they are the points separated at some angular distance over the same peripherial circular contour of radius unity. That is to mention that, the point C does not end on the point corresponding to a full complete turn of circular angle of $x = 2\pi$. It glides in excess of 2π, as observed for the various Nbic circles, with the exception for hyperbolic circle, where it is less than this value.

The appropriate terminology suggested for the Nbic circles is then **Overcomplete** or **Superior circles**, where as, that for hyperbolic circle is **Incomplete** or **Inferior circle**. The **full circle** or **complete circle** is the name earmarked for our year old Euclidean circle.

Consider that a point R on the Nbic circle (**Figure 1.3**), corresponding to a point Q for the circular angle x = a, is located at an Nbic angle N = gA = b. It is represented by:

$$Q]_{x=a} \Rightarrow R]_{S=b} \tag{1.9b}$$

where, the teN angle = tan N = tan (Y / X) = b.

Sweep angle

The sweep angle (sA) is defined as the angle measure of the teN angle of a point on the Nbic circle in excess of the corresponding circular angle x, and is measured by, gA – x = N. Since the angles b and a, always differ from each other in magnitude, the sweep angle sA at R, corresponding to the point Q, is the angle difference (b – a). That is

$$sA \text{ (sweep angle)} = b - a \tag{1.10}$$

If the circle is Euclidean circle, the teN angle reduces to circular angle, i.e., $b = a$, and the point Q does not glide over the circular path to reach to the point R. So in that case, $R]_{S=b} = Q]_{x=a}$. Judging from the angle of sweep, the **Euclidean circle, hyperbolic circle, and Nbic circle** can also be appropriately called as, **zero circle, positive circle** and **negative circle**, respectively.

Pythagorean and circularity criteria

The Pythagorean criterion as well as the Circularity criterion, are then applicable for the triangle OAP, drawn inside the corresponding Nbic circles.

However, the generalised terminology, gA (geometrical angle), is equally suitable for all types of teN angles N.

1.3 SYMBOLIC IDENTIFICATION OF Nbic FUNCTIONS

Within the format of the subject matter proposed and discussed, the various Nbic functions, their symbolic identification, and the corresponding teN angles N adopted, are:

Single Nbic function: N

Symbolically, N_A : N (x, y), N (x, x) [= N (x) = N x]

N_B : N (x, $-y$), N (x, $-x$)

teN angle : N

Double Nbic function (E type): N_{2e}

Symbolically, N_A : N_{2e} (x, y), N_{2e} (x, x) [= N_{2e} (x) = N_{2e} x]

N_B : N_{2e} (x, $-y$), N_{2e} (x, $-x$)

teN angle : N_{2e}

Double Nbic function (F type) : N_{2f}

Symbolically, $N_A : N_{2f}(x, y), N_{2f}(x, x)$ [$= N_{2f}(x) = N_{2f}\, x$]
$N_B : N_{2f}(x, -y), N_{2f}(x, -x)$

teN angle : N_{2f}

Triple Nbic function (E type) : N_{3e}

Symbolically, $N_A : N_{3e}(x, y), N_{3e}(x, x)$ [$= N_{3e}(x) = N_{3e}\, x$]
$N_B : N_{3e}(x, -y), N_{3e}(x, -x)$

teN angle : N_{3e}

Triple Nbic function (F type) : N_{3f}

Symbolically, $N_A : N_{3f}(x, y), N_{3f}(x, x)$ [$= N_{3f}(x) = N_{3f}\, x$]
$N_B : N_{3f}(x, -y), N_{3f}(x, -x)$

teN angle : N_{3f}

In Chapter 6, in connection with the topic: Circular representation of Nbic functions [Nbic circles], the concept of teN angles have been made use of.

If in a right triangle, x and y are the two adjacent sides which meet at right angle, and z is the hypotenuse, then Pythagorean criterian states that: $x^2 + y^2 = z^2$. The three quantities x, y and z are known as Pythagorean triplets. Let symbolically these triplets be shown as: [(x, y), z].

There is a very simple method to find all these triplets. The simplest one is, [(3, 4), 5]. Others are [(24, 7), 25], [(24, 70), 74], and so on.

Whether or not there exists any Pythagorean triplets where, x and y are the consecutive numbers, (i.e., the difference of x and y is 1), and if so, whether such triplets are limited or endless in number, and if so, how to find them directly, is the main inquisitiveness in mind needed to be probed and answered. In **Appendix A**, a very interesting, and at the same time, simple direct method for answering these quarries raised on this special variety of Pythagorean triplets, developed and proposed by Bairagi (author of this book), is presented.

1.3.1 Nbic Trigonometry

A closed space formed by three distinct lines, each pair of which intersects at distinct points called vertex, is called a triangle. So a triangle ABC possesses three vertices A, B and C, and three boundary contour lines AB, BC and CA, called sides. Two intersecting boundary lines (sides) meeting at each vertex, forms an angle. Thus a triangle ABC has three sides (AB, BC and CA), three vertices A, B and C, and three angles A, B and C at each vertex.

Trigonometry is that branch of mathematics with which the relationship among the sides and angles of triangles is studied. Geometry, on the other hand, is the study of geometrical shapes and their characteristics and associated aspects, in which the help of trigonometric relations in the form of rules, conditions and guidelines as the tool, are called for. Again, trigonometric relations, (or in short, trigonometry) are expressed as mathematical functions.

If the triangle is in the Euclidean plane (flat surface), the triangle ABC is a flat or plane triangle. The boundary lines AB, BC and CA are all segments of straight lines. The sum of the three internal angles formed at the three vertex, *viz.*, (angle A + angle B + angle C) = 180 degree. The corresponding study of the Euclidean triangle is taken care of through the trigonometry, which is better known as plane trigonometry.

A right triangle ABC (right angle at C, AB = c, as hypotenuse; BC = a, as altitude; AC = b, as base; x = angle formed by the intersecting lines AB and AC at A), as a geometrical figure in a flat plane, forms the basis of the plane trigonometric relations, from which the functions like sin$x$, cos$x$, tan$x$ etc., are derived as the proportion of the sides, namely:

$$\sin x = a/c,\ \cos x = b/c,\ \tan x = a/b$$

Alternatively (in terms of the angle formed at the vertex),

$$\sin B = b/c,\ \sin A = a/c \tag{1.11}$$

However, this is not the case for hyperbolic triangle. Two types of hyperbolic triangles (h-triangle), as the basic figure can be cited. They are, h-triangle in Euclidean plane, and h-triangle in hyperbolic plane (hyperbolic paraboloid).

The first type has been described as a triangle OAP, in **Figure 1.2a** [in connection with the parameter x as hyperbolic angle, for a unit hyperbola $X^2 - Y^2 = 1$, defined by the cartesian coordinates (X, Y) in 2-D flat plane]. This gives,

chx = X-projection of OP

shx = Y-projection of OP

thx = BC (where, the point B is the apex of the right side of the hyperbola, C is the meeting point of the perpendicular drawn at B, on OP) (1.12)

The second type is described in **Figure 1.2b**. The h-triangle ABC is on the hyperbolic paraboloid surface, which is not flat, but twisted. Since, hyperbolic geometry is developed as a break away departure from the Euclidean concept of geometry, on challenging the worthiness and necessity of the Euclid's parallel postulate, the plane trigonometry and hyperbolic trigonometry, utilizing circular functions and hyperbolic functions, respectively, in no case, have the basis same.

In fact, hyperbolic functions like, shx (= sinhx), chx (= coshx), thx (= tanhx) are usually expressed through series expansion. That is

$$chx = \sum_{n=0}^{\infty}\frac{x^{2n}}{(2n)!} = 1 + \frac{x^2}{(2)!} + \frac{x^4}{(4)!} + \frac{x^6}{(6)!} + \ldots$$

$$shx = \sum_{n=0}^{\infty}\frac{x^{2n+1}}{(2n+1)!} = 1 + \frac{x^3}{(3)!} + \frac{x^5}{(5)!} + \frac{x^7}{(7)!} + \ldots$$

$$thx = x - \frac{x^3}{3} + \frac{2x^5}{15} - \frac{17x^7}{315} + \ldots \tag{1.13}$$

For any h-triangle ABC in hyperbolic plane, (with k = 1):

Hyperbolic law of sines : $\dfrac{\sin A}{sha} = \dfrac{\sin B}{shb} = \dfrac{\sin C}{shc}$

Hyperbolic law of cosines : $chc = cha.chb - sha.shb.\cos C$

Dual hyperbolic law of cosines : $chc = \dfrac{\cos A.\cos B + \cos C}{\sin A.\sin B}$ (1.14a)

The equivalent analogous expressions using plane trigonometry for an arbitrary Euuclidean triangle ABC are:

Plane trigonometric law of sines : $\dfrac{\sin A}{a} = \dfrac{\sin B}{b} = \dfrac{\sin C}{c}$

Plane trigonometric law of cosines :

$$c^2 = a^2 + b^2 - 2ab \cos C$$
$$a^2 = b^2 + c^2 - 2bc \cos A$$
$$b^2 = c^2 + a^2 - 2ca \cos B \quad (1.14b)$$

For a right triangle ABC in hyperbolic plane (and, $k = 1$), with CB perpendicular on AB at C, (angle at C = $\pi / 2$), the angles and sides are related in hyperbolic geometry as

$$\sin A = sha / shc,\ \sin B = shb / shc$$
$$\cos A = thb / thc,\ \cos B = tha / thc$$
$$cha = \cos A / \sin B,\ chb = \cos B / \sin A$$
$$chc = cha.chb = \cot A \cot B \quad (1.15a)$$

The equivalent analogous expressions using plane trigonometry for an arbitrary Euclidean triangle ABC (angle at C = $\pi / 2$, so that, $\cos C = 0$), are:

law of sines : $\sin A = a / c,\ \sin B = b / c$
$\cos A = b / c,\ \cos B = a / c$

law of cosines : $\cos A = b / c,\ \cos B = a / c$

$$c^2 = a^2 + b^2 \quad (1.15b)$$

[The last one can be deduced from, $chc = cha.chb$, as follows ⇒ Expanding hyperbolic functions through power series, and cosidering that the *h*-triangle is very very small, so that higher powers of the sides a, b and c, can be neglected, this expression finally reduces to, $c^2 = a^2 + b^2$].

1.3.2 Interchangeability of Trigonometric and Hyperbolic Functions

Osborn's Rule [due to G. Osborn]

There exist a lot of structural similarities between the formulas derived for plane trigonometric functions (circular functions) and their corresponding hyperbolic counterparts.

The formulas include, single angle, two angle (addition and subtraction), multiple angle, half angle, powers of functions, sum, difference and product of functions, derivatives, integrals, law of sine and cosine, area formulas, etc.

It is possible to convert any trigonometric identity into the corresponding hyperbolic identity, by expanding it in terms of integral powers of sines and cosines, simply by replacing, **cos** to ***ch*.** When the expansion contains a product of 2, 6, 10, 14...***sh*-s,** the switching of sign of that term is necessary while replacing, **sin to *sh*.**

For example,

No change of sign, where the product of 2 *sh*-*s* does not appear:

$$\sin (x - y) = \sin x \cos y - \cos x \sin y \Rightarrow sh\ (x - y) = shx\ chy - chx\ shy$$
$$\sin 2x = 2 \sin x \cos x \Rightarrow sh\ 2x = 2\ shx\ chx$$

Change of sign, where the product of 2 *sh*-*s* appear:

$$\cos (x + y) = \cos x \cos y - \sin x \sin y \Rightarrow ch\ (x + y) = chx\ chy\ (+)\ shx\ shy$$
$$\cos 2x = \cos^2 x - \sin^2 x \Rightarrow ch\ 2x = ch^2 x\ (+)\ sh^2 x \quad (1.16a)$$

Gudermann Function [due to Christoph Gudermann]

Direct conversion from circular to hyperbolic function or *vice versa*, is possible with the application of Gudermann function, which by definition is:

$$\text{gd}\ (x) = \int_0^x \frac{dp}{ch(p)} = \sin^{-1}(th\ x) = \tan^{-1}(sh\ x) = 2\tan^{-1}(e^x) - \pi/2 \tag{1.16b}$$

From which,

$$\begin{aligned}\sin\ (\text{gd}(x)) &= \text{t}hx\\ \cos\ (\text{gd}(x)) &= \text{sec}hx\\ \tan\ (\text{gd}(x)) &= shx\\ \csc\ (\text{gd}(x)) &= \text{ct}hx\\ \sec\ (\text{gd}(x)) &= chx\\ \cot\ (\text{gd}(x)) &= \text{csc}hx\end{aligned} \tag{1.16c}$$

[**Note** that, in hyperbolic geometry, the quantity, $[\pi/2 - \text{gd}\ (x)]$ defines the angle of parallelism.]

1.3.3 Surface, Gaussian Curvature and Angle Sum

Again, a *h*-triangle ABC drawn on a hyperbolic surface (*viz.*, hyperbolic paraboloid surface as shown in **Figure 1.3**) naturally, will be distorted, since each of the boundary lines AB, BC and CA between the pair of vertices A-B, B-C and C-A, follows the contour of the said surface. The sum of the three internal angles at A, B and C in this case ceases to be equal to 180 degree, in fact it is always < 180 degree.

In contrast to this, if the circle ABC is considered on a spherical surface, the corresponding geometry of which is governed by spherical geometry, (which again, is the simplest model of elliptic geometry in which through a given point, a line has no parallels), the sum of the three internal angles at A, B and C will always > 180 degree.

The classification of the curved surfaces, are governed by the measure of the corresponding Gaussian curvature. Note that, the curvature at any point is the reciprocal of its radius. Thus, if the two radii of a given surface along the two principal directions are R_1 and R_2, and $K_1 = 1/R_1$ and $K_2 = 1/R_2$, are the two curvatures, (with centres of curvatures C_1 and C_2, respectively), along the two principal directions of the given surface, the Gaussian curvature K is defined as the product of K_1 and K_2, that is

$$K = K_1K_2 = 1/(R_1R_2).$$

In a **flat surface** (surface with no curvature):

$$R_1 = \infty,\ R_2 = \infty,$$
$$K_1 = 0,\ K_2 = 0,$$

So, **Gaussian curvature $\mathbf{K = K_1K_2 = 0}$** (1.16d)

In a **cylindrical surface** (which is a singly curved surface):

The surface is straight along one of the principal direction, (say) $K_1 = 0$, and curved along the other principal direction, i.e., K_2 = finite and (+ ve).

$$R_1 = \infty,\ R_2 = \text{finite and (+ ve)}$$

So, **Gaussian curvature $\mathbf{K = K_1K_2 = 0}$** (1.16e)

In a **spherical surface**, (which is a doubly curved surface):

There are two curvatures K_1 and K_2, along the two principal directions.

R_1 = finite and (+ ve), R_2 = finite and (+ ve)

K_1 = finite and (+ ve), K_2 = finite and (+ ve)

Here both the centres of curvatures C_1 and C_2 lie on same side of the surface.

So, **Gaussian curvature K = K_1K_2 = finite and (+ ve)** (1.16f)

The surfaces of positive Gaussian curvature are also known as **Synclastic** surfaces.

In a **hyperbolic surface**, i.e., Hyperbolic paraboloid (which is a doubly curved surface):

There are two curvatures K_1 and K_2, along the two principal directions.

either, R_1 = finite and (+ ve), and R_2 = finite and (– ve)

[or, R_1 = finite and (– ve), and R_2 = finite and (+ ve)]

K_1 = finite and (+ ve), and K_2 = finite and (– ve)

[or, K_1 = finite and (– ve), and K_2 = finite and (+ ve)] (1.16g)

Here one of the centres of curvatures, (say) C_1 lies on one side of the surface, while the other C_2 occupies the other side of the surface. If the curvature K_1 is (+ ve), K_2 will be (– ve), or *vice versa*.

So, **Gaussian curvature K = K_1K_2 = finite and (–ve)** (1.16h)

The surfaces of negative Gaussian curvature are also known as **Anticlastic** surfaces.

The measure of the Gaussian curvature K of a surface reflects on the angle sum of the three internal angles of an arbitrarily oriented triangle ABC drawn on that particular surface. It may then be inferred that:

for flat surface, K = 0, angle sum = 180 degree

for anticlastic surface, K = (– ve), angle sum < 180 degree

for synclastic surface, K = (+ ve), angle sum > 180 degree ...(1.17)

1.3.4 Nbic Functions and Nbic Trigonometric Relations

The terms Trigonometry and Geometry often appear to be synonymous, that is to say that, they carry the same meaning. The term geometry covers the study of various aspects and characteristics of different shapes, and establishes the interrelation and / or correlation of angles, sides, tangents, similarity, proportionality, congruency, etc. To perform these tasks, the tool in hand in the form of formula, or proposition, or reasoning, and likewise, must be available before hand.

Precisely in this sense, the study of a certain class of geometry turns out to be the exercise of the corresponding mathematical tool developed for that class governing the various relations through the corresponding functions.

The study of the geometrical relationships among the sides and angles of the hyperbolic triangle, for obvious reason, will have to be done using hyperbolic trigonometry, involving hyperbolic functions, where the applicability of the plane trigonometric relations fail to work. The same observation is true for the plane trigonometry where the use of circular functions is called for.

Viewing from this angle, the subject developed and presented in this book, namely, Nbic trigonometry, is no exception from the synonimousity of the proposed class of functions forwarded with suitable terminology as Nbic function. Precisely, this forms the subject matter which is dealt in this book.

1.4 COMPLEX Nbic FUNCTIONS

1.4.1 Some Basic Complex Functions

Circular function

Consider a complex number, $a + i\,b$. If it is represented through circular function, (*viz.*, plane trigonometric functions like, cos$x$, sin$x$, etc.), it would become,

$$a + i\,b = r_1\,(\cos x + i\,\sin x) = r_1\,e^{ix} \quad (1.18a)$$

Then, the system of representing the said number through circular function, is governed by the rule:

$$a = r_1\cos x,\; b = r_1\sin x,\; (b/a) = \tan x,\; r_1 = \sqrt{a^2+b^2}$$
$$\text{since, } \cos^2 x + \sin^2 x = 1 \quad (1.18b)$$

If we consider, r_1 = unity, the said complex function takes the form,

$$f(x) = (\cos x + i\,\sin x) = e^{ix},\ [\text{so also, } (\cos x - i\,\sin x) = e^{-ix}] \quad (1.18c)$$

Here, $(\cos x + i\,\sin x) = e^{ix}$, is a function of x only, *viz.*, $f(x)$, which is formed of trigonometric functions, cos$x$ and sin$x$. In this sense, this function, $f(x)$ better, is to be termed as Complex Circular Function.

Similarly, the same number, $a + i\,b$, can well be represented ideally through any function, and there is no restriction for such selection of functions for this purpose. These functions can well be $f(x)$, or, $g(y)$, etc., either used individually, or as a product.

The set of other circular functions which can be used for the same purpose are, (sec$y$, tan$y$), (cosec$u$, cot$u$).

When we use the set (sec$y$, tan$y$), the governing rule is likely to differ from those of the set (cos$x$, sin$x$). For example, if

$$a + i\,b = r_2\,(\sec y + i\,\tan y) \quad (1.19a)$$

$$a = r_2\sec y,\; b = r_2\tan y,\; (b/a) = \sin y,\; r_2 = \sqrt{a^2-b^2}$$

Since, $\sec^2 x - \tan^2 x = 1$ (1.19b)

However, as understood, proper care need to be taken, in this case, if the value of a, is less than b, while opting for the set of circular function, (sec$y$, tan$y$). Similar caution is to be exercised for the set, (cosec$u$, cot$u$).

For the sake of simple illustration, $(a + i\,b)$ may be represented individually by, $f(x)$, or, by the product of $f(x).g(x)$.

Hyperbolic function

Suppose, $g(x) = \text{R}(chx + i\,shx)$, is proposed to represent the complex number, $(c + i\,d)$, using hyperbolic functions, chx and shx.

$$c + i\,d = r_3(chx + i\,shx) \quad (1.20a)$$

The same situation arises, as encountered while adopting the set of circular function, (sec*y*, tan*y*), even though it is a valid proposition. Here, in this case

$$a = r_3\ chx,\ b = r_3\ shx,\ (b/a) = thx,\ r_3 = \sqrt{a^2 - b^2}$$

Since, $$ch^2 x - sh^2 x = 1 \qquad (1.20b)$$

The function, (*chx* + i *shx*), which is composed of hyperbolic functions, is to be termed as Complex hyperbolic function, and, analogous to Complex circular function, is identified symbolically as, f^{ix}.

That is

$$g(x) = (chx + i\ shx) = f^{ix},\ [\text{so also, } (chx - i\ shx) = f^{-ix}], \qquad (1.20c)$$

[The letter ***f***, preferred in f^{ix} has nothing to do with the base of natural logarithm, ***e***, which in e^{ix}, does.]

1.4.2 Generation of Single Nbic Function, N (*x*, *y*)

[As generated from Complex Single Nbic function, N / (*x*, *y*)]

[**Note**: (1) Symbolically, Single Nbic function should be written as N_1 (*x*, *y*). However, the suffix 1 does not carry any special meaning, and so, omitted.

(2) Additional slash sign (/), immediately after N, in both the forms of complex Nbic functions (*viz.*, in N / (*x*, *y*), and also in $e^{iN/(x,\ y)}$), are adopted throughout, in order to differentiate from the Nbic functions N (*x*, *y*), which are written without slash sign.].

Now consider, if the product of the functions *f*(*x*), and *g*(*y*), is proposed to represent a complex function, with the former one is a complex circular, and the later one, is a complex hyperbolic function.

$$f(x) = (\cos x + i\ \sin x) = e^{ix}$$
$$g(y) = (chy + i\ shy) = f^{iy} \qquad (1.21a)$$

so that, the complex function turns out to be

$$N/(x,\ y) = f(x)\ g(y) = e^{ix} f^{iy}$$
$$= (\cos x + i\ \sin x)\ (chy + i\ shy)$$
$$= (\cos x\ chy - \sin x\ shy) + i\ (\sin x\ chy + \cos x\ shy) \qquad (1.21b)$$

Each of the real and imaginary parts of this expression, when separated, is seen to have been formed as a product of trigonometric and hyperbolic terms. By adopting altogether new terminology as Nbic function, (1.21b), can be written as:

$$N/(x,\ y) = \cos N(x,\ y) + i\ \sin N(x,\ y) \qquad (1.21c)$$

In abbreated form, the complex expressions, N / (*x*, *y*), again may be:

N / (*x*, *y*):

$$\cos N(x,\ y) + i\ \sin N(x,\ y) = C_1^* + i\ S_1^* = e^{i\,N/(x,\ y)}$$
$$\cos N(x,\ y) - i\ \sin N(x,\ y) = C_1^* - i\ S_1^* = e^{-i\,N/(x,\ y)} \qquad (1.21d)$$

[The letter, *e* is **not** the base of natural logarithm, it is just a symbol.].

1.4.3 Single Nbic Function with Suffixes A and B

Identify the Nbic function N(*x*, *y*) with variables (*x*, *y*) or (*x*, –*y*), with the suffixes **A** or **B**, respectively. Then separating the real and imaginary parts from (1.21c or 1.21d), we get

N_A (*x*, *y*):

$$C_1^* = \cos N(x, y) = \cos x\ ch y - \sin x\ sh y$$
$$S_1^* = \sin N(x, y) = \sin x\ ch y + \cos x\ sh y \tag{1.21e}$$

If instead, if the selected functions are:

$$f(x) = (\cos x - i \sin x) = e^{-ix}$$
$$g(y) = (ch y - i\ sh y) = f^{-iy} \tag{1.21f}$$

the various combinations of the product in N / (*x*, *y*) = *f*(*x*) *g*(*y*), would yield the other Nbic functions, N(*x*, *y*), of the same category, as

N_B (*x*, *y*):

$$C_2^* = \cos N(x, -y) = \cos x\ ch y + \sin x\ sh y$$
$$S_2^* = \sin N(x, -y) = \sin x\ ch y - \cos x\ sh y \tag{1.21g}$$

The structural appearance of these Nbic functions, N(*x*, *y*) in (1.21e), and (1.21g), resembles identically with the well known trigonometric expansions of, cos(A + B), sin(A + B), cos(A – B) and sin(A – B). The similarity is

$$\cos(A + B) = \cos A \cos B - \sin A \sin B$$
$$\Leftrightarrow \cos N\ (x, y) = C1^* = \cos x\ ch y - \sin x\ sh y$$
$$\sin(A + B) = \sin A \cos B + \cos A \sin B$$
$$\Leftrightarrow \sin N\ (x, y) = S1^* = \sin x\ ch y + \cos x\ sh y$$
$$\cos(A - B) = \cos A \cos B + \sin A \sin B$$
$$\Leftrightarrow \cos N\ (x, -y) = C2^* = \cos x\ ch y + \sin x\ sh y$$
$$\sin(A - B) = \sin A \cos B - \cos A \sin B$$
$$\Leftrightarrow \sin N\ (x, -y) = S2^* = \sin x\ ch y - \cos x\ sh y \tag{1.22}$$

(1.22) describes the structure of Single Nbic functions, N(*x*, *y*).

1.4.4 Particular Case

For the case of single variable *x*, it may be sufficient to use only *x*, i.e., N(*x*, *x*) [= N (*x*) = N *x*]. They are:

N_A (*x*) [= N_A *x*]:

$$C_1 = \cos N(x, x) = \cos Nx = \cos x\ ch x - \sin x\ sh x$$
$$S_1 = \sin N(x, x) = \sin Nx = \sin x\ ch x + \cos x\ sh x \tag{1.23a}$$

N_B (*x*) [= N_B *x*]:

$$C_2 = \cos N(x, -x) = \cos x\ ch x + \sin x\ sh x$$
$$S_2 = \sin N(x, -x) = \sin x\ ch x - \cos x\ sh x \tag{1.23b}$$

[Refer to **Table D.1** and **Table D.2** for the numerical values of Trigonometric and Hyperbolic functions of (*x*) and (2*x*), respectively, in **Appendix D**.]

1.4.5 Complex Single Nbic Function with Suffixes A and B, [N_A / (*x*, *x*), N_B / (*x*, *x*)]

A particular case of N / (*x*, *y*), conforming to the single variable (*x*, *x*) and (*x*, –*x*), of N / (*x*, *x*) and N / (*x*, –*x*) types, are obtained similarly.

In abbreviated form, the complex expressions, N / (*x*, *x*) and N / (*x*, – *x*), for single variable *x*, again, may be represented symbolically similar to (1.21d) on putting *y* = *x*, as follows:

$N_A/(x, x)$ [$=N_A/x$]:

$$\cos Nx + i \sin Nx = C_1 + i\, S_1 = e^{i\, N/x}$$
$$\cos Nx - i \sin Nx = C_1 - i\, S_1 = e^{-i\, N/x}$$

$N_B/(x, -x)$:

$$\cos N(x, -x) + i \sin N(x, -x) = C_2 + i\, S_2 = e^{iN/(x,\,-x)}$$
$$\cos N(x, -x) - i \sin N(x, -x) = C_2 - i\, S_2 = e^{-iN/(x,\,-x)} \quad (1.24a)$$

In (1.24a), the Nbic functions, $N_A(x, x)$ and $N_B(x, -x)$, when collected from the complex Nbic functions, $N/(x, x)$ and $N/(x, -x)$, are:

Collecting Formulas Together

$N_A\ x$:

$$C_1 = \cos N(x, x) = \cos Nx = \cos x\, chx - \sin x\, shx$$
$$S_1 = \sin N(x, x) = \sin Nx = \sin x\, chx + \cos x\, shx$$

$N_B\ x$:

$$C_2 = \cos N(x, -x) = \cos x\, chx + \sin x\, shx$$
$$S_2 = \sin N(x, -x) = \sin x\, chx - \cos x\, shx \quad (1.24b)$$

[Refer to **Table D.3** for the numerical values of $N(x,x)$ and $N(x, -x)$, in **Appendix D**]

1.5 GENERATION OF DOUBLE Nbic FUNCTION, $N_2\ (x, y)$

1.5.1 As Generated from Complex Double Nbic Function, $N_2/(x, y)$

Double Nbic functions, $\mathbf{N_2(x, y)}$, can be thus, generated by multiplying, $N/(x, y)$ by either of the two, *viz.*,

$$f(y) = (\cos y \pm i \sin y) = e^{\pm iy},$$
$$g(y) = (chy \pm i\, shy) = f^{\pm iy} \quad (1.25a)$$

so that, two types of Double Nbic functions, in the same category, are formed. For example, they are:

Category 1 : (E type)

There are two E types *viz.*, **E_1 type** and **E_2 type,** i.e.,

$$E_1 \text{ type: } N_{2e}/(x, y) = N/(x, y).\, e^{\pm iy} = e^{\pm iN/(x,y)}.e^{\pm iy}$$
$$E_2 \text{ type: } N_{2e}/(x, y) = N/(x, x).\, e^{\pm iy} = e^{\pm iN/(x,x)}.e^{\pm iy} \quad (1.25b)$$

Category 2 : (F type)

There are two F types viz., **F_1 type** and **F_2 type,** i.e.,

$$F_1 \text{ type: } N_{2f}/(x, y) = N/(x, y).\, f^{\pm iy} = e^{\pm iN/(x,y)}.f^{\pm iy}$$
$$F_2 \text{ type: } N_{2f}/(x, y) = N/(x, x).\, f^{\pm iy} = e^{\pm iN/(x,x)}.f^{\pm iy} \quad (1.25c)$$

[**Note :** suffix, *e* or *f*, have been extensively used to identify functions related to, E type or, F type, as the case may be.]

This way, the number of Double Nbic functions generated will be **four**, *viz.*,

$[N/(x, y).e^{\pm iy}]$, $[N/(x, y).f^{\pm iy}]$, and

$[N/(x, x).e^{\pm iy}]$, $[N/(x, x).f^{\pm iy}]$.

1.5.2 Category 1 : (E type)

As for illustration, conforming to this category:

Using, N / (x, y) from (1.21d), and, e^{iy} as per Category 1, we have

E_1 type : N_{2e} / (x, y):

$$
\begin{aligned}
e^{iN/x}\,e^{iy} &= \cos N(x, y) + i \sin N(x, y) (\cos y + i \sin y)\\
&= [\cos N(x, y) \cos y - \sin N(x, y) \sin y]\\
&\quad + i\,[\sin N(x, y) \cos y + \cos N(x, y) \sin y]\\
&= \cos N_{2e}(x, y) + i \sin N_{2e}(x, y)\\
&= e^{iN_{2e}/(x,\,y)}
\end{aligned}
$$

From which we get

$$
\begin{aligned}
\cos N_{2e}(x, y) &= \cos N(x, y) \cos y - \sin N(x, y) \sin y\\
&= [\cos x\, ch y - \sin x\, sh y] \cos y\\
&\quad - [\sin x\, ch y + \cos x\, sh y] \sin y\\
&= \cos (x + y)\, ch y - \sin (x + y)\, sh y\\
&= \cos N[(x + y), y]\\
\sin N_{2e}(x, y) &= \sin N(x, y) \cos y + \cos N(x, y) \sin y\\
&= [\sin x\, ch y + \cos x\, sh y] \cos y\\
&\quad + [\cos x\, ch y - \sin x\, sh y] \sin y\\
&= \sin (x + y)\, ch y + \sin (x + y)\, sh y\\
&= \sin N[(x + y), y] \qquad (1.25d)
\end{aligned}
$$

Using, N / x from (1.24a), and, e^{iy}, as per Category 1, we have

E_2 type: N_{2e} / (x, y):

$$
\begin{aligned}
e^{iN/x}\,e^{iy} &= (\cos Nx + i \sin Nx) (\cos y + i \sin y)\\
&= [\cos Nx \cos y - \sin Nx \sin y]\\
&\quad + i\,[\sin Nx \cos y + \cos Nx \sin y]\\
&= \cos N_{2e}(x, y) + i \sin N_{2e}(x, y)\\
&= e^{iN_{2e}/(x,\,y)}
\end{aligned}
$$

From which we get

$$
\begin{aligned}
\cos N_{2e}(x, y) &= \cos Nx \cos y - \sin Nx \sin y\\
&= [\cos x\, ch x \sin x\, sh x] \cos y\\
&\quad - [\sin x\, ch x + \cos x\, sh x] \sin y\\
&= \cos (x + y)\, ch x - \sin (x + y)\, sh x\\
&= \cos N[(x + y), x]\\
\sin N_{2e}(x, y) &= \sin Nx \cos y + \cos Nx \sin y\\
&= [\sin x\, ch x + \cos x\, sh x] \cos y\\
&\quad + [\cos x\, ch x - \sin x\, sh x] \sin y\\
&= \sin (x + y)\, ch x + \cos (x + y)\, sh x\\
&= \sin N[(x + y), x] \qquad (1.25e)
\end{aligned}
$$

1.5.3 Particular Case

when $y = x$, we have both for **E_1 type** and **E_2 type**

N_{2e} / (x, x) = N_{2e} / (x) = N_{2e} / x:

$$
e^{iN/x}\,e^{iy} = [\cos N(x, x) + i \sin N(x, x)] (\cos x + i \sin x)
$$

$$= [\cos N(x, x) \cos x - \sin N(x, x) \sin x] + i\,[\sin N(x, x) \cos x + \cos N(x, x) \sin x]$$
$$= \cos N_{2e}(x, x) + i \sin N_{2e}(x, x)$$
$$= e^{i\,N_{2e}/(x,\,x)} = e^{i\,N_{2e}/x}$$

From which putting $y = x$ in (1.25e), we get

$$\begin{aligned}\cos N_{2e}(x, x) &= \cos N(x, x) \cos x - \sin N(x, x) \sin x\\ &= \cos N(2x, x)\\ \sin N_{2e}(x, x) &= \sin N(x, x) \cos x + \cos N(x, x) \sin x\\ &= \sin N(2x, x)\end{aligned} \qquad (1.26a)$$

[Refer to **Table D.4** for the numerical values of $N_{2e}(x, x)$, in **Appendix D**]

1.5.4 Category 2 : (F type)

As for illustration, conforming to this category, we have

F_1 type : $N_{2f}/(x, y)$:

$$\begin{aligned}e^{i\,N/x}\,f^{\,iy} &= [\cos N(x, y) + i \sin N(x, y)]\,(chy + i\,shy)\\ &= [\cos N(x, y)\,chy - \sin N(x, y)\,shy] + i\,[\sin N(x, y)\,chy + \cos N(x, y)\,shy]\\ &= \cos N_{2f}(x, y) + i \sin N_{2f}(x, y)\\ &= e^{i\,N_{2f}/(x,\,y)} = e^{i\,N_{2f}/(x,\,y)}\end{aligned}$$

From which we get

$$\begin{aligned}\cos N_{2f}(x, y) &= \cos N(x, y)\,chy - \sin N(x, y)\,shy\\ \sin N_{2f}(x, y) &= \sin N(x, y)\,chy + \cos N(x, y)\,shy\end{aligned} \qquad (1.26b)$$

F_2 type : $N_{2f}\ /(x, y)$:

$$\begin{aligned}e^{i\,N/x}\ f^{\,iy} &= (\cos Nx + i \sin Nx)\,(chy + i\,shy)\\ &= [\cos Nx\,chy - \sin Nx\,shy] + i\,[\sin Nx\,chy + \cos Nx\,shy]\\ &= \cos N_{2f}(x, y) + i \sin N_{2f}(x, y)\\ &= e^{i\,N_{2f}/(x,\,y)}\end{aligned}$$

From which we get

$$\begin{aligned}\cos N_{2f}(x, y) &= \cos Nx\,chy - \sin Nx\,shy\\ &= [\cos x\,chx - \sin x\,shx]\,chy - [\sin x\,chx + \cos x\,shx]\,shy\\ &= \cos N(x, y)\,chx - \sin N(x, y)\,shx\\ &= \cos N_{2f}[(x, y), x]\\ \sin N_{2f}(x, y) &= \sin Nx\,chy + \cos Nx\,shy\\ &= [\sin x\,chx + \cos x\,shx]\,chy + [\cos x\,chx - \sin x\,shx]\,shy\\ &= \sin N(x, y)\,chx + \sin N(x, y)\,shx\\ &= \sin N_{2f}[(x, y), x]\end{aligned} \qquad (1.26c)$$

1.5.5 Particular Case

When $y = x$, we have both for **F_1 type** and **F_2 type**

$N_{2f}/(x, x) = N_{2f}/(x) = N_{2f}/x$:

$$\begin{aligned} e^{iN/xfix} &= (\cos Nx + i\sin Nx)(chx + i\,shx) \\ &= [\cos Nx\, chx - \sin Nx\, shx] \\ &\quad + i\,[\sin Nx\, chx + \cos Nx\, shx] \\ &= \cos N_{2f}(x, x) + i\sin N_{2f}(x, x) \\ &= e^{i\,N_{2f}/(x,\,x)} = e^{i\,N_{2f}/x} \end{aligned}$$

From which we get

$$\begin{aligned} \cos N_{2f}(x, x) &= \cos Nx\, chx - \sin Nx\, shx \\ &= \cos x - \sin x\, sh\, 2x \\ \sin N_{2f}(x, x) &= \sin Nx\, chx + \cos Nx\, shx \\ &= \sin x + \cos x\, sh\, 2x \end{aligned} \tag{1.26d}$$

[Refer to **Table D.5** for the numerical values of $N_{2f}(x, x)$ and $N_{2f}(x, -x)$, in **Appendix D**]

1.5.6 Double Nbic Function with Suffixes A and B

Illustrated here, are the two groups of individual Double Nbic functions, $N_2(x, x)$ [$= N_2 x$], and $N_2(x, -x)$], each belonging to the category, namely, $N_{2e}(x, y)$, and, $N_{2f}(x, y)$.

Note: Both E_1 type and E_2 type, as well as F_1 type and F_2 type, belonging to E type and F type respectively, are further described with suffixes A and B to indicate for (x, y) or $(x, -y)$, (x, x) or $(x, -x)$, respectively, as the case may be. This way, we have the variations for E types as: E_{1A}, E_{2A}, E_{1B}, E_{2B}, and for F types as: F_{1A}, F_{2A}, F_{1B}, F_{2B}.

$N_2(x, x)$ [$= N_2 x$], and $N_2(x, -x)$, when compiled (belonging to Category 1 and Category 2), are:

as per Category 1, (1.26a), we have

E_{1A} type and E_{2A} type: $N_{2e}(x, x)$:

$$\begin{aligned} \cos N_{2e}(x, x) &= \cos N_{2e}(x) = \cos Nx \cos x - \sin Nx \sin x \\ &= \cos N(2x, x) \\ \sin N_{2e}(x, x) &= \sin N_{2e}(x) = \sin Nx \cos x + \cos Nx \sin x \\ &= \sin N(2x, x) \end{aligned} \tag{1.27a}$$

Similarly

E_{1B} type and E_{2B} type: $N_{2e}(x, x)$:

$$\begin{aligned} \cos N_{2e}(x, -x) &= \cos Nx \cos x + \sin Nx \sin x \\ \sin N_{2e}(x, -x) &= \sin Nx \cos x - \cos Nx \sin x \end{aligned} \tag{1.27b}$$

as per Category 2, (1.26b), we have

F_{1A} type and F_{2A} type: $N_{2f}(x, x)$:

$$\begin{aligned} \cos N_{2f}(x, x) &= \cos N_{2f}(x) = \cos Nx\, chx - \sin Nx\, shx \\ &= \cos x - \sin x\, sh\, 2x \\ \sin N_{2f}(x, x) &= \sin N_{2f}(x) = \sin Nx\, chx + \cos Nx\, shx \\ &= \sin x + \cos x\, sh\, 2x \end{aligned} \tag{1.27c}$$

Similarly

F_{1B} type and F_{2B} type: $N_{2f}(x, x)$:

$$\begin{aligned} \cos N_{2f}(x, -x) &= \cos Nx\, chx + \sin Nx\, shx \\ \sin N_{2f}(x, -x) &= \sin Nx\, chx - \cos Nx\, shx \end{aligned} \tag{1.27d}$$

Collecting Double Nbic Function Formulas Together

For $N_{2e}(x, y)$ [E_{1A} type from $N_{2e}/(x, y)$]:

$$\cos N_{2e}(x, y) = \cos N(x, y) \cos y - \sin N(x, y) \sin y$$
$$\sin N_{2e}(x, y) = \sin N(x, y) \cos y + \cos N(x, y) \sin y$$

For $N_{2e}(x, y)$ [E_{2A} type from $N_{2e}/(x, y)$]:

$$\cos N_{2e}(x, y) = \cos N(x, x) \cos y - \sin N(x, x) \sin y$$
$$\sin N_{2e}(x, y) = \sin N(x, x) \cos y + \cos N(x, x) \sin y$$

For $N_{2e}(x, x)$ [E_{1A} type and E_{2A} type from $N_{2e}/(x, x)$]:

$$\cos N_{2e}(x, x) = \cos N(x, x) \cos x - \sin N(x, x) \sin x$$
$$\sin N_{2e}(x, x) = \sin N(x, x) \cos x + \cos N(x, x) \sin x$$

For $N_{2e}(x, x)$ [E_{1B} type and E_{2B} type from $N_{2e}/(x, x)$]:

$$\cos N_{2e}(x, -x) = \cos Nx \cos x + \sin Nx \sin x$$
$$\sin N_{2e}(x, -x) = \sin Nx \cos x - \cos Nx \sin x$$

For $N_{2f}(x, y)$ [F_{1A} type from $N_{2f}/(x, y)$]:

$$\cos N_{2f}(x, y) = \cos N(x, y)\, chy - \sin N(x, y)\, shy$$
$$\sin N_{2f}(x, y) = \sin N(x, y)\, chy + \cos N(x, y)\, shy$$

For $N_{2f}(x, y)$ [F_{2A} type from $N_{2f}/(x, y)$]:

$$\cos N_{2f}(x, y) = \cos Nx\, chy - \sin Nx\, shy$$
$$\sin N_{2f}(x, y) = \sin Nx\, chy + \cos Nx\, shy$$

For $N_{2f}(x, x)$ [F_{1A} type and F_{2A} type from $N_{2f}/(x, x)$]:

$$\cos N_{2f}(x, x) = \cos Nx\, chx - \sin Nx\, shx$$
$$\sin N_{2f}(x, x) = \sin Nx\, chx + \cos Nx\, shx$$

For $N_{2f}(x, x)$ [F_{1B} type and F_{2B} type from $N_{2f}/(x, x)$]:

$$\cos N_{2f}(x, -x) = \cos Nx\, chx + \sin Nx\, shx$$
$$\sin N_{2f}(x, -x) = \sin Nx\, chx - \cos Nx\, shx \tag{1.27e}$$

[Suffixes 1 and 2 indicate that the corresponding function is generated using $N/(x, y)$ or $N/(x, x)$, respectively, and suffixes A and B here refer to the Nbic functions $N_2(x, x)$ or $N_2(x, -x)$, respectively.]

1.6 GENERATION OF TRIPLE Nbic FUNCTION, $N_3(x, y)$

1.6.1 As Generated from Complex Triple Nbic Function, $N_3/(x, y)$

The generation of Triple Nbic functions, $N_3/(x, y)$, as the name implies, can be simply done by multiplying any complex form of Double Nbic function, $N_2/(x, y)$, with either, $e^{\pm iy}$, or $f^{\pm iy}$ and are grouped into two categories, *viz.*,

Category 1 : (E type)

There are two E types *viz.*, **E_1 type** and **E_2 type,** i.e.,

$$\textbf{E}_1 \textbf{ type: } N_{3e}/(x, y) = N_{2e}/(x, y).\ e^{\pm iy} = e^{\pm iN_{2e}/(x,y)}.e^{\pm iy}$$
$$\textbf{E}_2 \textbf{ type: } N_{3e}/(x, y) = N_{2e}/(x, x).\ e^{\pm iy} = e^{\pm iN_{2e}/(x,x)}.e^{\pm iy}$$

Category 2 : (F type)

There are two F types *viz.*, **F_1 type** and **F_2 type,** i.e.,

$$\textbf{F}_1 \textbf{ type: } N_{3f}/(x, y) = N_{2f}/(x, y).\ f^{\pm iy} = e^{\pm iN_{2f}/(x,y)}.f^{\pm iy}$$
$$\textbf{F}_2 \textbf{ type: } N_{3f}/(x, y) = N_{2f}/(x, x).\ f^{\pm iy} = e^{\pm iN_{2f}/(x,x)}.f^{\pm iy} \tag{1.28a}$$

[**Note:** suffix, ***e*** or ***f***, have been extensively used to identify functions related to, E type or, F type, as the case may be.]

Again, the complex Nbic functions $N_2/(x, y)$ in (1.28a), are of four varieties, namely, $N_{2e}/(x, y)$ and $N_{2f}/(x, y)$. So the number of Triple Nbic functions, $N_3/(x, y)$, generated this way, will be **eight**. The combination of the products which produce the complex Triple Nbic functions $N_3/(x, y)$, are:

$$[N_{2e}/(x, y).e^{\pm iy}],\ [N_{2f}/(x, y).e^{\pm iy}],\ [N_{2e}/(x, y).f^{\pm iy}],\text{ and }[N_{2f}/(x, y).f^{\pm iy}],\text{ and}$$

$$[N_{2e}/(x, x).e^{\pm iy}],\ [N_{2f}/(x, x).e^{\pm iy}],\ [N_{2e}/(x, x).f^{\pm iy}],\text{ and }[N_{2f}/(x, x).f^{\pm iy}].$$

How the complex Triple Nbic functions $N_3/(x, y)$ [both $N_{3e}/(x, y)$, and $N_{3f}/(x, y)$], can be obtained, is as shown. Thus

1.6.2 Category 1 : (E type)

E_1 type : $N_{3e}/(x, y)$:

$$\begin{aligned} e^{i\,N_{2e}/(x,\,y)}.e^{iy} &= [\cos N_{2e}(x, y) + i\sin N_{2e}(x, y)](\cos y + i\sin y) \\ &= [\cos N_{2e}(x, y)\cos y - \sin N_{2e}(x, y)\sin y] \\ &\quad + i[\sin N_{2e}(x, y)\cos y + \cos N_{2e}(x, y)\sin y] \\ &= \cos N_{3e}(x, y) + i\sin N_{3e}(x, y) \\ &= e^{iN_{3e}/(x,\,y)} \end{aligned}$$

From which separating the real and imaginary parts, we get

$$\begin{aligned} \cos N_{3e}(x, y) &= \cos N_{2e}(x, y)\cos y - \sin N_{2e}(x, y)\sin y \\ &= [\cos(x + y)\,chy - \sin(x + y)\,shy]\cos y \\ &\quad - [\sin(x + y)\,chy + \cos(x + y)\,shy]\sin y \\ &= \cos N(y, y)\cos(x + y) - \sin N(y, y)\sin(x + y) \\ &= \cos N_{2e}[(y, y), (x + y)] \\ \sin N_{2e}(x, y) &= \sin N_{2e}(x, y)\cos y + \cos N_{2e}(x, y)\sin y \\ &= [\sin(x + y)\,chy + \cos(x + y)\,shy]\cos y \\ &\quad + [\cos(x + y)\,chy - \sin(x + y)\,shy]\sin y \\ &= \cos N(y, y)\sin(x + y) + \cos N(y, y)\sin(x + y) \\ &= \sin N_{2e}[(y, y), (x + y)] \end{aligned} \qquad (1.28b)$$

E_2 type : $N_{3e}/(x, y)$:

$$\begin{aligned} e^{i\,N_{2e}/(x,\,x)}.e^{iy} &= [\cos N_{2e}(x, x) + i\sin N_{2e}(x, x)](\cos y + i\sin y) \\ &= [\cos N_{2e}(x, x)\cos y - \sin N_{2e}(x, x)\sin y] \\ &\quad + i[\sin N_{2e}(x, x)\cos y + \cos N_{2e}(x, x)\sin y] \\ &= \cos N_{3e}/(x, y) + i\sin N_{3e}/(x, y) \\ &= e^{iN_{3e}/(x,\,y)} \end{aligned}$$

From which separating the real and imaginary parts, we get

$$\begin{aligned} \cos N_{3e}(x, y) &= \cos N_{2e}(x, x)\cos y - \sin N_{2e}(x, x)\sin y \\ &= [\cos N(x, x)\cos y - \sin N(x, x)\sin y]\cos y \\ &\quad - [\sin N(x, x)\cos y + \cos N(x, x)\sin y]\sin y \\ &= \cos N(x, x)\cos 2y - \sin N(x, x)\sin 2y \\ &= \cos N_{2e}[(x, x), 2y] \end{aligned}$$

$$\begin{aligned} \sin N_{3e}(x, y) &= \sin N_{2e}(x, x)\cos y + \cos N_{2e}(x, x)\sin y \\ &= [\sin N(x, x)\cos y + \cos N(x, x)\sin y]\cos y \\ &\quad + [\cos N(x, x)\cos y - \sin N(x, x)\sin y]\sin y \end{aligned}$$

$$= \sin N(y, x) \cos(x + y) + \cos N(y, x) \sin(x + y)$$
$$= \sin N_{2e}[(y, x), (x + y)] \qquad (1.28c)$$

1.6.3 Particular Case

When $y = x$, (for single variable, x), we have both for $\mathbf{E_1}$ **type** and $\mathbf{E_2}$ **type**

$\mathbf{N_{3e} / (x, x) = N_{3e} / (x) = N_{3e} / x}$:

$$\begin{aligned} e^{iN_{2e}/x}\, e^{ix} &= [\cos N_{2e}(x, x) + i \sin N_{2e}(x, x)] (\cos x + i \sin x) \\ &= [\cos N_{2e}(x, x) \cos x - \sin N_{2e}(x, x) \sin x] \\ &\quad + i [\sin N_{2e}(x, x) \cos x + \cos N_{2e}(x, x) \sin x] \\ &= \cos N_{3e}(x, x) + i \sin N_{3e}(x, x) \\ &= e^{iN_{3e}/(x, x)} = e^{iN_{3e}/x} \end{aligned}$$

From which separating the real and imaginary parts, we get

$$\begin{aligned} \cos N_{3e}(x, x) &= \cos N_{2e}(x, x) \cos x - \sin N_{2e}(x, x) \sin x \\ \sin N_{3e}(x, x) &= \sin N_{2e}(x, x) \cos x + \cos N_{2e}(x, x) \sin x \end{aligned} \qquad (1.28d)$$

[Refer to **Table D.6** for the numerical values of $N_{3e}(x, x)$, in **Appendix D**]

1.6.4 Category 2 : (F type)

$\mathbf{F_1}$ **type :** $\mathbf{N_{3f} / (x, y)}$:

$$\begin{aligned} e^{iN_{2f}/(x, y)}.f^{iy} &= [\cos N_{2f} / (x, y) + i \sin N_{2f} / (x, y)] (chy + i\, shy) \\ &= [\cos N_{2f}(x, y)\, chy - \sin N_{2f}(x, y)\, shy] \\ &\quad + i [\sin N_{2f}(x, y)\, chy + \cos N_{2f}(x, y)\, shy] \\ &= \cos N_{3f}(x, y) + i \sin N_{3f}(x, y) \\ &= e^{iN_{3f}/(x, y)} \end{aligned}$$

From which separating the real and imaginary parts, we get

$$\begin{aligned} \cos N_{3f}(x, y) &= \cos N_{2f}(x, y)\, chy - \sin N_{2f}(x, y)\, shy \\ &= [\cos N(x, y)\, chy - \sin N(x, y)\, shy]\, chy \\ &\quad - [\sin N(x, y)\, chy + \cos N(x, y)\, shy]\, shy \\ &= [\cos x - \sin x . sh\, 2y]\, chy - [\sin x + \cos x . sh\, 2y] \\ &= \cos N(x, y) - \sin N(x, y)\, sh\, 2y \\ \sin N_{3f}(x, y) &= \sin N_{2f}(x, y)\, chy + \cos N_{2f}(x, y)\, shy \\ &= [\sin N(x, y)\, chy + \cos N(x, y)\, shy]\, chy \\ &\quad + [\cos N(x, y)\, chy - \sin N(x, y)\, shy]\, shy \\ &= [\sin x + \cos x . sh\, 2y]\, chy + [\cos x - \sin x . sh\, 2y] \\ &= \sin N(x, y) + \cos N(x, y)\, sh\, 2y \end{aligned} \qquad (1.28e)$$

$\mathbf{F_2}$ **type :** $\mathbf{N_{3f} / (x, y)}$:

$$\begin{aligned} e^{iN_{2f}/x}.f^{iy} &= [\cos N_{2f} / (x, x) + i \sin N_{2f} / (x, x)] (chy + i\, shy) \\ &= [\cos N_{2f}(x, x)\, chy - \sin N_{2f}(x, x)\, shy] \\ &\quad + i [\sin N_{2f}(x, x)\, chy + \cos N_{2f}(x, x)\, shy] \\ &= \cos N_{3f}(x, y) + i \sin N_{3f}(x, y) \\ &= e^{iN_{3f}/(x, y)} \end{aligned}$$

From which separating the real and imaginary parts, we get

$$\begin{aligned} \cos N_{3f}(x, y) &= \cos N_{2f}(x, x)\, chy - \sin N_{2f}(x, x)\, shy \\ &= [\cos N(x, x)\, chy - \sin N(x, x)\, shy]\, chy \\ &\quad - [\sin N(x, x)\, chy + \cos N(x, x)\, shy]\, shy \end{aligned}$$

$$= [\cos x - \sin x.sh\ 2x]\ chy - [\sin x + \cos x.sh\ 2x]\ shy$$
$$= \cos N(x, x) - \sin N(x, x)\ sh\ 2y$$
$$\sin N_{3f}(x, y) = \sin N_{2f}(x, x)\ chy + \cos N_{2f}(x, x)\ shy$$
$$= [\sin N(x, x)\ chy + \cos N(x, x)\ shy]\ chy + [\cos N(x, x)\ chy - \sin N(x, x)\ shy]\ shy$$
$$= [\sin x + \cos x.sh\ 2x]\ chy - [\cos x - \sin x.sh\ 2x]\ shy$$
$$= \sin N(x, x) + \cos N(x, x)\ sh\ 2y \quad (1.28f)$$

1.6.5 Particular Case

When $y = x$, (for single variable, x), we have both for $\mathbf{F_1}$ **type** and $\mathbf{F_2}$ **type**

$\mathbf{N_{3f}/(x, x) = N_{3f}/(x) = N_{3f}/x}$:

$$e^{iN_{2f}/x}.\ f^{ix} = [\cos N_{2f}/(x, x) + i \sin N_{2f}/(x, x)]\ (chx + i\ shx)$$
$$= [\cos N_{2f}(x, x)\ chx - \sin N_{2f}(x, x)\ shx] + i\ [\sin N_{2f}(x, x)\ chx + \cos N_{2f}(x, x)\ shx]$$
$$= \cos N_{3f}(x, x) + i \sin N_{3f}(x, x)$$
$$= e^{iN_{3f}/(x, x)} = e^{iN_{2f}/x}$$

From which separating the real and imaginary parts, we get

$$\cos N_{3f}(x, x) = \cos N_{2f}(x, x)\ chx - \sin N_{2f}(x, x)\ shx$$
$$= \cos N(x, x) - \sin N(x, x)\ sh\ 2x$$
$$\sin N_{3f}(x, x) = \sin N_{2f}(x, x)\ chx + \cos N_{2f}(x, x)\ shx$$
$$= \sin N(x, x) + \cos N(x, x)\ sh\ 2x \quad (1.28g)$$

[Refer to **Table D.7** for the numerical values of $N_{3f}(x, x)$, in **Appendix D**]

1.6.6 Category M (Mixed Category)

Interchangeability of Nbic functions of third order is noticed in some forms, which can be identified with the functions written with two suffixes $\boldsymbol{e}$ as well as $\boldsymbol{f}$, *viz.*, $N_{2f,e}(x, x)$ and $N_{2e,f}(x, x)$, which has not been taken care of in either of the two Categories, *viz.*, Category 1 and Category 2. They are grouped under a separate category to be termed as:

Category M (**Mixed category** involving EF type and FE type).

FE type : $\mathbf{N_{2f,e}(x, x)}$:

$$\cos N_{2f,e}(x, x) = \cos N_{2f}(x, x) \cos x - \sin N_{2f}(x, x) \sin x$$
$$\sin N_{2f,e}(x, x) = \sin N_{2f}(x, x) \cos x - \cos N_{2f}(x, x) \sin x$$
$$\cos N_{2f,e}(x, -x) = \cos N(x, x) \cos N(x, -x) + \sin N(x, x) \sin N(x, -x)$$
$$\sin N_{2f,e}(x, -x) = \sin N(x, x) \cos N(x, -x) - \cos N(x, x) \sin N(x, -x) \quad (1.28h)$$

EF type : $\mathbf{N_{2e,f}(x, x)}$:

$$\cos N_{2e,f}(x, x) = \cos N_{2e}(x, x)\ chx - \sin N_{2e}(x, x)\ shx$$
$$\sin N_{2e,f}(x, x) = \sin N_{2e}(x, x)\ chx - \cos N_{2e}(x, x)\ shx$$
$$\cos N_{2e,f}(x, -x) = \cos N^2(x, x) + \sin N^2(x, x)$$
$$\sin N_{2e,f}(x, -x) = 2 \sin N(x, x) \cos N(x, x) \quad (1.28i)$$

It can be easily shown that $N_{2f,e}(x, x) = N_{2e,f}(x, x)$. That is:

$$\cos N_{2f,e}(x, x) = \cos N_{2e,f}(x, x) = \cos 2x - \sin 2x\ sh\ 2x$$
$$\sin N_{2f,e}(x, x) = \sin N_{2e,f}(x, x) = \sin 2x + \cos 2x\ sh\ 2x \quad (1.28j)$$

But, $N_{2f,e}(x, x) \neq N_{2e,f}(x, x)$.

$$\cos N_{2f,e}(x, -x) = 1$$
$$\cos N_{2e,f}(x, -x) = ch\ 2x$$

So, $\cos N_{2f,e}(x, -x) \neq \cos N_{2e,f}(x, -x)$

$$\sin N_{2f,e}(x, -x) = sh\ 2x$$
$$\sin N_{2e,f}(x, -x) = \sin 2x\ ch\ 2x$$

So, $\sin N_{2f,e}(x, -x) \neq \sin N_{2e,f}(x, -x)$ (1.28k)

Collecting Triple Nbic Function Formulas Together

For $N_{3e}(x, y)$ [E_{1A} type from $N_{3e} / (x, y)$]:

$$\cos N_{3e}(x, y) = \cos N_{2e}(x, y)\ chy - \sin N_{2e}(x, y)\ shy$$
$$\sin N_{2f}(x, y) = \sin N_{2e}(x, y)\ chy + \cos N_{2e}(x, y)\ shy$$

For $N_{3e}(x, y)$ [E_{2A} type from $N_{3e} / (x, y)$]:

$$\cos N_{3e}(x, y) = \cos N_{2e}(x, x) \cos y - \sin N_{2e}(x, x) \sin y$$
$$\sin N_{3e}(x, y) = \sin N_{2e}(x, x) \cos y + \cos N_{2e}(x, x) \sin y$$

For $N_{3e}(x, x)$ [E_{1A} type and E_{2A} type from $N_{3e} / (x, x)$]:

$$\cos N_{3e}(x, x) = \cos N_{2e}(x, x) \cos x - \sin N_{2e}(x, x) \sin x$$
$$\sin N_{3e}(x, x) = \sin N_{2e}(x, x) \cos x + \cos N_{2e}(x, x) \sin x$$

For $N_{3f}(x, y)$ [F_{1A} type $N_{3f} / (x, y)$]:

$$\cos N_{3f}(x, y) = \cos N_{2f}(x, y)\ chy - \sin N_{2f}(x, y)\ shy$$
$$\sin N_{3f}(x, y) = \sin N_{2f}(x, y)\ chy + \cos N_{2f}(x, y)\ shy$$

For $N_{3f}(x, y)$ [F_{2A} type from $N_{3f} / (x, y)$]:

$$\cos N_{3f}(x, y) = \cos N_{2f}(x, x)\ chy - \sin N_{2f}\ x, x)\ shy$$
$$\sin N_{3f}(x, y) = \sin N_{2f}(x, x)\ chy + \cos N_{2f}(x, x)\ shy$$

For $N_{3f}(x, x)$ [F_{1A} type and F_{2A} type from $N_{3f} / (x, x)$]:

$$\cos N_{3f}(x, x) = \cos N_{2f}(x, x)\ chx - \sin N_{2f}(x, x)\ shx$$
$$\sin N_{3f}(x, x) = \sin N_{2f}(x, x)\ chx + \cos N_{2f}(x, x)\ shx \quad (1.28l)$$

[Suffixes 1 and 2 indicate that the corresponding function is generated using $N_2 / (x, y)$ or $N_2 / (x, x)$, respectively, and suffixes A and B refer to the Nbic functions $N_3(x, -y)$ or $N_3(x, -x)$, respectively.]

In the structure of the Triple Nbic functions $\mathbf{N_3 (x, y)}$, the form of the Double Nbic functions *viz.*, $\mathbf{N_2 (x, y)}$ used, can be written with:

E type $N_{2e}(x, y)$, that is

as **E_1 type** with, $\cos N_{2e}(x, y)$, and $\sin N_{2e}(x, y)$, and
as **E_2 type** with, $\cos N_{2e}(x, x)$, and $\sin N_{2e}(x, x)$, and

F type $N_{2f}(x, y)$, that is

as **F_1 type** with, $\cos N_{2f}(x, y)$, and $\sin N_{2f}(x, y)$, and
as **F_2 type** with, $\cos N_{2f}(x, x)$, and $\sin N_{2f}(x, x)$

In the present discussion that follows, we prefer to use the function as a product of pure function of x, and pure function of y, that is, with the form as, $F(x, y) = f(x, x).g(y, y)$. The form, *viz.*, $F(x, y) = f(x, y).g(y)$, normally are not dealt with. This later form, if at all is intended to be used, will be specially mentioned there, to avoid any ambiguity. Further note that:

in $F(x, y) = f(x, y).g(y, y)$, the first function is $f(x, y)$.

So the function belongs to E_1 or F_1 type, whereas

in $F(x, y) = f(x, x).g(y, y)$, the first function is $f(x, x)$.

So the function belongs to E_2 or F_2 type.

1.6.7 Triple Nbic Function with Suffixes A and B

Now for example, using E_2 or F_2 forms, the third order Nbic functions appear as:

Category 1 (E_{2A} type, E_{2B} type) : $N_{3e}(x, y)$:

E_{2A} type:

$$\cos N_{3e}(x, y) = \cos N_{2f}(x) \cos y - \sin N_{2f}(x) \sin y$$
$$\sin N_{3e}(x, y) = \sin N_{2f}(x) \cos y + \cos N_{2f}(x) \sin y$$

E_{2B} type:

$$\cos N_{3e}(x, -y) = \cos N_{2f}(x) \cos y + \sin N_{2f}(x) \sin y$$
$$\sin N_{3e}(x, -y) = \sin N_{2f}(x)\, ch y - \cos N_{2f}(x) \sin y \quad (1.29a)$$

[**Note :** At a first glance, it appears that, use of $N_{2e}(x)$ in (1.29a) is logical. But, on the otherhand, in view of (1.27c), (1.29a) becomes:

$$\cos N_{3e}(x, y) = \cos (x + y) - \sin (x + y)\, sh\, 2x$$
$$\sin N_{3e}(x, y) = \sin (x + y) + \cos (x + y)\, sh\, 2x \quad (1.29a')$$

which for $y = x$, reduces to (4.1d).

Again, (1.29a) can also be defined in FE category. See later in (4.2a). Hence, (1.29a) with $N_{2f}(x)$ retained here.]

Category 2 (F_{2A} type, F_{2B} type) : $N_{3f}(x, y)$:

F_{2A} type:

$$\cos N_{3f}(x, y) = \cos N_{2f}(x)\, ch y - \sin N_{2f}(x)\, sh y$$
$$\sin N_{3f}(x, y) = \sin N_{2f}(x)\, ch y + \cos N_{2f}(x)\, sh y$$

F_{2B} type:

$$\cos N_{3f}(x, -y) = \cos N_{2f}(x)\, ch y + \sin N_{2f}(x)\, sh y$$
$$\sin N_{3f}(x, -y) = \sin N_{2f}(x)\, ch y - \cos N_{2f}(x)\, sh y \quad (1.29b)$$

where, $N_{2f}(x, x)$ is given by (1.27e).

1.6.8 Particular Case

Now taking help of (1.27e) for $N_{2f}(x, x)$ [which are derived from the complex form of $N_{2f}\ /\ (x, x)$], *viz.*,

$$\cos N_{2f}(x, x) = \cos Nx\, chx - \sin Nx\, shx$$
$$\sin N_{2f}(x, x) = \sin Nx\, chx + \cos Nx\, shx$$

let us write directly the third order Nbic function $N_3(x, x)$, belonging to Category 1, and Category 2. The generated corresponding Triple Nbic functions $N_{3e}(x, x)$ [$=N_{3e}(x)$], and $N_{3f}(x, x)$ [$=N_{3f}(x)$], and $N_{3e}(x, -x)$, and $N_{3f}(x, -x)$, are:

Category 1 (E_{2A} type, E_{2B} type) : $N_{3e}(x, x)$:

E_{2A} type:

$$\cos N_{3e}(x, x) = \cos N_{2f}(x, x) \cos x - \sin N_{2f}(x, x) \sin x$$
$$\sin N_{3e}(x, x) = \sin N_{2f}(x, x) \cos x + \cos N_{2f}(x, x) \sin x$$

E_{2B} type:

$$\begin{aligned} \cos N_{3e}(x, -x) &= \cos N_{2f}(x, x) \cos x + \sin N_{2f}(x, x) \sin x \\ \sin N_{3e}(x, -x) &= \sin N_{2f}(x, x) \cos x - \cos N_{2f}(x, x) \sin x \end{aligned} \tag{1.30a}$$

Category 2 (F_{2A} type, F_{2B} type) : $N_{3f}(x, x)$:

F_{2A} type:

$$\begin{aligned} \cos N_{3f}(x, x) &= \cos N_{2f}(x, x)\, chx - \sin N_{2f}(x, x)\, shx \\ \sin N_{3f}(x, x) &= \sin N_{2f}(x, x)\, chx + \cos N_{2f}(x, x)\, shx \end{aligned}$$

F_{2B} type:

$$\begin{aligned} \cos N_{3f}(x, -x) &= \cos N_{2f}(x, x)\, chx + \sin N_{2f}(x, x)\, shx \\ \sin N_{3f}(x, -x) &= \sin N_{2f}(x, x)\, chx - \cos N_{2f}(x, x)\, shx \end{aligned} \tag{1.30b}$$

and so on.

This way, other higher order Nbic functions can be generated on multiplying any function of a certain order, say, $N_n/(x, y)$, by either, $e^{\pm iy}$ or, $f^{\pm iy}$, to get the next higher order function, $N_{n+1}/(x, y)$. The corresponding Nbic functions can be picked up from these $N_{n+1}/(x, y)$, on separating the real and imaginary parts.

For example, $n = 3$, in $N_n/(x, y)$ means: 3rd order Nbic function, $N_3/(x, y)$, and, $N_{n+1}/(x, y)$ means, 4rd order Nbic function, $N_4/(x, y)$.

1.7 DEFINITION AND DEVELOPMENT OF Nbic FUNCTION

Define **sinN(*x*, *x*)** and **cosN(*x*, *x*),** similar to, sin(A+B) and cos(A+B) form, as:

$$\begin{aligned} S1 &= \sin N(x, x) = \sin N(x) = \sin N\, x = \sin x\, chx + \cos x\, shx \\ C1 &= \cos N(x, x) = \cos N(x) = \cos N x = \cos x\, chx - \sin x\, shx \end{aligned} \tag{1.31}$$

Note: The basic units or components of the single Nbic functions as a product of the functions with, (sin or cos) and with, (*sh* or *ch*), are not unknown functions. These are the rearranged form of the previously known functions of trigonometric-hyperbolic types, often used for solving **Beams on Elastic Foundation** [of single variable (*x*)], and **Cylindrical Shell Equations** [of two variables (*x*, *y*)].

Basically Nbic functions, particularly the single Nbic functions, identified by the additional symbol N, such as, sinN(*x*), cosN(*x*, *y*) etc., are composed of two parts, each part again, is a product of circular function, (like, sin*x*, or cos*x*), and hyperbolic function, (like, sin*hx*, or cos*hx*, which are abbreviated as *shx*, and *chx*, respectively), as described by the functions: A*, B*, C*, D* or A, B, C, D. Until and otherwise stated, Nbic function will mean Single Nbic function [N(*x*, *x*) or N(*x*, *y*)]. The other Nbic functions proposed and discussed are, Double Nbic function [like, $N_2(x, x)$ etc], and Triple Nbic function [like, $N_3(x, y)$ etc.], along with their corresponding complex forms [like, $N_2/(x, x)$, $N_3/(x, y)$].

This way, Nbic functions of higher order can be formed and designed as: $N_n(x, x)$, for $n = 3$, 4, 5…, and the process, as it appears, is unlimited.

1.7.1 Single Nbic Function with Variable (*x*, *y*) : N(*x*, *y*)

In introductory proposal, the generation of various categories of Nbic functions and their forms, have been given. Refer to $\mathbf{N_A}(x, y)$ and $\mathbf{N_B}(x, y)$ of (1.21e, 1.21g, 1.24b), and once again, let us define functions A*, B*, C*, D*, as functions of (x, y) as

$$\begin{aligned} A^* &= Axy = A\ (x, y) = \sin x\ chy \\ B^* &= Bxy = B\ (x, y) = \sin x\ shy \\ C^* &= Cxy = C\ (x, y) = \cos x\ shy \\ D^* &= Dxy = D\ (x, y) = \cos x\ chy \end{aligned} \quad (1.32)$$

Utilising these, the basic Nbic functions, S_1^*, S_2^*, C_1^* and C_2^*, and other derived forms are defined as:

$$\begin{aligned} S_1^* &= \sin N(x, y) = A^* + C^* = \sin x\ chy + \cos x\ shy \\ S_2^* &= \sin N(x,-y) = A^* - C^* = \sin x\ chy - \cos x\ shy \\ C_1^* &= \cos N(x, y) = D^* - B^* = \cos x\ chy - \sin x\ shy \\ C2^* &= \cos N(x,-y) = D^* + B^* = \cos x\ chy + \sin x\ shy \\ T_1^* &= S_1^* / C_1^* = [\sin N(x, y)] / [\cos N(x, y)] \\ &= \tan N\ (x, y) = [\tan x + thy] / [1 - \tan x.thy] \end{aligned} \quad (1.33a)$$

[Derivation is shown in (1.61). It is similar to, tan(A + B) formula of plane trigonometry]

$$\begin{aligned} T_2^* &= S_2^* / C_2^* = [\sin N(x, -y)] / [\cos N(x, -y)] \\ CC^* &= \text{cosec}N\ (x, y) = \csc N(x, y) = [\sec x.\text{sech}y] / [\tan x + thy] \\ SC^* &= \sec N(x, y) = [\sec x.\text{sech}y] / [1 - \tan x.thy] \\ CT^* &= \cot N(x, y) \end{aligned} \quad (1.33b)$$

1.7.2 Single Nbic Function with Variable of *x* Only : N(*x*, *x*)

When both the variables x and y are same, (that is, $y = x$), Nbic Functions A*, B*, C* and D*, reduce to Nbic functions A, B, C, and D of variables (x, x), respectively. Refer to $\mathbf{N_A(x, x)}$ and $\mathbf{N_B(x, x)}$ of (1.23a, 1.23b, 1.24b), as follows:

$$\begin{aligned} A &= Ax = A(x) = \sin x\ chx \\ B &= Bx = B(x) = \sin x\ shx \\ C &= Cx = C(x) = \cos x\ shx \\ D &= Dx = D(x) = \cos x\ chx \end{aligned} \quad (1.34)$$

$\mathbf{N_A(x, x)\ [=N_A(x)]}$:

$$\begin{aligned} S_1 &= \sin N(x, x) = A + C = \sin x\ chx + \cos x\ shx \\ C_1 &= \cos N(x, x) = D - B = \cos x\ chx - \sin x\ shx \\ T_1 &= S_1 / C_1 = [\sin N(x, x)] / [\cos N(x, x)] \\ &= \tan N(x, x) = [\tan x + thx] / [1 - \tan x.thx] \end{aligned}$$

[Derivation is shown in (1.55). It is similar to tan (A + B) formula of trigonometry]

$$\begin{aligned} CC &= \csc N(x, x) = [\sec x\ .\ \text{sech}x] / [\tan x + thx] \\ SC &= \sec N(x, x) = [\sec x\ .\ \text{sech}x] / [1 - \tan x\ .\ thx] \\ CT &= \cot N(x, x) \end{aligned} \quad (1.35a)$$

$\mathbf{N_B(x, x)\ [= N_B(x)]}$:

$$\begin{aligned} S_2 &= \sin N(x, -x) = A - C = \sin x\ chx - \cos x\ shx \\ C_2 &= \cos N(x, -x) = D + B = \cos x\ chx + \sin x\ shx \\ T_2 &= S_2 / C_2 = [\sin N(x, -x)] / [\cos N(x, -x)] \end{aligned} \quad (1.35b)$$

∇^4 of $(S_1, S_2, C_1, C_2) = 0$,

where, $\nabla^4 = \nabla^2\nabla^2$ = biharmonic operator.

[**Note:** When the function is N(*x*, *x*) or, N(– *x*, – *x*), it is enough to use only N(*x*) or, N(– *x*). But for, N(*x*, – *x*) or, N(–*x*, *x*), this abbreviation is not permissible, because the very structure of the expression becomes altogether different.]

In **Figure 1.4**, the variation of single Nbic function N(*x*) versus *x*(deg) is shown.

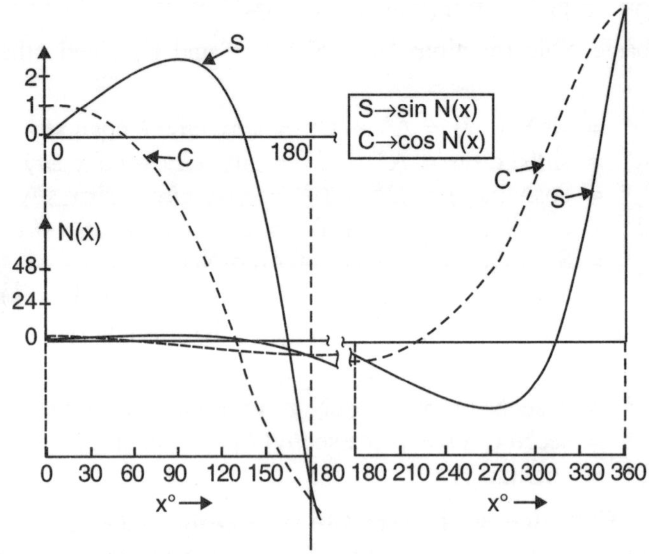

Fig. 1.4: N(x) vs. x

1.7.3 Graphical Determination of Single Nbic Functions

In **Figure 1.5(a)**, graphical determination of the single Nbic functions, $S_1 = \sin N(x, x)$, $C_1 = \cos N(x, x)$, $S_2 = \sin N(x, -x)$ and, $C_2 = \cos N(x, -x)$, is described.

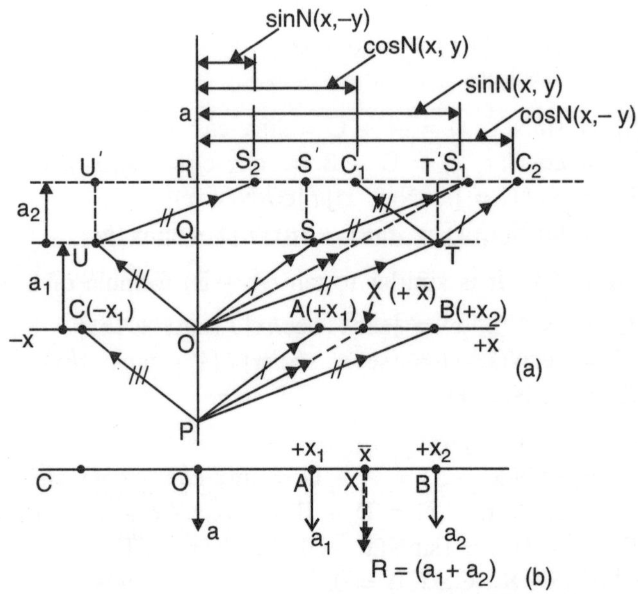

Fig. 1.5: Graphical determination of single Nbic functions

The Graphical determination of the Nbic functions in terms of their constituent functions can be done with ease. To illustrate that, we recall the expansion of (1.33a):

$$
\begin{aligned}
S_1^* &= \sin N(x, y) = A^* + C^* = \sin x\ chy + \cos x\ shy\\
S_2^* &= \sin N(x,-y) = A^* - C^* = \sin x\ chy - \cos x\ shy\\
C_1^* &= \cos N(x, y) = D^* - B^* = \cos x\ chy - \sin x\ shy\\
C_2^* &= \cos N(x,-y) = D^* + B^* = \cos x\ chy + \sin x\ shy,
\end{aligned}
$$

for demonstration, and to do that, we first demonstrate on the function:

$$F = a_1x_1 + a_2x_2,$$

where, (a_1, a_2) may be constants or functions, and (x_1, x_2) may be constants or functions, and possess numerical values (say, given). This is shown in **Figure 1.5**.

Procedure

Refer to **Figure 1.5(b)**

Get the rectangular axis system (x, a). On the a-axis, mark off OQ = + a_1 from O, and from Q mark off QR = + a_2. If a_2 is positive, it will be above Q, and if negative, it will be below Q, and so on.

Draw horizontals through Q and R.

On the a-axis, from O mark off OA = + x_1 and OB = + x_2. If x_2 is positive it will be towards positive x-axis, and *vice versa* [as shown, OC = $- x_1$].

On the other side of the a-axis, locate P at unit distance from the origin O. This is the pole distance OP = 1. Draw rays PA and PB.

Draw the ray OS parallel to OA, to get S on the horizontal line through Q.

Since, in the similar triangles POA and OQS,

$$QS / OQ = OA / OP, \text{ where } OQ = a_1,\ OA = x_1, \text{ and } OP = 1,$$

we have $$QS = a_1x_1$$

Draw the ray SS_1 parallel to PB, to get S_1 on the horizontal line through R.

Since, in the similar triangles POB and $SS'S_1$,

$$S'S_1 / SS' = OB / OP, \text{ where } SS' = QR = a_2,\ QS = RS',\ OB = x_2,$$

and $$OP = 1,\ QR = a_2x_2.$$

Hence, $$RS_1 = QS + S'S_1 = RS' + S'S_1 = a_1x_1 + a_2x_2.$$

Draw the ray PC. Similarly, parallel to the ray PB draw OT, to locate T on the horizontal line through Q, and from T draw TC_1 parallel to the ray PC, to locate C_1 on the horizontal line through R.

Since, in the similar triangles POC and $TT'C_1$,

$$QT = RT' = a_1\, x_2, \text{ and } T'\, C_1 = a_2\, (- x_1),$$

Hence, $$RC_1 = RT' - T'\, C_1 = a_1x_2 - a_2x_1.$$

As described with two terms here, this graphical method is an open ended one. That is to say that, there is absolutely no problem to represent the function like:

$$F = a_1x_1 + a_2x_2 + a_3x_3 + a_4x_4 + \ldots + a_nx_n.$$

Now to represent the functions

$$\begin{aligned} S_1^* &= \sin N(x, y) = \sin x\ chy + \cos x\ shy \\ S_2^* &= \sin N(x,-y) = \sin x\ chy - \cos x\ shy \\ C_1^* &= \cos N(x, y) = \cos x\ chy - \sin x\ shy \\ C2^* &= \cos N(x,-y) = \cos x\ chy + \sin x\ shy, \end{aligned}$$

we use the graphics of the function F with only two terms, i.e., $F = a_1x_1 + a_2x_2$,

Substitute,

$$\begin{aligned} OQ &= a_1 = + \cos x, \\ QR &= a_2 = + \sin x \\ OA &= + x_1 = shx, \\ OB &= + x_2 = chx \\ OC &= - x_1 = - shx \end{aligned}$$

so that, we read the following Nbic functions from **Figure 1.5(a)** as:

$$\begin{aligned} RS_1 &= QS + S'S_1 = RS' + S'S_1 = a_1x_1 + a_2x_2 = \sin N(x, y) \\ RS_2 &= - QU + U'S_2 = - U'R + U'S_2 = - a_1x_1 + a_2x_2 = \sin N(x, -y) \\ RC_1 &= QT - T'C_1 = RT' - T'C_1 = a_1x_2 - a_2x_1 = \cos N(x, y) \\ RC_2 &= QT + T'C_2 = a_1x_2 + a_2x_1 = \cos N(x, -y) \end{aligned}$$

We collect additional information from the construction for, $F = a_1 x_1 + a_2 x_2$. To do that, join OS_1, and from P, get the point X on the x-axis by drawing PX parallel to it. Now consider the similar triangles ORS_1 and POX (with OP = 1)

$$OX / OP = RS_1 / OR = [a_1x_1 + a_2x_2] / [a_1 + a_2] = \bar{x}$$

or, $$OX = \bar{x}$$

Interpretation of Statement

The RHS of this expression is that of the location of the centroid, for the case of like parallel force system, if a_1 and a_2 are taken as the weights or forces, and x_1 and x_2 are the corresponding locations measured from a certain reference system. In fact, if the forces a_1 and a_2 act at x_1 and x_2 , respectively, the line of action of the resultant of the two forces $R = (a_1 + a_2)$, will then pass through X, which is given by, $\bar{x}$

So,

$$OX = [a_1x_1 + a_2x_2] / [a_1 + a_2] = \bar{x}$$

The location of the centroid of a lamina of area a, when divided into component parts of areas $a_1, a_2, \ldots, a_n$, comes identically under this category of problem.

Derived from the idea of the interpretation of the like parallel forces system $a_i - s$ acting at the corresponding positions at $x = x_i - s$, and its line of action of the resultant $R = \sum_{i=0}^{n} a_i$, at location $x = \bar{x}$, a novel method based on the **Polynomial root analogy** applied to the parallel force systems, has been provided, elaborated and illustrated with problems and solved examples in **Appendix B**.

This procedure is equally valid for any combination of $a_i - s$ and $x_i - s$, $(i = 1, 2)$. For example,

(a_1, a_2) may be: $(\cos x, \sin x)$, (chx, shx),

(cosNx, sinNx), (cosN$_2x$, sinN$_2x$), and so on,

and

(x_1, x_2) may be: (cos$y$, sin$y$), (chy, shy),

(cosNy, sinNy), (cosN$_2y$, sinN$_2y$), and so on.

1.7.4 Single Nbic Function with Complex Variable of (*ix*) Only : N (*ix, ix*)

For complex variable (ix), the Single Nbic functions (1.35a, b) when recast, on replacing, x by (ix), and using these transformation, *viz.*,

$$\begin{aligned} sh(ix) &= i \sin x \\ ch(ix) &= \cos x \\ \sin(ix) &= i\, shx \\ \cos(ix) &= chx \end{aligned}$$

take the following form

$$\begin{aligned} \text{sinN}(ix, ix) &= \text{sinN}(ix) = \sin(ix)\, ch(ix) + \cos(ix)\, sh(ix) \\ &= i\,(\sin x\; chx + \cos x\; shx) \\ &= i\, \text{sinN}(x, x) \\ \text{cosN}(ix, ix) &= \cos(ix)\, ch(ix) - \sin(ix)\, sh(ix) \\ &= \cos x\; chx + \sin x\; shx \\ &= \text{cosN}(x, -x) \\ \text{sinN}(ix, -ix) &= \sin(ix)\, ch(ix) - \cos(ix)\, sh(ix) \\ &= -\, i\,(\sin x\; chx - \cos x\; shx) \\ &= -\, i\, \text{sinN}(x, -x) \\ \text{cosN}(ix, -ix) &= \cos(ix)\, ch(ix) + \sin(ix)\, sh(ix) \\ &= \cos x\; chx - \sin x\; shx \\ &= \text{cosN}(x, x) \end{aligned}$$

Collecting formulas in one place

$$\begin{aligned} \text{sinN}(ix, ix) &= i\, \text{sinN}(x, x) \\ \text{cosN}(ix, ix) &= \text{cosN}(x, -x) \\ \text{sinN}(ix, -ix) &= -\, i\, \text{sinN}(x, -x) \\ \text{cosN}(ix, -ix) &= \text{cosN}(x, x) \end{aligned} \tag{1.35c}$$

1.7.5 Comparison with Corresponding Circular and Hyperbolic Functions

We know that

$$\begin{aligned} \sin(-x) &= -\sin x,\ \cos(-x) = \cos x \\ sh(-x) &= -\, shx,\ ch(-x) = chx \end{aligned} \tag{1.36}$$

[where, shx = sinh$x$, chx = cosh$x$

thx = tanh$x$, cschx = cosech$x$, sechx = sech$x$, cthx = coth$x$]

With these

Compare with: [sin, *sh*] (–*x*) = – [sin, *sh*] (*x*), and
[cos, *ch*] (–*x*) = [cos, *ch*] (*x*)

$$\begin{aligned} \textbf{sinN}(-x, -x) = \textbf{sinN}\,(-x) &= \sin(-x)\, ch(-x) + \cos(-x)\, sh(-x) \\ &= -\sin x\; chx - \cos x\; shx \\ &= \boldsymbol{-\, \text{sinN}(x, x) = -\, \text{sinN}x} \end{aligned}$$

$$\mathbf{cosN(-x, -x) = cosN(-x)} = \cos x\ chx - (-\sin x)(-shx)$$
$$= \cos x\ chx - \sin x\ shx$$
$$= \mathbf{cosN(x, x) = cosNx} \tag{1.37}$$

Then

$$\text{sinN}[(x+\frac{p}{4}), x] = \sin(x+\frac{p}{4})\ chx + \cos(x+\frac{p}{4})\ shx$$
$$= \frac{1}{\sqrt{2}}\ [\mathbf{cosN(x, x) + sinN(x, x)}]$$
$$\cos \text{N}[(x+\frac{p}{4}), x] = \cos(x+\frac{p}{4})\ chx - \sin(x+\frac{p}{4})\ shx$$
$$= \frac{1}{\sqrt{2}}\ [\mathbf{cosN(x, x) - sinN(x, x)}] \tag{1.38}$$

Further

$$\text{sinN}[(x+\frac{\pi}{2}), x] = \sin(x+\frac{\pi}{2})\ chx + \cos(x+\frac{\pi}{2})\ shx = \cos x\ chx - \sin x\ shx$$
$$= \text{cosN}(x, x)$$
$$\text{sinN}[(x-\frac{\pi}{2}), x] = \sin(x-\frac{\pi}{2})\ chx + \cos(x-\frac{\pi}{2})\ shx = -\cos x\ chx + \sin x\ shx$$
$$= -\text{cosN}(x, x)$$
$$\text{cosN}[(x+\frac{\pi}{2}), x] = \cos(x+\frac{\pi}{2})\ chx - \sin(x+\frac{\pi}{2})\ shx = -\sin x\ chx - \cos x\ shx$$
$$= -\text{sinN}(x, x)$$
$$\text{cosN}[(x-\frac{\pi}{2}), x] = \cos(x-\frac{\pi}{2})\ chx + \sin(x-\frac{\pi}{2})\ shx = \sin x\ chx + \cos x\ shx$$
$$= \text{sinN}(x, x) \tag{1.39}$$

Collecting formulas in one place

$$\mathbf{sinN}[(x+\frac{\pi}{2}), x] = \mathbf{cosN(x, x)}$$
$$\mathbf{sinN}[(x-\frac{\pi}{2}), x] = -\mathbf{cosN(x, x)}$$
$$\mathbf{cosN}[(x+\frac{\pi}{2}), x] = -\mathbf{sinN(x, x)}$$
$$\mathbf{cosN}[(x-\frac{\pi}{2}), x] = \mathbf{sinN(x, x)} \tag{1.40}$$

Nbic functions of cosN and sinN in expressions (1.40) maintain striking similarity with the corresponding ones of the plane trigonometry. That is

$$\mathbf{sinN}[(x+\frac{\pi}{2}),x] = -\ \mathbf{sinN}[(x-\frac{\pi}{2}),x]$$

$$\mathbf{cosN}[(x+\frac{\pi}{2}),x] = -\ \mathbf{cosN}[(x-\frac{\pi}{2}),x] \qquad (1.41)$$

Nbic functions of **cos** and **sin** in expressions (1.40) and (1.41) maintain striking similarity with the corresponding ones of plane trigonometry. That is

$$\mathbf{sin}\,(x+\frac{\pi}{2}) = \mathbf{cos}x$$

$$\mathbf{sin}\,(x-\frac{\pi}{2}) = \mathbf{cos}x$$

$$\mathbf{cos}\,(x+\frac{\pi}{2}) = -\ \mathbf{sin}x$$

$$\mathbf{cos}\,(x-\frac{\pi}{2}) = \mathbf{sin}x$$

$$\mathbf{sin}\,(x+\frac{\pi}{2}) = -\ \mathbf{sin}\,(x-\frac{\pi}{2})$$

$$\mathbf{cos}\,(x+\frac{\pi}{2}) = -\ \mathbf{cos}\,(x-\frac{\pi}{2}) \qquad (1.42)$$

$$\begin{aligned}\mathbf{sinN}x + \mathbf{cosN}x &= chx\ (\sin x + \cos x) + shx\ (\cos x - \sin x)\\ &= chx\ (\sin x \cos\frac{\pi}{4} + \cos x \sin\frac{\pi}{4})\\ &\quad + shx\ (\cos x \cos\frac{\pi}{4} - \sin x \sin\frac{\pi}{4})\\ &= \sqrt{2}\,[chx \sin\,(x+\frac{\pi}{4}) + shx \cos\,(x+\frac{\pi}{4})]\\ &= \sqrt{2}\ \mathbf{sinN}\ [(x+\frac{\pi}{4}),x] \qquad (1.43)\end{aligned}$$

Similarly

$$\begin{aligned}\mathbf{sinN}x - \mathbf{cosN}x &= chx\ (\sin x - \cos x) + shx\ (\sin x + \cos x)\\ &= \sqrt{2}\,[chx\ (\sin x \sin\frac{\pi}{4} - \cos x \cos\frac{\pi}{4})\\ &\quad + shx\ (\cos x \sin\frac{\pi}{4} + \sin x \cos\frac{\pi}{4})]\\ &= \sqrt{2}\,[-\ chx \cos(x+\frac{\pi}{4}) + shx \sin\,(x+\frac{\pi}{4})]\end{aligned}$$

$$= -\sqrt{2}\ \cos N[(x+\frac{\pi}{4}), x] \tag{1.44}$$

Again

$$\begin{aligned}\mathbf{sinN}x \ .\ \mathbf{cosN}x &= (\sin x\ chx + \cos x\ shx)(\cos x\ chx - \sin x\ shx)\\ &= \sin x \cos x\ ch2x - \sin^2 x\ shx\ chx\\ &\quad + \cos^2 x\ shx\ chx - \cos x \sin x\ sh^2 x\\ &= \sin x \cos x\ (ch^2 x - sh^2 x) + chx\ shx\ (\cos^2 x - \sin^2 x)\\ &= \frac{1}{2} \mathbf{\sin 2x} + \frac{1}{2} \mathbf{\cos 2x\ sh\ 2x}\end{aligned}$$

$$S_1\ C_1 = \frac{1}{2}\ [\sin 2x + \cos 2x\ sh\ 2x] \tag{1.45}$$

In plane trigonometry, we have only the first term present. That is

$$\mathbf{\sin x \cos x = \frac{1}{2}\ \sin 2x}$$

Thus

$$\sin N x = \sqrt{2}\,[\sin N[(x+\frac{\pi}{4}), x] - \cos N[(x+\frac{\pi}{4}), x)]$$

$$\cos N x = \sqrt{2}\,[\sin N[(x+\frac{\pi}{4}), x] + \cos N[(x+\frac{\pi}{4}), x)] \tag{1.46}$$

Also

$$\begin{aligned}\mathbf{cosN^2}x + \mathbf{sinN^2}x &= \sin^2 x\ ch^2 x + \cos^2 x\ sh^2 x + 2 \sin x \cos x\ shx\ chx\\ &\quad + \cos^2 x\ ch^2 x + \sin^2 x\ sh^2 x - 2 \sin x \cos x\ shx\ chx\\ &= ch^2 x + sh^2 x\\ &= \mathbf{ch\ 2x}\end{aligned}$$

$$S_1^2 + C_1^2 = ch\ 2x \tag{1.47}$$

$$\begin{aligned}\mathbf{cosN^2}x - \mathbf{sinN^2}x &= \cos^2 x\ ch^2 x + \sin^2 x\ sh^2 x - 2 \sin x \cos x\ shx\ chx\\ &\quad - \sin^2 x\ ch^2 x - \cos^2 x\ sh^2 x - 2 \sin x \cos x\ shx\ chx\\ &= (\cos^2 x - \sin^2 x)(ch^2 x - sh^2 x) - 4 \sin x \cos x\ shx\ chx\\ &= \mathbf{\cos 2x - \sin 2x\ sh\ 2x}\end{aligned}$$

$$C_1^2 - S_1^2 = \cos 2x - \sin 2x\ sh\ 2x \tag{1.48}$$

In plane trigonometry, we have only the first term present. That is

$$\cos^2 x - \sin^2 x = \cos 2x$$

Note:

As discussed later separately, it will be seen that, (1.29) and (1.31), when written through E type Triple Nbic function, i.e., $N_{3e}(x)$, as discussed in Chapter 4, will become as:

$$\begin{aligned}\cos N^2 x - \sin N^2 x &= \cos 2x - \sin 2x\ sh\ 2x = \cos N_{3e}(x)\\ 2 \sin N x \cos N x &= \sin 2x + sh\ 2x \cos 2x = \sin N_{3e}(x)\end{aligned} \tag{1.49}$$

So that

$$\begin{aligned}\tan N_{3e}(x) &= [\sin N_{3e}(x)] / [\cos N_{3e}(x)]\\ &= [\cos N^2 x - \sin N^2 x] / [2 \sin Nx \cos Nx]\\ &= [\cos 2x - \sin 2x\ sh\ 2x] / [\sin 2x + sh\ 2x \cos 2x]\\ &= [\tan 2x + sh\ 2x] / [1 - \tan 2x\ sh\ 2x]\end{aligned}$$

This expression can also be shown as

$$\tan N_{3e}(x) = 2 \tan Nx / [1 - \tan N^2 x].$$

Again

$$\begin{aligned}(\sin Nx + \cos Nx)^2 &= (\sin N^2 x + \cos N^2 x) + 2 \sin Nx \cos Nx\\ &= \sin 2x + ch\ 2x + \cos 2x\ sh\ 2x\\ (\sin Nx - \cos Nx)^2 &= (\sin N^2 x + \cos N^2 x) - 2 \sin Nx \cos Nx\\ &= \sin 2x + ch\ 2x - \cos 2x\ sh\ 2x\end{aligned} \qquad (1.50)$$

Check: (1.50) satisfies the formula: $(a + b)^2 - (a - b)^2 = 4ab$

Further we have

$$\begin{aligned}[\sin N(x, x) &+ \sin N(x, -x)] / 2 = \sin x\ chx = A\\ [\sin N(x, x) &- \sin N(x, -x)] / 2 = \cos x\ shx = C\\ [\cos N(x, x) &+ \cos N(x, -x)] / 2 = -\cos x\ chx = -D\\ [\cos N(x, x) &- \cos N(x, -x)] / 2 = \sin x\ shx = B\end{aligned} \qquad (1.51)$$

From which get

$$\begin{aligned}&[(\sin N(x, x) + \sin N(x, -x))\ (\sin(x, x) - \sin N(x, -x))] /\\ &[(\cos N(x, x) + \cos N(x, -x))\ (\cos N(x, x) - \cos N(x, -x))] = -1\end{aligned}$$

That is, $A \cdot C / B \cdot D = 1$

Also

$$\mathbf{\sin N^2(x, x) + \cos N^2(x, x) = \sin N^2(x, -x) + \cos N^2(x, -x)} \qquad (1.52)$$

1.8 DERIVATION OF EXPRESSIONS OF OTHER BASIC Nbic FUNCTIONS

Recollect some already known formulae

$$\begin{aligned}sh(x + y) &= shx\ chy + chx\ shy\\ sh(x - y) &= shx\ chy - chx\ shy\\ ch(x + y) &= chx\ chy + shx\ shy\\ ch(x - y) &= chx\ chy - shx\ shy\\ th\ (x \pm y) &= [thx \pm thy] / [1 \pm thx\ thy]\\ sh(-x) &= -shx,\ ch(-x) = chx,\ th(-x) = -thx\\ sh\ ix &= i \sin x,\ ch\ ix = \cos x,\ th\ ix = i \tan x\\ shx &= -i \sin ix,\ chx = \cos ix,\ thx = i \tan ix\\ shx\ shy &= [ch(x + y) - ch(x - y)] / 2\\ chx\ chy &= [ch(x + y) + ch(x - y)] / 2\\ shx\ chy &= [sh(x + y) - sh(x - y)] / 2\\ thx\ thy &= (thx + thy) / (cthx + chy)\end{aligned}$$

$$shx \pm shy = 2\ sh\frac{x \pm y}{2}\ ch\ \frac{x \pm y}{2}$$

$$chx + chy = 2\ ch\frac{x + y}{2}\ ch\ \frac{x - y}{2}$$

$$chx - chy = 2\ sh\frac{x+y}{2}\ sh\ \frac{x-y}{2}$$

$$thx \pm thy = [sh\ \frac{x \pm y}{2}] / [chx.chy] \tag{1.53a}$$

1.8.1 To Find sinN*x* and cosN*x*, when only, tanN*x* is given

To find, sinNx and cosNx, from tanNx, take help of (1.47) and (1.49), i.e.,

$$\cos N^2x + \sin N^2x = ch\ 2x \tag{1.53b}$$

Let

$$\cos N^2x - \sin N^2x = a \tag{1.53c}$$

$$2 \sin Nx \cos Nx = b \tag{1.53d}$$

From (1.53d)

$$\sin N^2x = b^2 / (4 \cos N^2x)$$

Putting it to (1.53c)

$$\cos N^2x - \sin N^2x = \cos N^2x - b^2 / (4 \cos N^2x) = a$$

or, $4 [\cos N^2x]^2 - 4a [\cos N^2x] - b^2 = 0$

Solve for the quadratic equation in, cosN^2x

$$\cos N^2x = \frac{1}{2}\ [a \pm \sqrt{a^2 + b^2}\ ] = -\frac{1}{2} [a \pm ch\ 2x]$$

$$\cos Nx = \pm \sqrt{\frac{a \pm ch2x}{2}} \tag{1.53e}$$

Substitute (1.53e) in (1.53b)

$$\sin N^2x = ch\ 2x - \cos N^2x = ch\ 2x - \frac{1}{2} [a \pm ch\ 2x]$$

$$\sin Nx = \pm\ [ch\ 2x - \frac{1}{2} (a \pm ch\ 2x)]^{1/2} \tag{1.53f}$$

Thus, cosNx and sinNx, are obtained, following (1.53e, 1.53d).

To find [tanN, cosecN, secN and cotN], of (*x*, *x*):

For tanN:, cosecN:, secN: and cotN, of (*x*, *x*):

Already we have from (1.35a)

$$S_1 = \sin N(x, x) = \sin Nx = \sin x\ chx + \cos x\ shx$$
$$C_1 = \cos N(x, x) = \cos Nx = \cos x\ chx - \sin x\ shx \tag{1.54a}$$

From which

$$\cos Nx + \sin Nx = (\cos x + \sin x)\ chx + (\cos x - \sin x)\ shx$$
$$= (chx + shx) \cos x + (chx - shx) \sin x$$
$$\cos Nx - \sin Nx = (\cos x - \sin x)\ chx - (\cos x + \sin x)\ shx$$
$$= (chx - shx) \cos x - (chx + shx) \sin x \tag{1.54b}$$

Then from (1.54a)

$$T_1 = \tan N\ (x, x) = [\sin N(x, x)] / [\cos N(x, x)]$$
$$= (\sin x\ chx + \cos x\ shx) / (\cos x\ chx - \sin x\ shx)$$

$$= \frac{\sin x.chx + \cos x.shx}{\cos x.chx - \sin x.shx}$$

$$= \frac{\sin x.chx + \cos x.shx}{\cos x.chx - \sin x.shx} \cdot \frac{\sin x.chx + \cos x.shx}{\cos x.chx + \sin x.shx}$$

$$= \frac{\sin x.\cos x.ch^2x + \sin^2 x.chx.shx - \cos^2 x.shx.chx + \sin x.\cos x.sh^2x}{\cos^2 x.ch^2x - \sin^2 x.sh^2x}$$

$$= \frac{\sin x.\cos x.(ch^2x + sh^2x) + chx.shx(\sin^2 x + \cos^2 x)}{\cos^2 x.ch^2x - \sin^2 x.sh^2x}$$

$$= \frac{\sin x.\cos x.ch2x + shx.chx}{\cos^2 x.ch^2x - \sin^2 x.sh^2x}$$

Divide both the numerator and denominator of the RHS by, ($\cos^2 x\ .ch^2x$), to yield

$$= \frac{\tan x.(1 + th^2x) + thx.\dfrac{1}{\cos^2 x}}{1 - \tan^2 x.th^2x} = \frac{\tan x.(1 + th^2x) + thx.(1 + \tan^2 x)}{(1 + \tan x.thx)(1 - \tan x.thx)}$$

The numerator when further simplified becomes = (1 + tan$x$ thx) (tan$x$ + thx)

So that with it, finally, T_1 = tanN (x, x) is given by

$$\mathbf{T_1 = tanN}(x,\ x) = \frac{\tan x + thx}{1 - \tan x.thx} \tag{1.55}$$

It is similar to, tan (A+B) $= \dfrac{\tan A + \tan B}{1 - \tan A.\tan B}$, formula of plane trigonometry.

[**Note:** To arrive at (1.55), all these round about steps are unnecessary. To avoid it, simply divide the numerator and denominator by (cos$x$·chx) at the very beginning.]

Again

secN(x, x) = 1 / [cosN(x, x)] = 1 / (cos$x$ chx – sin$x$ shx)

Divide both the numerator and denominator of the RHS by, (cos$x$ chx + sin$x$ shx), to yield

$$\text{secN}(x,\ x) = \frac{\cos x.chx + \sin x.shx}{\cos^2 x.ch^2x - \sin^2 x.sh^2x}$$

Divide both the numerator and denominator of the RHS by ($\cos^2 x\ .ch^2x$) to yield

$$\text{secN}(x,\ x) = \frac{\dfrac{1}{\cos x.chx} + \tan x.\sec x.thx.\sec hx}{1 - \tan^2 x.th^2x}$$

$$= \frac{\sec x.\sec hx.(1 + \tan x.thx)}{1 - \tan^2 x.th^2x}$$

$$\mathbf{secN}(x,\ x) = \frac{\sec x.\sec hx}{1 - \tan x.thx} \tag{1.56}$$

Check

$$\text{sinN}(x, x) = \text{secN}(x, x) / \text{tanN}(x, x) = [\frac{\sec x.\sec hx}{1-\tan x.thx}] / [\frac{\tan x + thx}{1-\tan x.thx}]$$

$$= (\tan x + thx) / (\cos x\ chx) = \left(\frac{\sin x}{\cos x} + \frac{shx}{chx}\right) / (\cos x\ chx)$$

Thus, $\text{sinN}(x, x) = \sin x\ chx + \cos x\ shx$ **(Checked)** (1.57)

Again, $\text{cscN}(x, x) = \text{cosecN}(x, x) = 1 / [\text{sinN}(x, x)] = 1 / (\sin x\ chx + \cos x\ shx)$

Divide Numerator and Denominator by $(\cos x.shx)$, to yield

$$\text{cscN}(x, x) = \frac{\frac{1}{\cos x.shx}}{1+\tan x.\cot hx} = \frac{\sec x.\csc hx}{\frac{\tan x + thx}{thx}} = \frac{\sec x.\csc hx.thx}{\tan x + thx}$$

Thus, $$\text{cscN}(x, x) = \frac{\sec x.\sec hx}{\tan x + thx} \quad (1.58)$$

Collecting formulas together

$$S_1 = \textbf{sinN}(x, x) = \sin x\ chx + \cos x\ shx$$
$$C_1 = \textbf{cosN}(x, x) = \cos x\ chx - \sin x\ shx$$
$$T_1 = \textbf{tanN}(x, x) = \frac{\tan x + thx}{1-\tan x.thx}$$
$$SC = \textbf{secN}(x, x) = \frac{\sec x.\sec hx}{1-\tan x.thx}$$
$$CC = \textbf{cscN}(x, x) = \frac{\sec x.\sec hx}{\tan x + thx}$$
$$CT = \textbf{cotN}(x, x) = \frac{1-\tan x.thx}{\tan x + thx} \quad (1.59)$$

To find (tanN, cosecN, secN and cotN) of (*x*, *y*):

For tanN, cscN, secN and cotN), of (*x*, *y*):

Follow similar procedure, as done with (x, x), now with variable of (x, y), we get

$$S_1^* = \text{sinN}(x, y) = \sin x\ chy + \cos x\ shy$$
$$C_1^* = \text{cosN}(x, y) = \cos x\ chy - \sin x\ shy \quad (1.60)$$
$$T_1^* = \text{tanN}(x, y) = [\text{sinN}(x, y)] / [\text{cosN}(x, y)]$$
$$= (\sin x\ chy + \cos x\ shy) / (\cos x\ chy - \sin x\ shy)$$
$$= \frac{\sin x.chy + \cos x.shy}{\cos x.chy - \sin x.shy} \cdot \frac{\sin x.chy + \cos x.shy}{\cos x.chy + \sin x.shy}$$

$$= \frac{\sin x.\cos x.ch^2 y + \sin^2 x.chy.shy + \cos^2 x.shy.chy + \sin x.\cos x.sh^2 y}{\cos^2 x.ch^2 y - \sin^2 x.sh^2 y}$$

$$= \frac{\sin x.\cos x.(ch^2 y + sh^2 y) + chy.shy(\sin^2 x + \cos^2 x)}{\cos^2 x.ch^2 y - \sin^2 x.sh^2 y}$$

$$= \frac{\sin x.\cos x.ch2x + shx{\cdot}chx}{\cos^2 x.ch^2 x - \sin^2 x.sh^2 x}$$

Divide Numerator and Denominator by ($\cos^2 x$.$ch^2 y$) to yield

$$= \frac{\tan x.(1 + th^2 y) + thy.\dfrac{1}{\cos^2 x}}{1 - \tan^2 x.th^2 y} = \frac{\tan x.(1 + th^2 y) + thy.(1 + \tan^2 x)}{(1 + \tan x.thy)(1 - \tan x.thy)}$$

Numerator = (1 + tan$x$ thy) (tan$x$ + thy)

$$\mathbf{T_1^* = tanN(x, y)} = \frac{\tan x + thy}{1 - \tan x.thy} \quad (1.61)$$

It is similar to, $\tan(A+B) = \dfrac{\tan A + \tan B}{1 - \tan A.\tan B}$ formula of plane trigonometry.

[**Note:** To arrive at (1.61), all these round about stpeps are unnecessary. To avoid it, simply divide the numerator and denominator by (cos$x$.chy) at the very beginning.]

Again

$$\text{secN}(x, y) = 1 / \text{cosN}(x, y) = 1 / (\cos x\ chy - \sin x\ shy)$$

Divide Numerator and Denominator by, (cos$x$ chy + sin$x$ shy), to yield

$$\text{secN}(x, y) = \frac{\cos x.chy + \sin x.shy}{\cos^2 x.ch^2 y - \sin^2 x.sh^2 y}$$

Divide Numerator and Denominator by, ($\cos^2 x$.$ch^2 y$), to yield

$$\text{secN}(x, y) = \frac{\dfrac{1}{\cos x.chy} + \tan x.\sec x.thy.\sec hy}{1 - \tan^2 x.th^2 y} = \frac{\sec x.\sec hx.(1 + \tan x.thx)}{1 - \tan^2 x.th^2 x}$$

$$\mathbf{secN(x, y)} = \frac{\sec x.\sec hy}{1 - \tan x.thy} \quad (1.62)$$

Check:

$$\text{sinN}(x, y) = [\text{secN}(x, y)] / [\text{tanN}(x, y)] = \frac{\sec x.\sec hy}{1 - \tan x.thy} / \frac{\tan x + thy}{1 - \tan x.thy}$$

$$= (\tan x + thy) / (\cos x\ chy) = \left(\frac{\sin x}{\cos x} + \frac{shy}{chy}\right) / (\cos x\ chy)$$

Thus

$$\text{sinN}(x, y) = \sin x\ chy + \cos x\ shy \quad \textbf{(Checked)} \quad (1.63)$$

Again

$$\text{cscN}(x,\ y) = 1/[\text{sinN}(x,\ y)] = 1/(\sin x\ chy + \cos x\ shy)$$

Divide Numerator and Denominator by, (cos$x$ .shy), to yield

$$\text{cscN}(x,\ y) = \frac{\frac{1}{\cos x.shy}}{1+\tan x.\ \coth y} = \frac{\sec x.\csc hy}{\frac{\tan x+thy}{thy}} = \frac{\sec x.\csc hy.thy}{\tan x+thy}$$

Thus

$$\text{cscN}(x,\ y) = \frac{\sec x.\sec hy}{\tan x + thy} \tag{1.64}$$

Collecting formulas together

$$\mathbf{S_1^* = sinN(x,\ y)} = \sin x\ chy + \cos x\ shy$$

$$\mathbf{C_1^* = cosN(x,\ y)} = \cos x\ chy - \sin x\ shy$$

$$\mathbf{T_1^* = tanN(x,\ y)} = \frac{\tan x + thy}{1 - \tan x.thy}$$

$$\mathbf{SC^* = secN(x,\ y)} = \frac{\sec x.\sec hy}{1 - \tan x.thy}$$

$$\mathbf{CC^* = cscN(x,\ y)} = \frac{\sec x.\sec hy}{\tan x + thy}$$

$$\mathbf{CT^* = cotN(x,\ x)} = \frac{1 - \tan x.thy}{\tan x + thy} \tag{1.65}$$

1.8.2 Differentiation Rule for Single Nbic Functions [with ( )′ = d( ) / dx, and, ( )^0 = d( ) / dy, etc.]

[**Note:** actually, $df(x)/dx$ should be expressed as: $f'(x)$, or $[f(x)]'$, for example, $[\sin(-x)]'$ or $(chx)'$, or, $[\text{tanN}(x,\ x)]'$. Here they, for brevity, are expressed as: $\sin'(-x)$ or $ch'x$, or, $\text{tanN}'(x,\ x)$, etc., throughout, without losing the very meaning.]

Now remembering, $sh'x = chx$, $ch'x = shx$, and with

$$\begin{aligned} \text{sinN}x &= \text{sinN}(x,\ x) = \sin x\ chx + \cos x\ shx \\ \text{cosN}x &= \text{cosN}(x,\ x) = \cos x\ chx - \sin x\ shx \end{aligned} \tag{1.66}$$

we get

$$\begin{aligned} S_1' &= \text{sinN}'(x,\ x) = (\sin x\ shx + \cos x\ chx) + (\cos x\ chx - \sin x\ shx) \\ &= 2\cos x\ chx = 2D \\ C_1' &= \text{cosN}'/(x, x) = (\cos x\ shx - \sin x\ chx) - (\sin x\ chx + \cos x\ shx) \\ &= -2\sin x\ chx = -2A \end{aligned} \tag{1.67}$$

from which

$$\text{sinN}'(x,\ x) + \text{cosN}'(x,\ x)$$

$= \cos N(x, -x) + \cos N(x, x) - \sin N(x, -x) - \sin N(x, x)$
$= 2\,[\cos x\ chx - \sin x\ chx]$
$= 2\,(D - A)$

$\sin N'(x, x) - \cos N'(x, x)$
$= \cos N(x, -x) + \cos N(x, x) + \sin N(x, -x) + \sin N(x, x)$
$= 2\,[\cos x\ chx + \sin x\ chx]$
$= 2\,(D + A)$

So collecting

$$\mathbf{\sin N'(x, x) + \cos N'(x, x)}$$
$$\Rightarrow \quad S_1' + C_1' = 2\,(D - A)$$
$$\mathbf{\sin N'(x, x) - \cos N'(x, x)}$$
$$\Rightarrow \quad S_1' - C_1' = 2\,(D + A) \tag{1.68}$$

Again

$\cos N(x, x).\cos N(x, -x)$

$$= \frac{1}{4}\,[\sin N'(x, x) - \sin N'(x, -x)]\,[\sin N'(x, x) + \sin N(x, -x)]$$

$$C_1C_2 = \frac{1}{4}\,[S_1' + S_2']\,[S_1' - S_2'] = \frac{1}{4}(S_1'^2 - S_2'^2) = D^2 - B^2 \tag{1.69}$$

and

$$\frac{\cos N(x,x)}{\cos N(x,-x)} = \frac{\sin N'(x,x) - \sin N'(x,-x)}{\sin N'(x,x) + \sin N'(x,-x)}$$

$$\frac{C_1}{C_2} = \frac{S_1' - S_2'}{S_1' + S_2'} = \frac{D - B}{D + B} \tag{1.70}$$

Similarly

$$\mathbf{\sin N'(x, -x)} = (\sin x\ shx + \cos x\ chx) - (\cos x\ chx - \sin x\ shx)$$

$$\mathbf{= 2 \sin x\ shx = 2B}$$

$$\mathbf{\cos N'(x, -x)} = (\cos x\ shx - \sin x\ chx) + (\sin x\ chx + \cos x\ shx)$$

$$\mathbf{= 2 \cos x\ shx = 2C} \tag{1.71}$$

$$\mathbf{\tan N'(x, x)} = \frac{d}{dx}\left(\frac{\sin N(x,x)}{\cos N(x,x)}\right) = \frac{\cos N(x,x).\sin N'(x,x) - \sin N(x,x).\cos N'(x,x)}{\cos N^2(x,x)}$$

$$= \frac{\cos N(x,x).[\cos N(x,x) + \cos N(x,-x)] + \sin N(x,x).[\sin N(x,x) + \sin N(x,-x)]}{\cos N^2(x,x)}$$

$$= \frac{[\cos N^2(x,x) + \sin N^2(x,x)] + [\cos N(x,x).\cos N(x,-x) + \sin N(x,x).\sin N(x,-x)]}{\cos N^2(x,x)}$$

$= [1 + ch2x] / \cos N^2(x,x)$

$$\mathbf{tanN'(x,\ x) = (2\ ch^2\ x)\ secN^2(x,\ x)} \tag{1.72}$$

Collecting formulas together:

[Expressions (1.67) – (1.72)]

$$\begin{aligned}
\mathbf{S_1'} &= \mathbf{sinN'(x,\ x) = [2\ chx]\ (cos\ x) = 2D} \\
\mathbf{C_1'} &= \mathbf{cosN'(x,\ x) = [2\ chx]\ (-\ sinx) = -2A} \\
\mathbf{S_2'} &= \mathbf{sinN'(x,\ -x) = [2\ shx]\ (sinx) = 2B} \\
\mathbf{C_2'} &= \mathbf{cosN'(x,\ -x) = [2\ shx]\ (cosx) = 2C} \\
\mathbf{T_1'} &= \mathbf{tanN'((x,\ x)} = \frac{1+ch^2 x}{\cos N^2(x,x)} = \frac{2ch^2 x}{\cos N^2(x,x)} \\
&= \mathbf{(2\ ch^2x)\ secN^2(x,\ x)}
\end{aligned} \tag{1.73}$$

Again from (1.73)

$$\mathbf{tan}x = \frac{\sin N'(x,-x)}{\cos N'(x,-x)} = \frac{S_2'}{C_2'} \tag{1.74}$$

Using expressions from (1.68) – (1.73), we get

$$\frac{\sin N'(x,x)}{\cos N'(x,x)} = -\frac{\cos N(x,x)+\cos N(x,-x)}{\sin N(x,x)+\sin N(x,-x)}$$

$$\frac{\cos N(x,x)+\cos N(x,-x)}{\sin N(x,x)+\sin N(x,-x)} = \frac{\sin N'(x,x)}{-\cos N'(x,x)} \tag{1.75}$$

Use formula

$$\frac{a+c}{b+d} = \frac{e}{-f} = \frac{(a+c)+(b+d)}{(a+c)-(b+d)} = \frac{e+(-f)}{e-(-f)},$$

with actual functions in (1.75), to get

$$\frac{2\cdot chx\cdot(\cos x+\sin x)}{2\cdot chx\cdot(\cos x-\sin x)} = \frac{\sin N'(x,x)-\cos N'(x,x)}{\sin N'(x,x)+\cos N'(x,x)}$$

or,

$$\frac{\cos x+\sin x}{\cos x-\sin x} = \frac{\sin N'(x,x)-\cos N'(x,x)}{\sin N'(x,x)+\cos N'(x,x)} \tag{1.76}$$

Now simplifying the LHS

$$\text{LHS} = \frac{\sqrt{2}.(\cos x\cdot\cos\frac{\pi}{4}+\sin x\cdot\sin\frac{\pi}{4})}{\sqrt{2}.(\sin x.\cos\frac{\pi}{4}-\cos x\cdot\sin\frac{\pi}{4})}$$

$$= \frac{\cos(x-\frac{\pi}{4})}{\sin(x-\frac{\pi}{4})} = \cot(x-\frac{\pi}{4})$$

So, taking reciprocal of both the sides

$$\tan(x-\frac{\pi}{4})=\frac{\cos N'(x,x)+\sin N'(x,x)}{\cos N'(x,x)-\sin N'(x,x)} \tag{1.77}$$

Again substitute, tan$x$ from (1.74), and simplify the LHS of (1.77), to get

$$\text{LHS}=\tan(x-\frac{\pi}{4})=\frac{\tan x-1}{\tan x+1}$$

$$=\frac{\sin N'(x,-x)-\cos N'(x,-x)}{\sin N'(x,-x)+\cos N'(x,-x)} \tag{1.78}$$

Again simplify the LHS of (1.77), in view of, tan$x$ from (1.74), and equate (1.77) and (1.78), to get

$$\frac{\sin N'(x,-x)-\cos N'(x,-x)}{\sin N'(x,-x)+\cos N'(x,-x)}=\frac{\cos N'(x,x)+\sin N'(x,x)}{\cos N'(x,x)-\sin N'(x,x)} \tag{1.79}$$

Expand and simplify

$$\sin N'(x,\ x)\ \sin N'(x,\ -x) = -\cos N'(x,\ x)\ \cos N'(x,\ -x) \tag{1.80}$$

$$\frac{\sin N'(x,x)}{\cos N'(x,x)}=-\frac{\cos N'(x,-x)}{\sin N'(x,-x)}$$ [This statement can be directly obtained.]

$$\frac{S_1'}{C_1'}=-\frac{C_2'}{S_2'}$$

$$\Rightarrow S_1'\,S_2' + C_1'\,C_2' = 0 \tag{1.81}$$

Further

$$\sin N(x,\ x)\ \sin N(x,\ -x) = \sin^2 x\ ch^2 x - \cos^2 x\ sh^2 x$$
$$\cos N(x,\ x)\ \cos N(x,\ -x) = \cos^2 x\ ch^2 x - \sin^2 x\ sh^2 x \tag{1.82}$$

Expanding, arranging and adding, to get

$$\sin N(x,\ x)\ \sin N(x,\ -x) + \cos N(x,\ x)\ \cos N(x,\ -x) = ch^2 x - sh^2 x = 1$$

$$\Rightarrow S_1S_2 + C_1C_2 = 1 \tag{1.83}$$

Also

$$\sin N(x,\ x)\ \cos N\ (x,\ -x)$$
$$= (\sin x\ chx + \cos x\ shx)\ (\cos x\ chx + \sin x\ shx)$$
$$\cos N(x,\ x)\ \sin\ (x,\ -x) \tag{1.84}$$
$$= (\cos x\ chx - \sin x\ shx)\ (\sin x\ chx - \cos x\ shx)$$

Expanding, arranging and adding, to get

$$\sin N(x,\ x)\ \cos N(x,\ -x) + \cos N(x,\ x)\ \sin N(x,\ -x)$$
$$= \sin\ 2x\ (ch^2 x + sh^2 x) = \sin\ 2x\ ch\ 2x$$
$$\Rightarrow S_1C_2 + C_1S_2 = \sin\ 2x\ ch\ 2x \tag{1.85}$$

1.8.3 Numerical Verification of Expressions

The following expressions are verified with some assumed numerical data:

Basic data for Examples 1.1–1.12

$$x = 18.45 \text{ deg} = 0.322 \text{ rad},\ \sin x = 0.3165,\ \cos x = 0.9486$$

$$shx = 0.3276,\ chx = 1.0523$$

Example 1.1a (Equation 1.35a)

$$\begin{aligned} \sin N(x, x) &= \sin x\ chx + \cos x\ shx \\ &= 0.3165 \times 1.0523 + 0.9486 \times 0.3276 \\ &= 0.3331 + 0.3108 = 0.6439 \\ \sin N(x, -x) &= \sin x\ chx - \cos x\ shx \\ &= 0.3165 \times 1.0523 - 0.9486 \times 0.3276 \\ &= 0.3331 - 0.3108 = 0.0223 \end{aligned}$$

Example 1.1b (Equation 1.35b)

$$\begin{aligned} \cos N(x, x) &= \cos x\ chx - \sin x\ shx \\ &= 0.9486 \times 1.0523 - 0.3165 \times 0.3276 \\ &= 0.9992 - 0.1039 = 0.8943 \\ \cos N(x, -x) &= \cos x\ chx + \sin x\ shx \\ &= 0.9486 \times 1.0523 + 0.3165 \times 0.3276 \\ &= 0.9992 + 0.1039 = 1.1021 \end{aligned}$$

Example 1.2 (Equation 1.47)

Given, the numerical values of, sinNx and cosNx, obtain the value of x. Take the basic data from Examples 1.1a and 1.1b.

Solution:

Given that

$$\sin Nx = 0.6439,\ \cos Nx = 0.8943$$

From (1.47), that is

$$\begin{aligned} ch\ 2x &= \cos N^2 x + \sin N^2 x = 1.214380 \\ 2x &= 0.643630 \\ x &= 0.321815 \text{ rad } (=18.438641 \text{ degree}) \end{aligned}$$

Then with reference to the results of the Examples 1.1a and 1.1b, proceed for the following Examples.

Example 1.3 (Equation 1.67, 1.73)

$$\begin{aligned} \sin N'(x, x) &= [\cos N(x, -x) + \cos N(x, x)] \\ &= 1.1021 + 0.8943 = 1.9964 \\ \cos N'(x, x) &= -\ [\sin N(x, -x) + \sin N(x, x)] \\ &= -\ 0.0223 - 0.6439 = -\ 0.6662 \end{aligned}$$

Similarly

Example 1.4 (Equation 1.71, 1.73)

$$\begin{aligned} \sin N'(x, -x) &= 2 \sin x\ shx \\ &= 2 \times 0.3165 \times 0.3276 = 0.2074 \\ \cos N'(x, -x) &= 2 \cos x\ shx \\ &= 2 \times 0.9486 \times 0.3276 = 0.6215 \end{aligned}$$

Example 1.5 (Equation 1.74)

$$\tan N(x, x) = [\sin N'(x, -x)] / [\cos N'(x, -x)]$$

$$= 0.2074 / 0.6215 = 0.3337$$

Now

$$2x = 36.9 \text{ deg} = 0.644 \text{ rad}$$

$$\sin 2x = 0.6004, \cos 2x = 0.7997, sh\ 2x = 0.6895, ch\ 2x = 1.2146$$

With this

Example 1.6 (Equation 1.45)

$$\sin Nx.\cos Nx = \frac{1}{2}[\sin 2x + sh\ 2x \cos 2x]$$

$$\text{LHS} = 0.6439 \times 0.8943 = \mathbf{0.5758}$$

$$\text{RHS} = \frac{1}{2}[0.6004 + 0.7997 \times 0.6895] = \mathbf{0.5759} \qquad \textbf{(Checked)}$$

Example 1.7 (Equation 1.55)

To check $\tan N(x, x) = \dfrac{\tan x + thx}{1 - \tan x.thx}$

Reference (the data is borrowed from Example 2.1, based equation 2.39a–f), *viz.*:

$$8 + 6i = R\,(chx + shx) = 5.292\ (1.5119 + 1.1339\ i)$$

$$R\ chx = 8,\ R\ shx = 6,\ R = 5.292, \text{ so we get}$$

$$chx = 1.5119,\ shx = 1.1339$$

$$thx = shx / chx = 1.1339 / 1.5119 = 0.75 \text{ rad } (= 42.9118 \text{ deg})$$

so, $$x = 0.97296 \text{ rad } (= 55.7468 \text{ deg}),\ \sin x = 0.8266,\ \cos x = 0.5628,$$

$$\tan x = 1.46852$$

$$chx = 1.5119,\ shx = 1.1339,\ thx = 0.75 \text{ rad } (= 42.9118 \text{ deg})$$

$$\tan x + thx = 1.46852 + 0.75 = 2.2185$$

$$1 - \tan x\ .\ thx = 1 - 1.4685 \times 0.75 = -0.101375$$

RHS $= 2.2185 / -0.101375 = -\mathbf{21.884}$

$$\sin N(x, x) = \sin x\ chx + \cos x\ shx$$

$$= 0.8266 \times 1.5119 + 0.5628 \times 1.1339 = 1.8879$$

$$\cos N(x, x) = \cos x\ chx - \sin x\ shx$$

$$= 0.5628 \times 1.5119 - 0.8266 \times 1.1339 = -0.08638$$

LHS $= \tan N(x, x) = 1.8879 / 0.08638 = \mathbf{-21.8558}$ **(Checked)**

Example 1.8 (Equation 1.61)

To check $\tan N(x, y) = \dfrac{\tan x + thy}{1 - \tan x.thy}$

Reference:

[for x**]** : $8 + 6i = R(chx + i\ shx) = 5.292\ (1.5119 + 1.1339\ i)$

$$R\ chx = 8,\ R\ shx = 6,\ R = 5.292, \text{ so that}$$

$$chx = 1.5119, \; shx = 1.1339$$
$$thx = shx / chx = 1.13399 / 1.5119 = 0.75 \text{ rad } (= 42.9118 \text{ deg})$$

so, $x = 0.97296$ rad ($= 55.7468$ deg), $\sin x = 0.8266$, $\cos x = 0.5628$,

$$\tan x = 1.46852$$
$$chx = 1.5119, \; shx = 1.1339, \; thx = th(0.97296) = 0.7500 \text{ rad}$$

[for y]: $12 + 5i = R(chy + i\, shy) = 10.9087\,(1.1 + 0.4583\, i)$

$$R\, chy = 12, \; R\, shy = 6, \; R = 10.9087, \text{ so that}$$
$$chy = 1.1, \; shy = 0.4583$$
$$thy = shy / chy = 0.4583 / 1.1 = 0.4166 \text{ rad } (= 23.8694 \text{ deg}),$$

so, $y = 0.4436$ rad ($= 25.4164$ deg), $\sin y = 0.4292$, $\cos y = 0.9032$,

$$\tan y = 0.4752$$
$$chy = 1.1, \; shy = 0.4583, \; thy = th(0.4466) = 0.4166 \text{ rad } (= 23.8694 \text{ deg})$$

$$\sin N(x, y) = \sin x\, chy + \cos x\, shy = 1.1672$$
$$\cos N(x, y) = \cos x\, chy - \sin x\, shy = 0.2403$$

LHS $= \tan N(x, y) = [\sin N(x, y)] / [\cos N(x, y)]$

$= 1.1672 / 0.2403 = \mathbf{4.8573}$

RHS $= (1.4685 + 0.4166) / (1 - 1.4685 \times 0.4166)$

$= \mathbf{4.8557}$ **(Checked)**

Example 1.9 (Equation 1.47)

$$\sin N^2 x + \cos N^2 x = ch 2x$$

LHS $= 0.6439^2 + 0.8943^2 = 0.4146 + 0.7997$

$= \mathbf{1.2146} =$ RHS **(Checked)**

Example 1.10 (Equation 1.48)

$$\cos N^2 x - \sin N^2 x = \cos 2x - \sin 2x\; sh\, 2x$$

LHS $= 0.7997 - 0.4146 = \mathbf{0.3857}$

RHS $= 0.7997 - 0.6004 \times 0.6895 = \mathbf{0.3851}$ **(Checked)**

Example 1.11 (Equation 1.50)

$$(\sin Nx + \cos Nx)^2 = \sin 2x + ch\, 2x + \cos 2x\; sh\, 2x$$

LHS $= (0.6439 + 0.8943)^2 = \mathbf{2.3664}$

RHS $= 0.6004 + 1.2146 + 0.7997 \times 0.6895 = \mathbf{2.3664}$ **(Checked)**

Example 1.12 (Equation 1.50)

$$(\sin Nx - \cos Nx)^2 = ch\, 2x - \sin 2x - \cos 2x\; sh\, 2x$$

LHS $= (0.6439 - 0.8943)^2 = \mathbf{0.0629}$

RHS $= 1.2146 - 0.6004 - 0.7997 \times 0.6895 = \mathbf{0.0623}$ **(Checked)**

Data for Example 1.13

$$x + \frac{\pi}{4} = 63.45 \text{ deg} = 1.1074 \text{ rad}$$

$$\sin x = 0.8945, \; \cos x = 0.4470, \; shx = 1.3480, \; chx = 1.6784$$

Example 1.13a (Equation 1.39)

$$\begin{aligned}\text{sinN}[(x+\frac{\pi}{4}),x] &= \sin(x+\frac{\pi}{4})\,chx + \cos(x+\frac{\pi}{4})\,shx\\ &= 0.8945 \times 1.0523 + 0.4470 \times 0.3276\\ &= 0.9413 + 0.1464 = 1.0877\end{aligned}$$

Example 1.13b (Equation 1.39)

$$\begin{aligned}\text{cosN}[(x+\frac{\pi}{4}),x] &= \cos(x+\frac{\pi}{4})\,chx - \sin(x+\frac{\pi}{4})\,shx\\ &= 0.4470 \times 1.0523 - 0.8945 \times 0.3276\\ &= 0.4704 - 0.2930 = 0.1774\end{aligned}$$

Example 1.13c (Equation 1.43, 1.44)

$$\text{sinN}(x, x) + \text{cosN}(x, x) = \sqrt{2}\,\text{sinN}\,[(x+\frac{\pi}{4}),x]$$

LHS = 0.6439 + 0.8943 = **1.5382**

RHS = $\sqrt{2}$ × 0.8945 = **1.5382** **(Checked)**

Example 1.13d (Equation 1.44)

$$\text{sinN}x - \text{cosN}x = -\sqrt{2}\,\text{cosN}\,[(x+\frac{\pi}{4}),x]$$

LHS = 0.6439 – 0.8943 = **– 0.2504**

RHS = – $\sqrt{2}$ × 0.1774 = **– 0.2508** **(Checked)**

Data for Example 1.14

$$x+\frac{\pi}{2} = 108.45 \text{ deg} = 1.893 \text{ rad}$$

sin$x$ = 0.9486, cos$x$ = – 0.3165, sh$x$ = 3.2443, ch$x$ =3.3949

Example 1.14a (Equation 1.39)

$$\begin{aligned}\text{sinN}\,[(x+\frac{\pi}{2}),x] &= \sin(x+\frac{\pi}{2})\,chx + \cos(x+\frac{\pi}{2})\,shx\\ &= 0.9486\,x1.0523 - 0.3165\,x\,0.3276\\ &= 0.9982 - 0.1037 = \mathbf{0.8945} = \text{cosN}(x, x)\end{aligned}$$ **(Checked)**

Example 1.14b (Equation 1.39)

$$\begin{aligned}\text{cosN}[(x+\frac{\pi}{2}),x] &= \cos(x+\frac{\pi}{2})\,chx - \sin(x+\frac{\pi}{2})\,shx\\ &= -0.3165 \times 1.0523 - 0.9486 \times 0.3276\\ &= -0.3331 - 0.3108 = \mathbf{-0.6439} = -\text{sinN}(x, x)\end{aligned}$$ **(Checked)**

Data for Example 1.15a-d

$$x-\frac{\pi}{2} = -71.55 \text{ deg} = -1.249 \text{ rad}$$

sin$x$ = – 0.9486, cos$x$ = 0.3165, sh$x$ = – 1.60003, ch$x$ = 1.88682

Example 1.15a (Equation 1.40)

$$\sin N[(x-\frac{\pi}{2}),x] = \sin\ (x-\frac{\pi}{2})\ chx + \cos\ (x-\frac{\pi}{2})\ shx$$
$$= -0.9486 \times 1.0523 + 0.3165 \times 0.3276$$
$$= -0.9982 + 0.1037 = -0.8945 = -\cos N(x,\ x)\ \textbf{(Checked)}$$

Example 1.15b (Equation 1.40)

$$\cos N[(x-\frac{\pi}{2}),x] = \cos(x-\frac{\pi}{2})\ chx + \sin(x-\frac{\pi}{2})\ shx$$
$$= 0.3165 \times 1.0523 + 0.9486 \times 0.3276$$
$$= 0.3331 + 0.3108 = 0.6439 = \sin N(x,\ x) \qquad \textbf{(Checked)}$$

Example 1.15c (Equation 1.40)

$$\sin N[(x-\frac{\pi}{2}),x] = -\cos N(x,\ x) = -0.8945 \qquad \textbf{(Checked)}$$

Example 1.15d (Equation 1.40)

$$\cos N[(x-\frac{\pi}{2}),x] = \sin N(x,\ x) = 0.6439 \qquad \textbf{(Checked)}$$

Example 1.15e (Equation 1.59)

Given, x = 120 deg (= 2.094395 rad), calculate the values of, sinNx and cosNx using (1.59). Check the values thus obtained.

Solution:

Call the necessary formulas (1.59).

$$\sec N(x,\ x) = \frac{\sec x.\sec hx}{1-\tan x.thx}$$
$$\csc N(x,\ x) = \frac{\sec x.\sec hx}{\tan x + thx}$$
$$x = 120 \text{ deg } (= 2.094395 \text{ rad})$$
$$\cos x = -0.5,\ \sec x = -2.0,\ \tan x = -1.732051$$
$$thx = 0.970124,\ chx = 4.121836,\ \sec hx = 0.242610$$
$$ch\ 2x = 32.97906$$

Then

$$\sin Nx = 1/[\csc Nx] = \frac{-1.732051+0.970124}{-2\times 0.242610} = \mathbf{1.570271}\ (1.5702691)$$
$$\cos Nx = 1/[\sec Nx] = \frac{1-0.970124\times(-1.732051)}{-2\times 0.242610} = \mathbf{-5.523895}\ (-5.523886)$$

The values within ( ) are those obtained using (1.23a).

Example 1.15f (Equations 1.69, 1.70, 1.81 1.85)

Given, x = 18.45 deg (= 0.322 rad), verify the corresponding equations.

Solution:

We have from the given data:

$2x$ = 36.9 deg (= 0.644 rad), sin $2x$ = 0.60042, *ch* $2x$ = 1.21463

$S_1 = 0.6439$, $S_2 = 0.0233$, $C_1 = 0.8943$, $C_2 = 1.1021$

$S'_1 = 1.9964$, $S'_2 = 0.2074$, $C'_1 = -0.6662$, $C'_2 = 0.6215$

With these

$$(1.69) \Rightarrow C_1C_2 = \frac{1}{4}(S_1'^2 - S_2'^2)$$

LHS = (0.8943) (1.1021) = **0.9856, RHS** = $(1.9964^2 - 0.2074^2) / 4$ = **1.10072**

Hence, **LHS = RHS.**

$$(1.70) \Rightarrow \frac{C_1}{C_2} = \frac{S_1' - S_2'}{S_1' + S_2'}$$

LHS = 0.81145, RHS = 1.789 / 2.2038 = **0.81148**

Hence, **LHS = RHS**

$$(1.81) \Rightarrow S_1' S_2' + C_1' C_2' = 0$$

LHS = (1.9964) (0.2074) + (–0.6662) (0.6215) = **0 = RHS.**

$$(1.83) \Rightarrow S_1S_2 + C_1C_2 = 1$$

LHS = (0.6439) (0.0233) + (0.8943) (1.1021) = 1.0006 = **1 = RHS.**

$$(1.85) \Rightarrow S_1C_2 + C_1S_2 = \sin 2x \; ch\; 2x$$

LHS = (0.6934) (1.1021) + (0.0233) (0.8943) = **0.73048**

RHS = (90.60042) (1.21463) = **0.7293**

Hence, LHS = RHS

1.8.4 Basic Nbic Functions and their Derivatives

Orthogonality of Single Nbic functions:

For Nbic Function as function of (x, x):

$$\begin{aligned} S_1 &= \sin N(x, x) = A + C = \sin x \; chx + \cos x \; shx \\ S_2 &= \sin N(x, -x) = A - C = \sin x \; chx - \cos x \; shx \\ C_1 &= \cos N(x, x) = D - B = \cos x \; chx - \sin x \; shx \\ C_2 &= \cos N(x, -x) = D + B = \cos x \; chx + \sin x \; shx \end{aligned} \qquad (1.86)$$

Remembering the differential rule of the functions A, B, C and D as functions of (x, x)

$$A = Axx = A\,(x, x) = \sin x \; chx$$

$$\begin{aligned}
B &= Bxx = B\ (x, x) = \sin x\ shx \\
C &= Cxx = C\ (x, x) = \cos x\ shx \\
D &= Dxx = D\ (x, x) = \cos x\ chx
\end{aligned} \tag{1.87}$$

$$\begin{aligned}
A' &= D + B = C_2 \\
B' &= C + A = S_1 \\
C' &= D - B = C_1 \\
D' &= C - A = - S_2
\end{aligned} \tag{1.88}$$

Thus with these

$$\begin{aligned}
S_1' &= \sin N'(x, x) = \cos N(x, -x) + \cos N(x, x) = C_1 + C_2 = A' + C' = 2\ D \\
S_2' &= \sin N'(x, -x) = 2\ \sin x\ shx = 2\ B \\
C_1' &= \cos N'(x, x) = -\ \sin N(x, -x) - \sin N(x, x) = -\ S_1 - S_2 = D' - B' = -\ 2\ A \\
C_2' &= \cos N'(x, -x) = 2\ \cos x\ shx = D' + B' = 2\ C
\end{aligned} \tag{1.89a}$$

$$\begin{aligned}
S_1'' &= 2\ \cos x\ shx - \sin N(x, -x) - \sin N(x, x) = 2\ D/ = 2\ (C - A) = -\ 2\ S_2 \\
S_2'' &= 2\ B' = 2\ S_1 \\
C_1'' &= -\ 2\ A' = -\ 2\ C_2 \\
C_2'' &= 2\ C' = 2\ C_1
\end{aligned} \tag{1.89b}$$

$$\begin{aligned}
S_1''' &= -\ 2\ S_2' = 2\ (C' - A') = -\ 4\ B \\
S_2''' &= 2\ S_1' = 4\ D \\
C_1''' &= -\ 2\ C_2' = -\ 4\ C \\
C_2''' &= 2\ C_1'' = -\ 4\ A
\end{aligned} \tag{1.90}$$

$$\begin{aligned}
S_1'''' &= -\ 4\ B' = -\ 4\ \sin N(x, x) = -\ 4\ S_1 \\
S_2'''' &= 4\ D' = -\ 4\ \sin N(x, -x) = -\ 4\ S_2 \\
C_1'''' &= -\ 4\ C' = -\ 4\ \cos N(x, x) = -\ 4\ C_1 \\
C_2'''' &= -\ 4\ A' = -\ 4\ \cos N(x, -x) = -\ 4\ C_2
\end{aligned} \tag{1.91}$$

Collecting from these

$$\begin{aligned}
S_1'' &= -\ 2\ S_2,\ S_2'' = 2\ S_1,\ C_1'' = -\ 2\ C_2,\ C_2'' = 2\ C_1 \\
S_1'''' &+ 4\ S_1 = 0,\ S_2'''' + 4\ S_2 = 0,\ C_1'''' + 4\ C_1 = 0,\ C_2'''' + 4\ C_2 = 0
\end{aligned} \tag{1.92}$$

Thus

$$\begin{aligned}
\sin N''(x, x) + 2\ \sin N(x, -x) = 0 &\Rightarrow S_1'' + 2\ S_2 = 0 \\
\cos N''(x, x) + 2\ \cos N(x, -x) = 0 &\Rightarrow C_1'' + 2\ C_2 = 0
\end{aligned} \tag{1.93}$$

Thus the following basic Nbic functions **follow orthogonality**

$$\begin{aligned}
\sin N''''(x, x) + 4\ \sin N(x, x) = 0 &\Rightarrow S_1'''' + 4\ S_1 = 0 \\
\cos N''''(x, x) + 4\ \cos N(x, x) = 0 &\Rightarrow C_1'''' + 4\ C_1 = 0 \\
\sin N''''\ (x, -\ x) + 4\ \sin N(x, -\ x) = 0 &\Rightarrow S_2'''' + 4\ S_2 = 0 \\
\cos N''''(x, -\ x) + 4\ \cos N(x, -\ x) = 0 &\Rightarrow C_2'''' + 4\ C_2 = 0
\end{aligned} \tag{1.94}$$

For Nbic Function as function of (*x*, *y*)

Taking help of the functions A*, B*, C* and D* as functions of (*x*, *y*) from (1.32), and the Nbic Functions from (1.33b), namely

$$\begin{aligned} S_1^* &= \text{sinN}(x, y) = \sin x\ chy + \cos x\ shy = A^* + C^* \\ S_2^* &= \text{sinN}(x, -y) = \sin x\ chy - \cos x\ shy = A^* - C^* \\ C_1^* &= \text{cosN}(x, y) = \cos x\ chy - \sin x\ shy = D^* - B^* \\ C_2^* &= \text{cosN}(x, -y) = \cos x\ chy + \sin x\ shy = D^* + B^* \end{aligned} \quad (1.95)$$

the following expressions can be obtained as:

$$\mathbf{S_1^* + C_1^* = sinN(x, y) + cosN(x, y) = \sqrt{2}\ cosN[(x - \frac{\pi}{4}), y]}$$

$$\mathbf{S_1^* - C_1^* = sinN(x, y) - cosN(x, y) = \sqrt{2}\ cosN[(x - \frac{\pi}{4}), -y]} \quad (1.96)$$

$$\begin{aligned} \mathbf{C_1^{*2} + S_1^{*2}} &= \mathbf{cosN^2(x, y) + sinN^2(x, y) = ch\ 2y} \\ \mathbf{C_1^{*2} - S_1^{*2}} &= \mathbf{cosN^2(x, y) - sinN^2(x, y) = cos\ 2x - sin\ 2x\ sh\ 2y} \end{aligned} \quad (1.97)$$

For differentiation, these are made use of

$$\begin{aligned} A^* &= \sin x\ chy \\ (A^*)'' &= -A^*,\ (A^*)'''' = A^*,\ (A^*)^{0000} = A^*,\ (A^*)''^{00} = -A^*,\ \nabla^4 A^* = 0 \\ B^* &= \sin x\ shy \\ (B^*)'' &= -B^*,\ (B^*)'''' = B^*,\ (B^*)^{0000} = B^*,\ (B^*)''^{00} = -B^*,\ \nabla^4 B^* = 0 \\ C^* &= \cos x\ shy \\ (C^*)'' &= -C^*,\ (C^*)'''' = C^*,\ (C^*)^{0000} = C^*,\ (C^*)''^{00} = -C^*,\ \nabla^4 C^* = 0 \\ D^* &= \cos x\ chy \\ (D^*)'' &= -D^*,\ (D^*)'''' = D^*,\ (D^*)^{0000} = D^*,\ (D^*)''^{00} = -D^*,\ \nabla^4 D^* = 0 \end{aligned} \quad (1.98a)$$

Recall (1.95)

$$\begin{aligned} S_1^* &= \text{sinN}(x, y) = A^* + C^* \\ S_2^* &= \text{sinN}(x, -y) = A^* - C^* \\ C_1^* &= \text{cosN}(x, y) = D^* - B^* \\ C_2^* &= \text{cosN}(x, y) = D^* + B^* \end{aligned} \quad (1.98b)$$

∇^4 Criterion

In view of (1.98a), expressions (1.98b), provides the criteria

$$\nabla^4[S_1^*, S_2^*, C_1^*, C_2^*] = 0 \quad (1.99)$$

where, $\nabla^4 = \dfrac{\delta^4}{\delta x^4} + 2\dfrac{d^4}{dx^2 dy^2} + \dfrac{d^4}{dy^4} = (\,)'''' + 2(\,)''^{00} + (\,)^{0000}$, is the biharmonic operator.

Thus the basic single Nbic functions N(*x*, *y*) are biharmonic. That is $\nabla^4 N(x, y) = 0$.

1.8.5 Integration Rule for Single Nbic Functions

[symbol, $\int$ or **I** represents integration symbol]:

The integration of the following Nbic Functions referred from (1.88) when performed, can be shown to reduce to the basic function (1.87) as:

$$\int \sin N(x, x)\, dx = \sin x\ shx, \qquad \text{or, } I(S_1)\, dx = B$$
$$\int \sin N(x, -x)\, dx = \cos x\ chx, \qquad \text{or, } I(S_2)\, dx = D$$
$$\int \cos N(x, x)\, dx = \cos x\ shx, \qquad \text{or, } I(C_1)\, dx = C$$
$$\int \cos N(x, -x)\, dx = \sin x\ chx, \qquad \text{or, } I(C_2)\, dx = A \qquad (1.100)$$

Taking help of (1.89a), we arrive at these integration results in terms of the basic functions (1.87) as

$$\int \sin x\ chx\, dx = -\frac{1}{2}\cos N(x, x)$$
or, $$I\ (A)\ dx = -C_1/2$$
$$\int \cos x\ shx\, dx = \frac{1}{2}\cos N(x, -x)$$
or, $$I\ (C)\ dx = C_2/2$$
$$\int \cos x\ chx\, dx = \frac{1}{2}\sin N(x, x)$$
or, $$I\ (D)\ dx = S_1/2$$
$$\int \sin x\ shx\, dx = \sin N(x, -x)$$
or, $$I\ (B)\ dx = S_2/2 \qquad (1.101)$$

1.8.6 Related Expressions Involving Differentiation and Integration

The basic Nbic Functions are represented in terms of the integration of the sum or difference of the other basic Nbic Functions, as shown

$$\cos N(x, x) = \int [\sin N(x, -x) - \sin N(x, x)]dx$$
or, $$C_1 = I\ (S_2 - S_1)\ dx$$
$$\cos N(x, -x) = \int [\sin N(x, -x) + \sin N(x, x)]dx$$
or, $$C_2 = I\ (S_2 + S_1)\ dx$$
$$\sin N(x, x) = \int [\cos N(x, -x) + \cos N(x, x)]dx$$
or, $$S_1 = I\ (C_2 + C_1)\ dx$$
$$\sin N(x, -x) = \int [\cos N(x, -x) - \cos N(x, x)]dx$$
or, $$S_2 = I\ (C_2 - C_1)\ dx \qquad (1.102)$$

The expressions of S_2' and C_2' from (1.89) as well as integrations of single Nbic functions (1.100) on suitable division, yield

$$\tan x = \frac{\sin N'(x,-x)}{\cos N'(x,-x)} = -\frac{\cos N'(x,x)}{\sin N'(x,x)}$$

$$= \frac{\int \cos N(x,-x)dx}{\int \sin N(x,-x)dx} = \frac{\int \sin N(x,x)dx}{\int \cos N(x,x)dx}$$

or, $\quad T_1 = S_2'/C_2' = -C_1'/S_1'$

$$= I\,(C_2)\,dx\,/\,I\,(S_2)\,dx = I\,(S_1)\,dx\,/\,I\,(C_1)\,dx \qquad (1.103)$$

Similarly the expressions of (C_2' and S_1'), and (S_2' and C_1'), from (1.89a) on suitable division, yields

$$thx = \frac{\cos N'(x,-x)}{\sin N'(x,x)} = -\frac{\sin N'(x,-x)}{\cos N'(x,x)}$$

or, $\quad thx = C_2'/S_1' = -S_2'/C_1' \qquad (1.104)$

Further, combinations of the various differentiation and integration terms provide the following identities:

$$\sin N'(x,x).\int \sin N(x,x)\,dx + \cos N'(x,x).\int \cos N(x,x)\,dx = 0$$

or, $\quad S_1'\,I\,(S_1)\,dx + C_1'\,I\,(C_1)\,dx = 0 \qquad (1.105)$

$$\int \sin N(x,x)\,dx\,.\int \sin N(x,-x)\,dx = \int \cos N(x,x)\,dx\,.\int \cos N(x,-x)\,dx$$

or, $\quad I\,(S_1)\,dx\;I\,(S_2)\,dx = I\,(C_1)\,dx\;I\,(C_2)\,dx \qquad (1.106)$

From (1.87) and (1.88)

$$\frac{\tan x}{thx} = -\frac{\cos N'(x,x)}{\cos N'(x,-x)} \qquad (1.107)$$

$$\tan x\, th\, x = \frac{\sin N'(x,-x)}{\sin N'(x,x)} \qquad (1.108)$$

Circular functions (*viz.*, sin*x* and cos*x*), can well be written from, tan*x*, and thus, from (1.103), after some simplification,
we get, in terms of S_2' and C_2' as:

$$\sin x = \frac{\sin N'(x,-x)}{\sqrt{\sin N'^2(x,-x) + \cos N'^2(x,-x)}}$$

or, $\quad S = S_2'/\,\text{sqrt}\;[(S_2')^2 + (C_2')^2]$

$$\cos x = \frac{\cos N'(x,-x)}{\sqrt{\sin N'^2(x,-x) + \cos N'^2(x,-x)}}$$

or, $\quad C = C_2'/\,\text{sqrt}\;[(S_2')^2 + (C_2')^2] \qquad (1.109a)$

Alternatively, in terms of S_1' and C_1' as:

$$\sin x = -\frac{\cos N'(x,x)}{\sqrt{\sin N'^2(x,x) + \cos N'^2(x,x)}}$$

or, $\quad S = -C_1'/\,\text{sqrt}\;[(S_1')^2 + (C_1')^2]$

$$\cos x = \frac{\sin \mathrm{N}'(x,x)}{\sqrt{\sin \mathrm{N}'^2(x,x) + \cos \mathrm{N}'^2(x,x)}}$$

or, $$C = S_1' / \text{sqrt}\ [(S_1')^2 + (C_1')^2] \tag{1.109b}$$

(Here, S and C represent, sin$x$ and cos$x$, respectively, and **sqrt** is understood as: **square root of**)

Equations (1.109a, b) are the expressions of circular functions written in terms of the first derivatives of the Nbic functions.

1.8.7 Interpretation and Representation in Terms of Circular Functions

[for (1.109a) and (1.109b)]

The very pattern of the derived expressions (1.109a) or (1.109b) suggests that, each of them follows the similar pattern which forms the basis of the formulation of plane trigonometry. This is elaborated as follows:

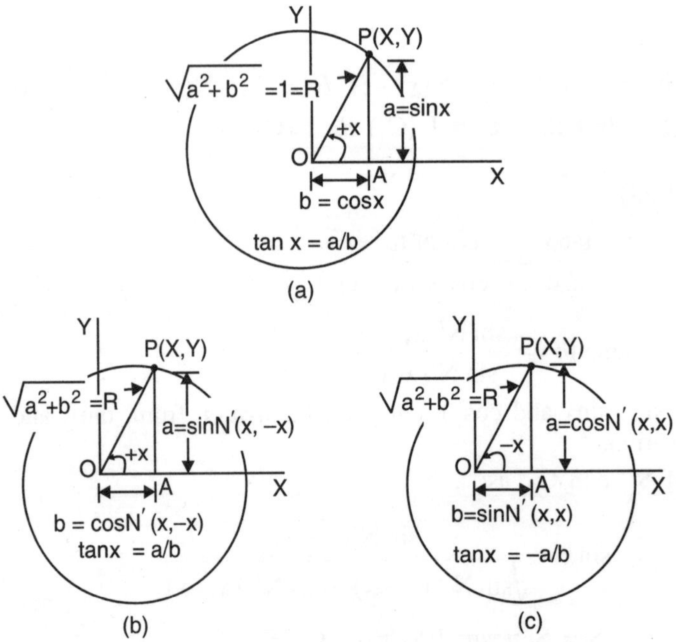

$$\begin{bmatrix} \sin x,\ \sin\mathrm{N}'(x,-x),\ \cos\mathrm{N}'(x,x) = a/\mathrm{R} \\ \cos x,\ \cos\mathrm{N}'(x,-x),\ \sin\mathrm{N}'(x,x) = b/\mathrm{R} \end{bmatrix}$$

Fig. 1.6: Tan-equivalence of the first derivative of tanN expressions

In plane trigonometry, consider a point P(X, Y) on the circumference of a circle having center O at (0, 0), and radius R = OP $= \sqrt{X^2 + Y^2}$, of unit magnitude, under the coordinate system (X, Y). The radius OP makes an angle x (deg) with the X-axis **(Figure 1.6a)**.

The X-projection and Y-projection of the radius, are cos$x$ and sin$x$, respectively. Since, the X-projection, Y-projection and the radius R follow the Pythagorean triangle, and $X^2 + Y^2 = \cos^2 x + \sin^2 x = R^2 = 1$, the functions, cos$x$ and sin$x$, thus are interpreted as the circular functions.

The same plane trigonometric functions (*viz.*, cos$x$ and sin$x$), are assured in the similar form of circular representation of the radius R = OP $=\sqrt{X^2+Y^2}$, when the first derivatives of the single Nbic functions N(x, –x), are represented suitably by X and Y (**Figure 1.6b**).

In (1.109a), if we represent,

X-axis as, $C_2' = \cos N'(x, -x)$; and Y-axis as, $S_2' = \sin N'(x, -x)$,

and the radius $R = OP = \sqrt{[\sin N'(x,-x)]^2 + [\cos N'(x,-x)]^2}$ (1.110a)

X-component of OP = $\cos x = \cos N'(x, -x) / OP$

Y-component of OP = $\sin x = \sin N'(x, -x) / OP$ (1.110b)

so that

$\tan x = Y / X = \sin N'(x, -x) / \cos N'(x, -x)$ (1.110c)

The same plane trigonometric functions (viz., cos$x$ and sin$x$), are assured in the similar form of circular representation of the radius R = OP $=\sqrt{X^2+Y^2}$, when the first derivatives of the single Nbic functions N(x, x), are represented suitably by X and Y (**Figure 1.6c**).

In (1.109b), if we represent,

X-axis as, $S_1' = \sin N'(x, x)$; and Y-axis as, $C_1' = \cos N'(x, x)$,

and, the radius $OP = \sqrt{[\sin N'(x,x)]^2 + [\cos N'(x,x)]^2}$ (1.110d)

X-component of OP = $\cos x = [\sin N'(x, x)] / OP$

Y-component of OP = $-\sin x = -[\cos N'(x, x)] / OP$ (1.110e)

so that

$\tan x = Y / X = -[\cos N'(x, x)] / [\sin N'(x, x)]$ (1.110f)

In (1.109a) and (1.109b), the triangle formed by the corresponding X-projection,Y-projection and the radius R, is a right angled triangle.

1.9 Nbic FUNCTIONS WITH VARIABLE (2x, ± 2x) AND (2x, ± x)

1.9.1 Similarity of Forms

The expanded expressions conform well to the pattern of the corresponding formulas involving terms as product of trigonometric terms, [like, sin(A + B), or, cos(A + B)], and hyperbolic terms [like, *shx* or, *chy*]. The final expressions again, appear nicely as that of Nbic function itself. Remembering that, the abbreviated form like, sinN(x + y) to mean sinN[(x + y), (x + y)], etc., when expanded, the following expressions of single Nbic function involving (2x, ± 2x) and (2x, ± x), namely,

$\sin N(2x, 2x)$, $\sin N(2x, -2x)$, $\sin N(2x, x)$, $\sin N(2x, -x)$
$\cos N(2x, 2x)$, $\cos N(2x, -2x)$, $\cos N(2x, x)$, $\cos N(2x, -x)$
$\tan N(2x, 2x)$, $\tan N(2x, -2x)$, $\tan N(2x, x)$, $\tan N(2x, -x)$

are derived as follows, and for that, first we elaborate the following expansions:

Expand **sinN(x + y)**

$$\begin{aligned}\sin N(x + y) &= \sin(x + y)\, ch(x + y) + \cos(x + y)\, sh(x + y)\\ &= [\sin x \cos y + \cos x \sin y]\,[chx\, chy + shx\, shy]\\ &\quad + [\cos x \cos y - \sin x \sin y]\,[shx\, chy + chx\, shy]\end{aligned}$$

$= \sin x\ chx\ (\cos y\ chy - \sin y\ shy)$
$+ \cos x\ shx\ (\cos y\ chy + \sin y\ shy)$
$+ \cos x\ chx\ (\sin y\ chy + \cos y\ shy)$
$- \sin x\ shx\ (- \cos y\ shy + \sin y\ chy)$
$= \sin x\ chx\ \cos N(y, y) + \cos x\ shx\ \cos N(y, -y)$
$+ \cos x\ chx\ \sin N(y, y) - \sin x\ shx\ \sin N(y, -y)$
$= [\sin x\ \cos N(y, y) + \cos x\ \sin N(y, y)]chx$
$+ [\cos x\ \cos N(y, -y) - \sin x\ \sin N(y, -y)]shx$

Note: According to the definition of Double Nbic function, this expression can be written as

$= \cos N_{2e}[(y, -y), x]shx + \sin N_{2e}[(y, y), x]chx$

The form is

$\Rightarrow$ [sin cos + cos sin]*ch* + [cos cos – sin sin]*sh*
$\Rightarrow$ sin(A + B)*chx* +cos(A + B)*shx*
$\Rightarrow$ sin *ch* + cos *sh* (1.111)

Expand **sinN(*x* –*y*)**

$\sin N(x - y) = \sin(x - y)\ ch(x - y) + \cos(x - y)\ sh(x - y)$
$= [\sin x \cos y - \cos x \sin y]\ [chx\ chy - shx\ shy]$
$+ [\cos x \cos y + \sin x \sin y][shx\ chy - chx\ shy]$
$= \sin x\ chx\ (\cos y\ chy - \sin y\ shy)$
$+ \cos x\ shx\ (\cos y\ chy + \sin y\ shy)$
$- \cos x\ chx\ (\sin y\ chy + \cos y\ shy)\ )$
$+ \sin x\ shx\ (- \cos y\ shy + \sin y\ chy)$
$= \sin x\ chx\ \cos N(y, y) + \cos x\ shx\ \cos N(y, -y)$
$- \cos x\ chx\ \sin N(y, y) + \sin x\ shx\ \sin N(y, -y)$
$= [\sin x\ \cos N(y, y) - \cos x\ \sin N(y, y)]chx$
$+ [\cos x\ \cos N(y, -y) + \sin x\ \sin N(y, -y)]shx$

Note: According to the definition of Double Nbic function, this expression can be written as

$= - \sin N_{2e}[(y, y), -x]\ chx + \cos N_{2e}[(y, -y), -x]\ shx$

The form is

$\Rightarrow$ [sin cos – cos sin] *ch* + [cos cos + sin sin] *sh*
$\Rightarrow$ sin (A – B) *chx* + cos (A – B) *shx*
$\Rightarrow$ sin *ch* – cos *sh* (1.112)

Expand **cosN(*x* + *y*)**

$\cos N(x + y) = \cos(x + y)\ ch(x + y) - \sin(x + y)\ sh(x + y)$
$= [\cos x \cos y - \sin x \sin y]\ [chx\ chy + shx\ shy]$
$- [\sin x \cos y + \cos x \sin y]\ [shx\ chy + chx\ shy]$
$= \cos x\ chx\ (\cos y\ chy - \sin y\ shy)$
$- \sin x\ shx\ (\cos y\ chy + \sin y\ shy)$
$- \sin x\ chx\ (\sin y\ chy + \cos y\ shy)$
$- \cos x\ shx\ (- \cos y\ shy + \sin y\ chy)$

$$= \cos x\ chx\ \cos N(y, y) - \sin x\ shx\ \cos N(y, -y)$$
$$- \sin x\ chx\ \sin N(y, y) - \cos x\ shx\ \sin N(y, -y)$$
$$= [\cos x\ \cos N(y, y) - \sin x\ \sin N(y, y)]chx$$
$$- [\sin x\ \cos N(y, -y) + \cos x\ \sin N(y, -y)]shx$$

Note: According to the definition of Double Nbic function, this expression can be written as

$$= \cos N_{2e}[(y, y), x]chx - \sin N_{2e}[(y, -y), x]shx$$

The form is

$$\Rightarrow [\cos\ \cos - \sin\ \sin]ch - [\sin\ \cos + \cos\ \sin]sh$$
$$\Rightarrow \cos(A + B)chx - \sin(A + B)shx$$
$$\Rightarrow \cos\ ch - \sin\ sh \qquad (1.113)$$

Expand **cosN($x - y$)**

$$\cos N(x - y) = \cos(x - y)\ ch(x - y) + \sin(x - y)\ sh(x - y)$$
$$= [\cos x\ \cos y + \sin x\ \sin y]\ [chx\ chy - shx\ shy]$$
$$+ [\sin x\ \cos y - \cos x\ \sin y]\ [shx\ chy - chx\ shy]$$
$$= \cos x\ chx\ (\cos y\ chy + \sin y\ shy)$$
$$- \sin x\ shx\ (\cos y\ chy + \sin y\ shy)$$
$$+ \sin x\ chx\ (\sin y\ chy - \cos y\ shy)$$
$$- \cos x\ shx\ (\cos y\ shy + \sin y\ chy)$$
$$= [\cos x\ \cos N(y, -y) + \sin x\ \sin N(y, -y)]chx$$
$$+ [\sin x\ \cos N(y,y) - \cos x\ \sin N(y, y)]shx$$

Note: According to the definition of Double Nbic function, this expression can be written as

$$= \cos N_{2e}[(y, -y), -x]chx + \sin N_{2e}[(y, y), -x]\ shx$$

The form is

$$\Rightarrow [\cos\ \cos + \sin\ \sin]\ ch + [\sin\ \cos - \cos\ \sin]\ sh$$
$$\Rightarrow \cos\ (A - B)\ chx + \sin\ (A + B)\ shx$$
$$\Rightarrow \cos\ ch + \sin\ sh \qquad (1.114)$$

Collecting formulas together

$$\mathbf{sinN(x + y)} = \sin(x + y)\ ch(x + y) + \cos(x + y)\ sh(x + y)$$
$$\mathbf{= [\sin x\ \cos N(y, y) + \cos x\ \sin N(y, y)]\ chx}$$
$$\mathbf{+ [\cos x\ \cos N(y, - y) - \sin x\ \sin N(y, - y)]shx}$$
$$= \cos N_{2e}\ [(y, -y), x]\ shx + \sin N_{2e}\ [(y, y), x]chx$$
$$\mathbf{\Rightarrow \sin(A + B)chx + \cos(A + B)shx}$$

The form is $\Rightarrow$ **sin.*ch* + cos.*sh*** $\Rightarrow [S_1* - \text{form}]$ (1.115)

$$\mathbf{sinN(x - y)} = \sin(x - y)\ ch(x - y) + \cos(x - y)\ sh(x - y)$$
$$\mathbf{= [\sin x\ \cos N(y, y) - \cos x\ \sin N(y, y)]chx}$$
$$\mathbf{+ [\cos x\ \cos N(y, - y) + \sin x\ \sin N(y, - y)]shx}$$

$= -\sin N_{2e}[(y, y), -x]chx + \cos N_{2e}[(y, -y), -x]shx$

$\Rightarrow \mathbf{\sin(A - B)}\boldsymbol{chx} + \mathbf{\cos(A - B)}\boldsymbol{shx}$

The form is $\Rightarrow$ **sin.*ch* – cos.*sh*** $\Rightarrow [S_2^* - \text{form}]$ (1.116)

$$\mathbf{cosN}(\boldsymbol{x + y}) = \cos(x + y)\, ch(x + y) - \sin(x + y)\, sh(x + y)$$

$$= \mathbf{[\cos x\ \cos N(y, y) - \sin x\ \sin N(y, y)]} \boldsymbol{chx}$$

$$\mathbf{-[\sin x\ \cos N(y, -y) + \cos x\ \sin N(y, -y)]} \boldsymbol{shx}$$

$$= \cos N_{2e}\ [(y, y), x]\ chx - \sin N_{2e}\ [(y, -y), x] shx$$

$$\Rightarrow \mathbf{\cos(A + B)}\boldsymbol{chx} - \mathbf{\sin(A + B)}\boldsymbol{shx}$$

The form is $\Rightarrow$ **cos.*ch* – sin.*sh*** $\Rightarrow [C_1^* - \text{form}]$ (1.117)

$$\mathbf{cosN}(\boldsymbol{x - y}) = \cos(x - y)\, ch(x - y) + \sin(x - y)\, sh(x - y)$$

$$= \mathbf{[\cos x\ \cos N(y, -y) + \sin x\ \sin N(y, -y)]} \boldsymbol{chx}$$

$$\mathbf{+ [\sin x\ \cos N(y, y) - \cos x\ \sin N(y, y)]} \boldsymbol{shx}$$

$$= \cos N_{2e}[(y, -y), -x]chx + \sin N_{2e}[(y, y), -x]shx$$

$$\Rightarrow \mathbf{\cos(A - B)}\boldsymbol{chx} + \mathbf{\sin(A + B)}\boldsymbol{shx}$$

The form is $\Rightarrow$ **cos.*ch* + sin.*sh*** $\Rightarrow [C_2^* - \text{form}]$ (1.118)

1.9.2 Single Nbic Function with Double Angle, N(2*x*, 2*x*) in Terms of, N(2*x*, *x*)

In hyperbolic functions, *shx*, *chx*, *thx*, etc., the parameter *x* is not an angle, but analogous to the Euclidean system, it is termed as hyperbolic angle. In this sense, following the similar logic, the parameters in all types of Nbic functions are preferred to be called by the name, **Nbic angle**.

Now for sinN(2*x*, 2*x*):

Putting $y = x$, on actual expansion of (1.115), that is

$$\sin N[(x + y), (x + y)] = \sin(x + y)\ ch(x + y) + \cos(x + y)\ sh(x + y)$$

we have

$$\sin N(2x, 2x) = [\sin x\ \cos N(y, y) + \cos x\ \sin N(y, y)]chx$$

$$+ [\cos x\ \cos N(y, -y) - \sin x\ \sin N(y, -y)]shx$$

$$= \sin x\ chx[\cos x\ chx - \sin x\ shx] + \cos x\ chx[\sin x\ chx + \cos x\ shx]$$

$$+ \cos x\ chx[\cos x\ chx + \sin x\ shx] - \sin x\ shx[\sin x\ chx - \cos x\ shx]$$

$$= \sin 2x\ [ch^2x + sh^2x] + sh\ 2x\ [\cos^2 x - \sin^2 x]$$

$= \sin 2x\ ch\ 2x + \cos 2x\ sh\ 2x$ **[by definition, this expression comes directly]**

$$= \sin 2x\ (ch^2x + sh^2x) + \cos 2x\ (shx\ chx + shx\ chx)$$

$$= [\sin 2x\ chx + \cos 2x\ shx]chx + [\sin 2x\ shx + \cos 2x\ chx]shx$$

$$= \sin N(2x, x)chx + \cos N(2x, -x)shx$$

Thus, on similar lines, we get by substituting $y = x$, in (1.115) through (1.118),

$$\begin{aligned}
\sin N(2x, 2x) &= \sin N(2x, x)\ chx + \cos N(2x, -x)\ shx \\
\cos N(2x, 2x) &= \cos N(2x, x)\ chx - \sin N(2x, -x)\ shx \\
\sin N(2x, -2x) &= \sin N(2x, -x)\ chx - \cos N(2x, x)\ shx \\
\cos N(2x, -2x) &= \cos N(2x, -x)\ chx + \sin N(2x, x)\ shx
\end{aligned} \quad (1.119)$$

where again, on actual expansion on similar lines and rearrangement

$$\begin{aligned}
\sin N(2x, x) &= \sin N(x, x) \cos x + \cos N(x, x) \sin x \\
\cos N(2x, x) &= \cos N(x, x) \cos x - \sin N(x, x) \sin x \\
\sin N(2x, -x) &= \sin N(x, -x) \cos x + \cos N(x, -x) \sin x \\
\cos N(2x, -x) &= \cos N(x, -x) \cos x - \sin N(x, -x) \sin x
\end{aligned} \tag{1.120}$$

Further

$$\begin{aligned}
\tan N(2x, x) &= [\sin N(2x, x)] / [\cos N(2x, x)] \\
&= \frac{\tan N(x, x) + \tan x}{1 - \tan N(x, x).\tan x} \\
\tan N(2x, -x) &= [\sin N(2x, -x)] / [\cos N(2x, -x)] \\
&= \frac{\tan N(x, -x) + \tan x}{1 - \tan N(x, -x).\tan x}
\end{aligned} \tag{1.121}$$

$$\sin N(2x, x).\cos N(2x, x) = \frac{1}{2}[\sin 4x + \cos 4x \; sh\, 2x] \tag{1.122}$$

Striking similarity is noticed when this expression (1.122) is compared with (1.45), namely

$$\sin Nx.\cos Nx = \frac{1}{2} [\sin 2x + \cos 2x \; sh\, 2x]$$

[Note: the RHS when written in terms of E type Nbic function N_{3e}, gives $\sin N_{3e}(x, x)/2$]

Now addition and subtraction among suitable expressions of (1.121), yields

$$\begin{aligned}
\sin N(2x, x) + \cos N(2x, x) &= (\cos 2x + \sin 2x)\, chx + (\cos 2x - \sin 2x)\, shx \\
\sin N(2x, -x) + \cos N(2x, -x) &= (\cos 2x + \sin 2x)\, chx - (\cos 2x - \sin 2x)\, shx
\end{aligned} \tag{1.123}$$

1.9.3 Some Examples Related to Nbic Functions with Variable (2*x*, ± 2*x*) and (2*x*, ± *x*)

These are demonstrated through Examples 1.16a – 1.16c.

Example 1.16a

Prove that

$$\sin N\, 2x = \frac{B_{2x}}{2B_x} \sin Nx - 2B_x \sin N(x, -x)$$

where, A_x, B_x, C_x, D_x are as in (1.34), and A_{2x}, B_{2x}, C_{2x}, D_{2x} are of the same form as in (1.34), but with functions of (2*x*), instead of (*x*), as follows:

$$\begin{aligned}
A &= Ax = A(x) = \sin x \; chx \\
B &= Bx = B(x) = \sin x \; shx \\
C &= Cx = C(x) = \cos x \; shx
\end{aligned}$$

$$D = Dx = D(x) = \cos x \; chx \tag{1.124a}$$

$$A_{2x} = A(2x) = \sin 2x \; ch \; 2x$$
$$B_{2x} = B(2x) = \sin 2x \; sh \; 2x$$
$$C_{2x} = C(2x) = \cos 2x \; sh \; 2x$$
$$D_{2x} = D(2x) = \cos 2x \; ch \; 2x \tag{1.124b}$$

Solution 1.16a

From (1.124a) and (1.124b)

$$\tan x = \frac{B_x}{C_x}$$

$$\tan 2x = \frac{B_{2x}}{C_{2x}} = \frac{2\tan x}{1-\tan^2 x} = \frac{2\frac{B_x}{C_x}}{1-(\frac{B_x}{C_x})^2} = \frac{2B_xC_x}{C_x{}^2 - B_x{}^2} \tag{1.124c}$$

$$th\; 2x = \frac{B_{2x}}{A_{2x}} = \frac{2thx}{1+th^2x} = \frac{2\frac{B_x}{A_x}}{1-(\frac{B_x}{A_x})^2} = \frac{2A_xB_x}{A_x{}^2 + B_x{}^2} \tag{1.124d}$$

Collecting together from (1.124c) and (1.124d)

$$\frac{B_{2x}}{C_{2x}} = \frac{2B_xC_x}{C_x{}^2 - B_x{}^2}, \text{ and, } \frac{B_{2x}}{A_{2x}} = \frac{2A_xB_x}{A_x{}^2 + B_x{}^2} \tag{1.124e}$$

Taking reciprocal of both the expressions, adding and simplifying

$$\frac{A_{2x}+C_{2x}}{B_{2x}} = \frac{A_x + C_x}{2B_x} + \frac{B_x}{2}\left(\frac{1}{A_x} - \frac{1}{C_x}\right)$$

or,
$$\frac{\sin N(2x)}{B_{2x}} = \frac{\sin N(x)}{2B_x} - \frac{B_x}{2}\left(\frac{A_x - C_x}{A_x.C_x}\right) = \frac{\sin N(x)}{2B_x} - \frac{B_x}{2}.\frac{\sin N(x,-x)}{A_x.C_x}$$

$$= \frac{\sin N(x)}{2B_x} - \frac{B_x}{2}.\frac{\sin N(x,-x)}{(\sin 2x.sh2x/4)} = \frac{\sin N(x)}{2B_x} - B_x.\frac{2\sin N(x,-x)}{B_{2x}}$$

So,
$$\sin N(2x) = \frac{B_{2x}}{2B_x} \sin N(x) - 2\, B_x \sin N(x, -x),$$

where, $B_x = \sin x \; shx$ (1.124f)

tan-representation of $\frac{\text{sinN}(2x)}{\text{sinN}(x)}$ and $\frac{\text{cosN}(2x)}{\text{cosN}(x)}$

Example 1.16b

Express $\frac{\text{sinN}(2x)}{\text{sinN}(x)}$ in terms of, tan$x$ and thx.

Solution 1.16b

We have by definition

$$\begin{aligned}\text{sinN } 2x &= \sin 2x\ ch\ 2x + \cos 2x\ sh\ 2x \\ &= 2 \sin x \cos x\ (ch^2x + sh^2x) + 2\ shx\ chx\ (\cos^2 x - \sin^2 x) \\ \text{sinN}x &= \sin x\ chx + \cos x\ shx \qquad (1.125a)\end{aligned}$$

Dividing, sinN $2x$, by sinNx, and again, dividing the numerator and denominator of the RHS by ($\cos^2 x\ ch^2x$),

$$\frac{\sin \text{N}(2x)}{\sin \text{N}(x)} = \frac{2\tan x.(1+th^2x)+2thx.(1-\tan^2 x)}{\dfrac{\tan x + thx}{\cos x.chx}}$$

This gives on simplification

$$\frac{\sin \text{N}(2x)}{\sin \text{N}(x)} = 2\text{D}_x.[\frac{\tan x.(1+th^2x)+thx.(1-\tan^2 x)}{\tan x + thx}],$$

(where $\text{D}_x = \cos x\ chx$)

(1.125b)

Example 1.16c

Express $\frac{\cos \text{N}(2x)}{\cos \text{N}(x)}$ in terms of, tan$x$ and thx.

Solution 1.16c

We have by definition

$$\begin{aligned}\text{cosN } 2x &= \cos 2x\ ch\ 2x - \sin 2x\ sh\ 2x \\ &= (ch^2x + sh^2x)(\cos^2 x - \sin^2 x) - 4 \sin x \cos x\ shx\ chx \\ &= \cos^2 x\ ch^2x - \sin^2 x\ ch^2x + \cos^2 x\ sh^2x - \sin^2 x\ sh^2x \\ &\quad - 4 \sin x \cos x\ shx\ chx \\ \text{cosN}x &= \cos x\ chx - \sin x\ shx \qquad (1.126a)\end{aligned}$$

Dividing, cosN $2x$, by cosNx, and again, dividing the numerator and denominator of the RHS by ($\cos^2 x\ ch^2x$),

$$\frac{\text{cosN}(2x)}{\text{cosN}(x)} = \frac{1-\tan^2 x + th^2x - \tan^2 x.th^2x - 4\tan x.thx}{\dfrac{\cos x.chx - \sin x.shx}{\cos^2 x.ch^2x}}$$

Simplifying further

$$\frac{\cos N(2x)}{\cos N(x)} = D_x \frac{(1-\tan^2 x)(1+th^2x) - 4\tan x.thx}{1-\tan x.thx},$$

(where, D_x = cos$x$ ch$x$) (1.126b)

(1.125b) and (1.126b) are the expressions which relate, sinN(2x, 2x) with sinN(x, x), and cosN(2x, 2x) with cosN(x, x), using tan$x$ and th$x$. These two expressions will be tested with the help of numerical data as follows:

Numerical Verification of tan-representation

Example 1.17 (Equation 1.125b, and 1.126b)

Common data

x	= 60 deg (=1.0472 rad), 2x = 120 deg (= 2.0944 rad)
sin$x$	= 0.867, cos$x$ = 0.5, tan$x$ =1.7321
sin 2x	= 0.867, cos 2x = – 0.5, tan 2x = – 1.7321
sh$x$	= 1.2494, ch$x$ = 1.6, th$x$ = 0.7809
sh 2x	= 3.9987, ch 2x = 4.1218, th 2x = 0.9701
sinNx	= sin$x$ ch$x$ + cos$x$ sh$x$ = 2.0106
cosNx	= cos$x$ ch$x$– sin$x$ sh$x$ = – 0.2818
D_x	= cos$x$ ch$x$ = 0.5 × 1.6 = 0.8
sinN 2x	= sin 2x ch 2x + cos 2x sh 2x
	= 0.867 × 4.1218 – 0.5 × 3.9987 = 1.5742
cosN 2x	= cos 2x ch 2x – sin 2x sh 2x
	= – 0.5 × 4.1218 – 0.867 × 3.9987 = – 5.5278

$$\frac{\sin N(2x)}{\sin N(x)} = \frac{1.5742}{2.0106} = 0.7830$$

$$\frac{\cos N(2x)}{\cos N(x)} = \frac{-5.5278}{-0.2818} = 19.6160$$

$$\frac{\tan N(2x)}{\tan N(x)} = [\frac{\sin N(2x)}{\sin N(x)}] / [\frac{\cos N(2x)}{\cos N(x)}] = \frac{(1.5742)(-0.2818)}{(-0.5278)(2.0106)} = 0.0399$$

Example 1.18 (Equation 1.125b)

Verify that

$$\frac{\sin N(2x)}{\sin N(x)} = 2D_x.[\frac{\tan x.(1+th^2x)+thx.(1-\tan^2 x)}{\tan x + thx}]$$

Solution 1.18

LHS = 0.7830

$$\textbf{RHS} = (2)(0.8)[\frac{1.7321.(1+0.7809^2)+0.7809.(1-1.7321^2)}{1.7321+0.7809}]$$

$$= (1.6).\frac{2.7883 - 1.5619}{2.5130} = 0.7808 = \textbf{LHS} \qquad \textbf{(Verified)}$$

Example 1.19 (Equation 1.126b)

Verify that

$$\frac{\cos N(2x)}{\cos N(x)} = D_x \frac{(1-\tan^2 x)(1+th^2 x) - 4\tan x.thx}{1-\tan x.thx}$$

Solution 1.19

LHS = 19.6160

$$\textbf{RHS} = (0.8)[\frac{(1+0.7809^2)(1-1.7321^2) - 4(1.7321)(0.7809)}{1-(1.7321)(0.7809)}]$$

$$= (0.8).\frac{(-2.00)(1.6098) - 5.5104}{-0.3526} = 19.5802 = \textbf{LHS} \qquad \textbf{(Verified)}$$

❑❑❑

2

COMPLEX Nbic FUNCTION AND ASSOCIATED TOPICS

2.1 De MOIVRE'S FORM EXTENDED IN Nbic FUNCTION

Remember that

$$
\begin{aligned}
sh(ix) &= i\ \sin x,\ ch(ix) = \cos x \\
sh(-x) &= -\ shx,\ ch(-x) = chx \\
\sin(ix) &= i\ shx,\ \cos(ix) = chx
\end{aligned}
\tag{2.1}
$$

2.1.1 Complex Circular Function (Coci-function)

The form of complex function made up of circular functions, (*viz.*, plane trigonometric functions, like, cos$x$, and sin$x$), namely

$$
\begin{aligned}
\cos x + i\ \sin x &= e^{ix} = ch(ix) + sh(ix) \\
\cos x - i\ \sin x &= e^{-ix} = ch(ix) - sh(ix)
\end{aligned}
\tag{2.2}
$$

be renamed as **Complex Circular Function** or in short, **Coci-functions.**

where, $\cos^2 x + \sin^2 x = e^{ix} \cdot e^{-ix} = e^0 = 1$

and, in general, $e^{nix}\ e^{-nix} = e^0 = 1$, and, $e^{mix}\ e^{nix} = e^{(m+n)ix}$ (2.3a)

In Coci-functions, the representations of the functions,

$$
\begin{aligned}
&e^{nix}\ [= (e^{ix})^n], \text{ and, } e^{inx}\ [= e^{i(nx)}], \text{ need to be identified.} \\
&e^{nix} = (\cos x + i\ \sin x)^n \\
&e^{inx}\ (= e^{i(nx)}) = \cos nx + i\ \sin nx
\end{aligned}
\tag{2.3b}
$$

Here in Coci-functions,

$$\cos nx + i\ \sin nx = (\cos x + i\ \sin x)^n$$

So, $$e^{nix} = (e^{ix})^n = e^{inx} = e^{i(nx)} \tag{2.3c}$$

Complex circular function ratio, R_{nci}

The complex function ratio, R_{nci}, for Coci-functions, is given by

$$R_{nci} = [\cos nx + i\ \sin nx]\ /\ (\cos x + i\ \sin x)^n$$

$$= [\cos nx + i \sin nx]/[\cos nx + i \sin nx]$$
$$= e^{inx}/e^{nix} = 1 \qquad (2.3d)$$

Alternatively, (2.3a) can be presented as follows:

We have the formula

$$(a + b)^2 - (a - b)^2 = 4ab$$

or,
$$ab = [(a + b)/2]^2 - [(a - b)/2]^2 \qquad (2.3e)$$

Put $a = e^{nix}$, $b = e^{-nix}$

so that,

$$(a + b)/2 = \cos nx,\ (a - b)/2 = i \sin nx$$
$$e^{nix}\, e^{-nix} = [\cos nx]^2 - [i \sin nx]^2 = \cos^2 nx + \sin^2 nx = 1 \qquad (2.3f)$$

Note: If the Coci-function involves one variable, (say, x only), it will be termed as **Coci 1 function**, *e.g.*, $e^{ix} = e^{i\,(x,\,x)} = \cos x + i \sin x$, or, $e^{-iy} = e^{-i(y,\,y)} = \cos y - i \sin y$. Otherwise, it will be **Coci 2 function**, where it is of two variables, namely, (x, y). For example,

$$e^{i(-x,\,y)} = \cos(-x) + i \sin(y), \text{ etc.}$$

2.1.2 Complex Hyperbolic Function (Cohy-function)

Let us define further, functions with hyperbolic terms like, chx and shx, on the similar lines of complex circular functions, as Complex Hyperbolic Function, or, in short, Cohy-function, as Cohy 1 function [with variable (x, x)], as well as, Cohy 2 function [with variable (x, y)].

Cohy 1 function [with variable (x, x)]

Similar to the Coci-function (2.2), let us frame functions with hyperbolic terms as

$$f^{\,ix} = (chx + i\, shx)$$
$$f^{\,-ix} = (chx - i\, shx) \qquad (2.4a)$$

where, unlike the relation with circular functions, *viz.*, $\cos^2 x + \sin^2 x = e^{ix}\, e^{-ix} = e^0 = 1$,

the corresponding governing relations are

$$ch^2 x - sh^2 x = 1,$$

and
$$ch^2 x + sh^2 x = ch\ 2x \qquad (2.4b)$$

Some products of two Cohy 1 functions

$$f^{\,ix} f^{\,-ix} = (chx + i\, shx)\,(chx - i\, shx)$$
$$= ch^2 x + sh^2 x = ch\ 2x, \text{ [and, } \neq 1] \qquad (2.4c)$$
$$f^{\,ix} f^{\,ix} = (chx + i\, shx)^2$$
$$= ch^2 x - sh^2 x + 2\, i\, chx\, shx = 1 + i\, sh\ 2x$$

With these

$$f^{\,ix} = chx + i\, shx$$
$$f^{\,-ix} = (chx - i\, shx)$$
$$f^{\,ix} f^{\,ix} = (f^{\,ix})^2 = f^{\,2ix} = 1 + i\, sh\ 2x, \text{ [it is } \neq f^{\,i\,2x}]$$
$$f^{\,ix} f^{\,-ix} = (chx + i\, shx)\,(chx - i\, shx)$$
$$= ch^2 x + sh^2 x = ch\ 2x, \text{ [it is } \neq (f^{\,0} = 1)]$$

$$
\begin{aligned}
f^{-ix} f^{-ix} &= (chx - i\ shx)(chx - i\ shx) = 1 - i\ sh\ 2x \\
(f^{ix})^2 (f^{-ix})^2 &= f^{2ix} f^{-2ix} = (1 + i\ sh\ 2x)(1 - i\ sh\ 2x) \\
&= 1 + sh^2\ 2x = ch^2\ 2x \\
f^{2ix} &= f^{ix} f^{ix} = (f^{ix})^2 \\
&= ch^2x - sh^2x + 2i\ shx\ chx = 1 + i\ sh\ 2x \qquad (2.4d) \\
f^{-2ix} &= f^{-ix} f^{-ix} = (f^{-ix})^2 \\
&= ch^2x - sh^2x - 2i\ shx\ chx = 1 - i\ sh\ 2x \\
f^{i\,2x} &= ch\ 2x + i\ sh\ 2x, \text{ [it is } \neq f^{2ix}] \\
f^{i\,2x} f^{-i\,2x} &= (ch\ 2x + i\ sh\ 2x)(ch\ 2x - i\ sh\ 2x) \\
&= ch^2\ 2x + sh^2\ 2x = ch\ 4x
\end{aligned}
$$

$$
\begin{aligned}
f^{ix} f^{2ix} &= (f^{ix})^3 = f^{3ix} = chx\ (1 - 2\ sh^2x) + i\ shx\ (1 + 2\ ch^2x) \\
f^{-ix} f^{-2ix} &= (f^{-ix})^3 = f^{-3ix} = chx\ (1 - 2\ sh^2x) - i\ shx\ (1 + 2\ ch^2x) \\
f^{3ix} &= (f^{ix})^3 = (chx + i\ shx)^3 \\
&= (ch^3x - 3\ chx\ sh^2x) + i\ (3ch^2x\ shx - sh^3x) \\
&= chx\ (1 - 2\ sh^2x) + i\ shx\ (1 + 2\ ch^2x) \\
f^{i\,3x} &= ch\ 3x + i\ sh\ 3x, \text{ [it is } \neq f^{3ix}] \\
f^{-3ix} &= (f^{-ix})^3 = (chx - i\ shx)^3 \\
&= (ch^3x - 3\ chx\ sh^2x) + i\ (3ch^2x\ shx - sh^3x) \\
&= chx\ (1 - 2\ sh^2x) - i\ shx\ (1 + 2\ ch^2x)
\end{aligned}
$$

$$
\begin{aligned}
f^{2ix} f^{2ix} &= (f^{2ix})^2 = [(f^{ix})^2]^2 = (f^{ix})^4 = f^{4ix} \\
f^{4ix} &= f^{2ix} f^{2ix} = (f^{2ix})^2 = [(f^{ix})^2]^2 = (f^{ix})^4 \\
&= (1 + i\ sh\ 2x)^2 = (1 - sh^2\ 2x) + 2i\ sh\ 2x \\
f^{i\,4x} &= ch\ 4x + i\ sh\ 4x, \text{ [it is } \neq f^{4ix}] \\
f^{-4ix} &= f^{-2ix} f^{-2ix} = (f^{-2ix})^2 = [(f^{-ix})^2]^2 = (f^{-ix})^4 \\
&= (1 - i\ sh\ 2x)^2 = (1 - sh^2\ 2x) - 2i\ sh\ 2x
\end{aligned}
$$

$$
\begin{aligned}
f^{3ix} f^{2ix} &= chx\ (1 - 2\ sh^2x - 2\ shx\ sh2x) \\
&\quad + i\ [shx\ (1 - sh^2\ 2x) + 2\ chx\ sh\ 2x] \\
&= (1 - sh^2\ 2x + i\ sh\ 2x)(chx + i\ shx) = f^{4ix} f^{ix} \\
f^{5ix} &= f^{4ix} f^{ix} = f^{3ix} f^{2ix} = (f^{ix})^5 \\
f^{i\,5x} &= ch\ 5x + i\ sh\ 5x, \text{ [it is } \neq f^{5ix}] \\
f^{-5ix} &= f^{-4ix} f^{-ix} = f^{-3ix} f^{-2ix} = (f^{-ix})^5 \qquad (2.4e)
\end{aligned}
$$

and so on.

When two Cohy-functions are in multiplication, powers either, both positive or, both negative, are additive. However, this is not the case, when one is negative with respect to the other. This is clearly noticed in the following:

$$
\begin{aligned}
f^{ix} f^{-ix} &= ch\ 2x\ , \text{ [it is } \neq f^0, \textit{ i.e.}, \neq 1] \\
f^{2ix} f^{-2ix} &= (f^{ix} f^{-ix})(f^{ix} f^{-ix}) = ch^2\ 2x, \text{ [it is } \neq (f^0 = 1)] \\
f^{2ix} f^{-ix} &= (1 + 2\ i\ shx\ chx)(chx - i\ shx) \\
&= chx\ (1 + 2\ sh^2x) - i\ shx\ (1 - 2\ ch^2x) \\
&= (ch^2x + sh^2x)(chx + i\ shx) \\
&= f^{ix}.ch\ 2x, \text{ [it is } \neq (f^{ix})] \\
f^{-2ix} f^{ix} &= f^{-ix}.ch\ 2x, \text{ [it is } \neq (f^{-ix})]
\end{aligned}
$$

$$f^{4ix} f^{-ix} = f^{3ix} (f^{ix} f^{-ix}) = f^{3ix}.ch\ 2x, \text{ [it is } \neq (f^{3ix})]$$
$$f^{3ix} f^{-3ix} = (f^{ix} f^{-ix})(f^{2ix} f^{-2ix}) = ch^2\ 2x\ .ch\ 2x = ch^3\ 2x$$

This statement can be directly obtained from the formula (2.3e) as shown.

$$\begin{aligned} f^{3ix} f^{-3ix} &= [(f^{3ix} + f^{-3ix})/2]^2 - [(f^{3ix} - f^{-3ix})/2]^2 \\ &= ch^2x\ (1 + 2\ sh^2x)^2 + sh^2x\ (1 + 2\ ch^2x)^2 \\ &= (ch^2x + sh^2x) + 4\ sh^2x\ ch^2x\ (ch^2x + sh^2x) \\ &= ch\ 2x\ (1 + 4\ sh^2x\ ch^2x) = ch\ 2x\ (1 + sh\ 2x)^2\ ch\ 2x \\ &= ch\ 2x\ ch^2\ 2x = ch^3\ 2x \end{aligned} \tag{2.4f}$$

and so on.

We know that

$e^{-ix} = 1/e^{ix}$, but, f^{-ix} is not the reciprocal of f^{ix}, *i.e.*, $f^{-ix} \neq (1/f^{ix})$.

$$f^{-ix} = ch\ 2x/f^{ix}$$

also, in other words

$$f^{-ix} = \sqrt{2-(f^{ix})^2} \tag{2.4g}$$

Alternative derivation for f^{-ix} using fexponential function

[*the proposed terminology fexponential is explained shortly*]

Expression (2.4g) can be proved in an alternative way, as follows.

From (2.6d), [which is shown later]

$$f^{x} = \cos x + \sin x$$
$$f^{-x} = \cos x - \sin x$$

so that

$$f^{x} f^{-x} = \cos\ 2x$$

Replace, x by, (ix), to get

$$f^{ix} f^{-ix} = \cos(i\ 2x) = ch\ 2x$$
$$f^{-ix} = ch\ 2x/f^{ix}$$

Again since

$$(\cos x - \sin x) = \sqrt{2(\cos^2 x + \sin^2 x) - (\cos x + \sin x)^2}$$

We get

$$f^{-x} = \sqrt{2-(f^{x})^2}$$

Replace, x by, (ix), to get

$$f^{-ix} = \sqrt{2-(f^{ix})^2}$$

The reason for, $f^{i\,nx} f^{-i\,nx} \neq 1$, is implied with the very nature of the grammar of hyperbolic functions. Unlike, $\cos^2x + \sin^2x = 1$, as in the case of circular functions, hyperbolic functions exibit, $ch^2x + sh^2x = ch\ 2x$, (and not $= 1$). This is evident as shown.

Call formula (2.3e), and put

$$a = f^{imx} = (ch\ mx + i\ sh\ mx)$$

CHAPTER 2

$$b = f^{-i\,mx} = (ch\ mx - i\ sh\ mx)$$

so that, $(a + b)/2 = ch\ mx$, $(a - b)/2 = i\ sh\ mx$ (2.5a)

$$f^{i\,mx} f^{-i\,mx} = [ch\ mx]^2 - [i\ sh\ mx]^2$$
$$= ch^2\ mx + sh^2\ mx = ch\ 2mx \quad (2.5b)$$

Then

$$f^{i\,2x} f^{-i\,2x} = (ch\ 2x + i\ sh\ 2x)\ (ch\ 2x - i\ sh\ 2x)$$
$$= ch^2\ 2x + sh^2\ 2x = ch\ 4x$$
$$f^{i\,4x} f^{-i\,4x} = (ch\ 4x + i\ sh\ 4x)\ (ch\ 4x - i\ sh\ 4x)$$
$$= ch^2\ 4x + sh^2\ 4x = ch\ 8x \quad (2.5c)$$

and so on.

Now

$$(f^{ix} f^{-ix}) + (f^{ix} f^{ix}) = ch\ 2x + 1 + i\ sh\ 2x = 1 + f^{i\,2x}$$
$$\Rightarrow f^{i\,2x} = f^{ix} f^{-ix} + f^{ix} f^{ix} - 1 \quad (2.5d)$$

Now referring to (2.4d)

$$f^{i\,2x} - f^{2ix} = (ch\ 2x + i\ sh\ 2x) - (1 + i\ sh\ 2x)$$
$$= ch\ 2x - 1$$
$$= 2\ sh^2 x$$

Also

$$f^{i\,2x} - f^{2ix} = f^{-i\,2x} - f^{-2ix} \quad (2.5e)$$

The first of (2.5e) can be arrived at, using a different way, as follows:

$$f^{2x} = \cos\ 2x + \sin\ 2x$$
$$= \cos^2 x - \sin^2 x + 2\ \sin x\ \cos x$$
$$= (\cos x + \sin x)^2 - 2\ \sin^2 x$$

As shown later in (2.6d) that, $f^x = \cos x + \sin x$. With this, we get

$$f^{2x} = (f^x)^2 - 2\ \sin^2 x$$

Replace, x by $(i\ x)$, in it, to yield

$$f^{i\,2x} = (f^{ix})^2 - 2\ [\sin\ (ix)]^2$$
$$f^{i\,2x} - f^{2ix} = 2\ sh^2 x$$

Additionaly we have

$$sh\ 2x + ch\ 2x = (shx + chx)^2$$
$$sh\ 2x - ch\ 2x = (shx - chx)^2 \quad (2.5f)$$

Generalising (2.5f)

$$sh\ nx + ch\ nx = (shx + chx)^n$$
$$sh\ nx - ch\ nx = (shx - chx)^n \quad (2.5g)$$

Generalized form of the product of two Cohy-functions

For m and $n = 1, 2, 3, \ldots$

$$f^{mix} f^{nix} = f^{(m+n)\ ix}$$
$$f^{-mix} f^{-nix} = f^{-(m+n)\ ix}$$

$$f^{\,mix} f^{\,-mix} = ch^m\, 2x, \text{ [it is } \neq 1]$$
$$f^{\,i\,mx} f^{\,-i\,mx} = ch\; 2mx, \text{ [it is } \neq 1]$$

Additionally, from (2.4e)

$$f^{\,mix} = f^{\,(m-1)ix}.ch\; 2x \tag{2.5h}$$

Complex hyperbolic function ratio, R_{nhy}

In Coci-functions (2.3d), $e^{inx} = e^{nix}$, so that $R_{nci} = e^{inx}/e^{nix} = 1$. But the functions, $f^{\,inx}$ and $f^{\,nix}$ are different in Cohy-functions, so their ratio, R_{nhy} is bound to be different from unity. In fact, it is a certain function of x.

The complex function ratio R_{nhy}, for Cohy-functions, is given by

$$R_{nhy} = f^{\,inx}/f^{\,nix} = [ch\; nx + i\; sh\; nx]/(chx + i\; shx)^n \tag{2.5i}$$

Multiply numerator and denominator by, $f^{\,-nix}$

$$\begin{aligned} R_{nhy} &= f^{\,inx}/f^{\,nix} \\ &= [f^{\,inx} f^{\,-nix}]/[f^{\,-nix} f^{\,-nix}] \\ &= [f^{\,inx} f^{\,-nix}]/ch^n\; 2x \end{aligned} \tag{2.5j}$$

For $n = 1$, $R_{nhy} = [f^{\,ix} f^{\,-ix}]/ch\; 2x = 1$

[Only for $n = 1$, R_{nhy} for Cohy-function (2.5j), is identical with R_{nci} of Coci-function (2.3d).]

2.1.3 Assessment of the Value of *e* and *f*, in Coci and Cohy-functions

We want to assess whether the symbol **f**, which is used to represent the Cohy-functions (2.4a), *viz.*,

$$\begin{aligned} f^{\,ix} &= (chx + i\; shx) \\ f^{\,-ix} &= (chx - i\; shx) \end{aligned} \tag{2.6a}$$

has any similar bearing as that of the symbol **e**, (which is nothing but the logarithmic base), used to represent the Coci-functions (2.2), *viz.*,

$$\begin{aligned} e^{ix} &= (\cos x + i\; \sin x) \\ e^{-ix} &= (\cos x - i\; \sin x) \end{aligned} \tag{2.6b}$$

Exponential function

Replace, (x) by (ix), in (2.6b), to arrive at the **exponential function**

$$\begin{aligned} e^{x} &= chx + shx \\ e^{-x} &= chx - shx \end{aligned} \tag{2.6c}$$

Take the first one, (2.4a), and replace, (x) by (ix)

$$\begin{aligned} f^{\,ix} &= (chx + i\; shx) \\ f^{\,i\,(ix)} &= [ch(i\,x) + i\; sh(i\,x)] \end{aligned}$$

So, $f^{\,-x} = \cos x - \sin x$

Take the second one, (2.4a), and replace, (x) by (ix)

$$\begin{aligned} f^{\,-ix} &= (chx - i\; shx) \\ f^{\,i\,(-\,ix)} &= [ch(ix) - i\; sh(ix)] \end{aligned}$$

So, $f^{\,x} = \cos x + \sin x$

Fexponential function

Collecting in one place

$$f^{x} = \cos x + \sin x$$
$$f^{-x} = \cos x - \sin x$$

so that,

$$f^{x} f^{-x} = \cos^2 x - \sin^2 x = \cos 2x \; (\neq f^0, \; i.e., \neq 1) \quad (2.6d)$$

Compare (2.6d) with (2.6c), which when recalled is

$$e^{x} = chx + shx$$
$$e^{-x} = chx - shx$$

so that,

$$e^x e^{-x} = ch^2 x - sh^2 x = 1 \; (= e^0) \quad (2.6e)$$

Then, analogous to the definition of the hyperbolic functions, *viz.*,

$$(e^x + e^{-x})/2 = chx$$
$$(e^x - e^{-x})/2 = shx \quad (2.6f)$$

when recast, in terms of (2.6d), appears as

$$(f^{x} + f^{-x})/2 = \cos x$$
$$(f^{x} - f^{-x})/2 = \sin x \quad (2.6g)$$

The very well known identity of plane trigonometry that, $\cos^2 x + \sin^2 x = 1$, can be seen to be satisfied by the form of the symbolic use of, ***f***. Refer to (2.6g).

$$\text{RHS} = \cos^2 x + \sin^2 x$$
$$\text{LHS} = [(f^{x})^2 + (f^{-x})^2]/2$$
$$= [(\cos x + \sin x)^2 + (\cos x - \sin x)^2]$$
$$= 2\,(\cos^2 x + \sin^2 x)/2 = 1 \quad \textbf{(Satisfied)} \quad (2.6h)$$

Note: Thus, it is possible to represent the circular functions, (*viz.*, cos*x*, and sin*x*), through the symbol ***f***, (2.6g), exatly in the same way, as is done for the hyberbolic functions, (*viz.*, *chx*, and *shx*), by the symbol **e**, as shown by (2.6f).

The function, e^x is the well known **exponential function**, which does not need any introduction. But the function f^{x}, proposed here, having similar mathematical appearance, is to be identified by the proposed terminology, **fexponential function**.

Further

$$f^{x}.\, f^{x} = (\cos x + \sin x)^2 = 1 + \sin 2x$$
$$f^{x}.\, f^{-x} = (\cos x + \sin x)(\cos x - \sin x) = \cos 2x \quad (2.6i)$$

Replacing, (*x*) by (*ix*), in (2.6i), we arrive at the same (2.4d), which is shown as

$$f^{ix} f^{ix} = 1 + \sin(2ix) = 1 + i\; sh\, 2x$$
$$f^{ix} f^{-ix} = \cos(2ix) = \cos(i2x) = ch\, 2x \quad (2.6i')$$

Now multiply the RHS of (2.6g), and view it by replacing, (*x*) by (*ix*), with the help of (2.4d)

$$\sin 2x = 2 \sin x \cos x = [f^{x} f^{x} - f^{-x} f^{-x}]/2$$

so that

$$\sin(2ix) = \sin(i2x) = [f^{ix} f^{ix} - f^{-ix} f^{-ix}]/2$$

$$= [1 + i\ sh\ 2x - 1 + i\ sh\ 2x]/2 = i\ sh\ 2x \qquad (2.6i'')$$

However, the product

$$f^{x}. f^{x} = (f^{x})^2 \text{ , is not equal to, } f^{2x} \qquad (2.6j)$$

This can be shown as

$$\text{LHS} = f^{x}. f^{x} = 1 + \sin 2x$$
$$\text{RHS} = f^{2x} = \cos 2x + \sin 2x$$

This confirms that, LHS ≠ RHS.

In other words, it means that, powers do not add up in the multiplication of, *f–s*.

On the contrary, in the multiplication of, *e–s*, powers are additive, *e.g.*,

$$e^x. e^x = (e^x)^2 = e^{2x}$$
$$e^{-x}. e^{-x} = (e^{-x})^2 = e^{-2x}$$

The proof is obvious, in accordance with (2.5f), or, in general, with (2.5g).

2.1.4 Value of the Parameter, *e*

Recall (2.6e). That is

$$e^x = ch\ x + shx$$

We get directly

$$e = (chx + shx)^{1/x}$$

The same relation can be arrived at on taking, log ( ), on both sides, as

$$x \log e = \log(chx + shx)$$

$$\log e = \frac{1}{x}\log(chx + shx) = \log (chx + shx)^{1/x}$$

$$e = (chx + shx)^{1/x}$$

Numerical Example

The value of ***e*** can be checked with numerical values of *x*, as shown in the following example:

Example 2.1a

Find the value of ***e*** with any three numerical data, say, $x = [\frac{\pi}{4}, \frac{\pi}{2}$ and 0.2].

Solution 2.1a

Put, $x = \frac{\pi}{4}$

so that, $\frac{1}{x} = 1.2732395$, $[ch, sh]\ [\frac{\pi}{4}] = [1.3246091, 0.86867]$

$$e = [1.3246091 + 0.86867]^{1.2732395} = \mathbf{2.7182813}, \textbf{ (call it, } e_1\textbf{)}$$

Put $x = \frac{\pi}{2}$

so that, $\frac{1}{x} = 0.6366198$, $[ch, sh]\ [\frac{\pi}{2}] = [2.5091777, 2.3012981]$

$$e = [2.5091777 + 2.3012981]^{0.6366198} = \mathbf{2.7182814, (call\ it,\ } e_2\mathbf{)}$$

Put, $x = 0.2$

so that, $\frac{1}{x} = 5$, $[ch\ ,\ sh]\ [0.2] = [1.0200668,\ 0.2013360]$

$$e = [1.0200668 + 0.2013360]^5 = \mathbf{2.7182823, (call\ it,\ } e_3\mathbf{)}$$

The value of, $e\ (= e_1 = e_2 = e_3) =$ **2.7182813,** (the value of the base of logarithm).

2.1.5 Value of the Parameter, *f*

But this is not the case with the parameter, *f*. It is demonstrated as follows.

Call (2.6d). That is

$$f^x = \cos x + \sin x$$

We get directly

$$f = (\cos x + \sin x)^{1/x}$$

Operating, log ( ), on both sides

$$x \log f = \log(\cos x + \sin x)$$

$$\log f = \frac{1}{x}\log(\cos x + \sin x) = \log(\cos x + \sin x)^{1/x}$$

$$f = (\cos x + \sin x)^{1/x}$$

Let us attempt to find the value of, *f*, similar to the process as adopted for *e*, with numerical values of *x*, as shown in the following example:

Numerical Example

Example 2.1b

Find the value of, *f* with any two numerical data, say, $x = [\frac{\pi}{4}$, and 0.2]

Solution 2.1b

Put, $x = \frac{\pi}{4}$

so that, $\frac{1}{x} = 1.2732395$, $[\cos x\ ,\ \sin x]\ [\frac{\pi}{4}] = [0.7071068,\ 0.7071068]$

$$f = [0.7071068 + 0.7071068]^{1.2732395} = \mathbf{1.55468223, (call\ it,\ } f_1\mathbf{)}$$

Put, say, $x = 0.2$ (= 11.459156 deg)

so that, $\frac{1}{x} = 5$, [cos , sin] [11.459156 deg] = [0.98006658, 0.19866933]

$$f = [0.98006658 + 0.19866933]^5 = \mathbf{2.275530043, (call\ it,\ } f_2\mathbf{)}$$

Note: Since, $f_1 \neq f_2$, it can be inferred that, the value of, *f* cannot be not uniquely obtained, as it is done for, e. This observation confirms that the character, *f*, is just a symbol, and does not carry any special status as enjoined by, *e*.

2.1.6 Alternative Formula for tanN(*x*, *x*) using Exponential and Fexponential Functions

Refer to (1.54a) here, where each of the expressions of the sum and difference of, cosN*x* and sinN*x*, are arranged in two ways.

$$\begin{aligned}\cos Nx + \sin Nx &= (\cos x + \sin x)\, chx + (\cos x - \sin x)\, shx \\ &= f^{x}\, chx + f^{-x}\, shx \\ &= (chx + shx)\cos x + (chx - shx)\sin x \\ &= e^{x}\cos x + e^{-x}\sin x \\ \cos Nx - \sin Nx &= (\cos x - \sin x)\, chx - (\cos x + \sin x)\, shx \\ &= f^{-x}\, chx - f^{x}\, shx \\ &= (chx - shx)\cos x - (chx + shx)\sin x \\ &= e^{-x}\cos x - e^{x}\sin x \end{aligned} \tag{2.6k}$$

In (2.6k), the use of the substitution (2.6d), is made. That is

$$\begin{aligned}\cos x + \sin x &= f^{x} \\ \cos x - \sin x &= f^{-x} \\ chx + shx &= e^{x} \\ chx - shx &= e^{-x}\end{aligned}$$

Thus, we get from (2.6k)

$$\frac{\cos N(x) - \sin N(x)}{\cos N(x) + \sin N(x)} = \frac{e^{-x}\cos x - e^{x}\sin x}{e^{x}\cos x + e^{-x}\sin x} = \frac{f^{-x}\, chx - f^{x}\, sh\, x}{f^{x}\, chx + f^{-x}\, sh\, x}$$

Dividing the numerator and denominator of the first one by, cosN*x*, second one by, $e^{-x}\cos x$, and, third one by, $f^{-x}\, chx$, and, remembering that, $e^{x}/e^{-x} = e^{2x}$, (but, $f^{x}/f^{-x} \neq f^{2x}$), we get

$$\frac{1 - \tan N(x)}{1 + \tan N(x)} = \frac{1 - e^{2x}\tan x}{e^{2x} + \tan x} = \frac{1 - \dfrac{f^{x}}{f^{-x}}\, th\, x}{\dfrac{f^{x}}{f^{-x}} + th\, x} = p \tag{2.61}$$

The tan-formula for Single Nbic function, tanN*x*, can now be written in two ways from (2.61), namely,

either, in terms of exponential function, e^{x} and circular function, tan*x*,

or, in terms of fexponential function, f^{x} and hyperbolic function, th*x*, as

$$\tan Nx = \frac{1-p}{1+p} \tag{2.6m}$$

in which, *p* is as given in (2.6l).

The tanN*x*, formula has been provided earlier, and the numerical verification of which is done in Example 1.7. The numerical verification of the present alternative formula (2.6m), is now shown in the following example 2.1c.

Numerical Example

Example 2.1c (Equation 2.6m)

Calculate the values of tanNx, using: (e^x and tan$x$), and ($f^{\,x}$ and thx) forms, for x = –1.0472 (= – 60 deg).

Solution

$$\begin{aligned} x &= -1.0472 \ (= -60 \text{ deg}) \\ \sin x &= -0.866025, \ \cos x = 0.5, \ \tan x = -1.732053 \\ shx &= -1.249371, \ chx = 1.6002899, \ thx = -0.78071 \\ e^{2x} &= (chx + shx)/(chx - shx) \\ &= (1.6002899 - 1.249371)/(1.6002899 + 1.249371) = 0.123144 \end{aligned}$$

[This value can be directly obtained.]

$$p = \frac{1 - e^{2x}\tan x}{e^{2x} + \tan x} = \frac{1 - (0.123144)(-1.732053)}{0.123144 - 1.732053} = 0.754108 \text{ (call it, } p_1)$$

$$\begin{aligned} f^{\,x}/f^{\,-x} &= (\cos x + \sin x)/(\cos x - \sin x) \\ &= (0.5 - 0.866025)/(0.5 + 0.866025) = -0.267949 \end{aligned}$$

$$p = \frac{1 - \dfrac{f^x}{f^{-x}}\, thx}{\dfrac{f^x}{f^{-x}} + thx} = \frac{1 - (-0.267949)(-0.78071)}{-0.267949 - 0.78071} = -0.754115 \text{ (call it, } p_2)$$

Since $p_1 = p_2$, (and it has to be), take any one of the values.

$$\tan Nx = \frac{1-p}{1+p} = \frac{1+0.754108}{1-0.754108} = \mathbf{7.13365217}$$

2.1.7 Representation of Expansion for c*h nx* and s*h nx*

The coefficients of the various terms involving chx and shx, in the expansion of (chx + shx)^n, binomially, of course, are the well known binomial coefficients, $nCr = (n)!/[(n-r)!\,(r)!]$, ($n$ = 0, 1, 2, …n). Since the binomial coefficients, arranged properly in triangular form, are the well known Pascal's triangular numbers, the coefficients in the expansion of the RHS of the functions for (2.6a), namely,

$$\begin{aligned} sh\ nx + ch\ nx &= (shx + chx)^n \\ sh\ nx - ch\ nx &= (shx - chx)^n, \end{aligned}$$

will then fit to the corresponding ones of the Pascal's number triangle.

Interesting to observe that, as will be seen shortly that, in each of the expansion of, ch nx, and, sh nx, for (n = 0, 1, 2, …n), the associated coefficients occupy suitably along the identified rows of the array of numbers of the Pascal's triangle. The task is to pick them simply from a certain row, and to use directly for the required expansion.

Coefficients for expansion of *ch nx* and *sh nx* through Pascal's triangle of numbers

We use the formulas from (1.53a)

$$sh(x + y) = shx\ chy + chx\ shy$$
$$sh(x - y) = shx\ chy - chx\ shy$$
$$ch(x + y) = chx\ chy + shx\ shy$$
$$ch(x - y) = chx\ chy - shx\ shy$$

For *ch nx* coefficients:

By adjusting *y* and *x*, in them on the actual expansion, we have

for **ch *nx* , (*n* = 0, 1, 2, 3, ...*n*), that is, for *chx*, *ch* 2*x*, *ch* 3*x*, *ch* 4*x*, ..., *ch nx*:**

Coefficients of **ch *n*x** are shown within [ ].

$$ch\ 0x = [1]$$
$$chx = [1]chx$$
$$ch\ 2x = [1]ch^2x + [1]sh^2x$$
$$ch\ 3x = [1]ch^3x + [3]chx\ sh^2x$$
$$ch\ 4x = [1]ch^4x + [6]ch^2x\ sh^2x + [1]sh^4x$$
$$ch\ 5x = [1]ch^5x + [10]ch^3x\ sh^2x + [5]chx\ sh^4x$$
$$ch\ 6x = [1]ch^6x + [15]ch^4x\ sh^2x + [15]ch^2x\ sh^4x + [1]sh^6x$$
$$ch\ 7x = [1]ch^7x + [21]ch^5x\ sh^2x + [35]ch^3x\ sh^4x + [7]chx\ sh^6x$$
$$ch\ 8x = [1]ch^8x + [28]ch^6x\ sh^2x + [70]ch^4x\ sh^4x + [28]ch^2x\ sh^6x + [1]sh^8x$$

and so on. (2.7a)

For *sh n*x coefficients:

Similarly, on actual expansion, we have

for **sh *nx* , (*n* = 0, 1, 2, 3, ...*n*), that is for *shx*, *sh* 2*x*, *sh* 3*x*, *sh* 4*x*, ..., *sh nx*:**

Coefficients of **sh *nx*** are shown within ( ).

$$sh\ 0x = 0 = (\text{nil})$$
$$shx = (1)shx$$
$$sh\ 2x = (2)chx\ shx$$
$$sh\ 3x = (3)ch^2x\ shx + (1)sh^3x$$
$$sh\ 4x = (4)ch^3x\ shx + (4)chx\ sh^3x$$
$$sh\ 5x = (5)ch^4x\ shx + (10)ch^2x\ sh^3x + (1)sh^5x$$
$$\sin 6x = (6)ch^5x\ shx + (20)ch^3x\ sh^3x + (6)chx\ sh^5x$$
$$\sin 7x = (7)ch^6x\ shx + (35)ch^4x\ sh^3x + (21)ch^2x\ sh^5x + (1)sh^7x$$
$$\sin 8x = (8)ch^7x\ shx + (56)ch^5x\ sh^3x + (56)ch^3x\ sh^5x + (8)chx\ sh^7x$$

and so on. (2.7b)

In **Figure 2.1**, the coefficients of the various terms of *ch nx*, as well as of *sh nx*, when expanded, are shown after identifying them within [ ] and ( ), respectively. In fact, it can be easily identified that, the numbers in **Figure 2.1** actually belong to the same ones of Pascal's triangle, in which the pattern is symmetric and regular.

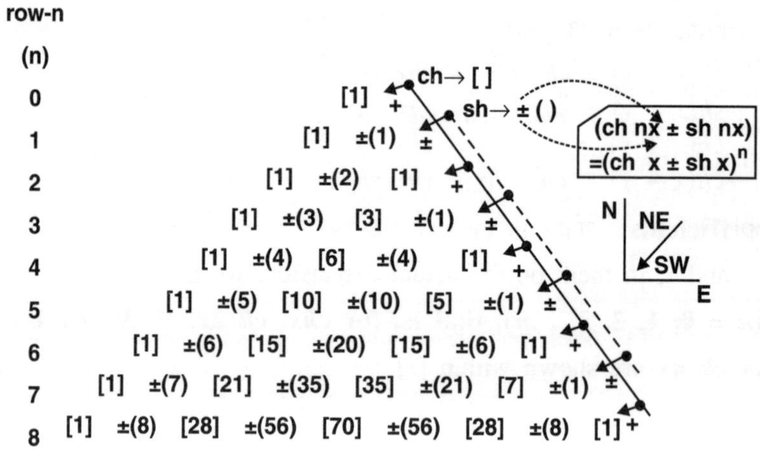

Fig. 2.1: Coefficents of hyperbolic functions of multiple angles through Pascal's triangle

The numbers lying on the leftmost inclined line, which decend from the N–E corner at the apex point towards S–W below, is the Line 1. The next one parallel to it is the Line 2, and so on. Similarly, the horizontal rows of numbers starting from top are demarcated as, row 0, row 1, row 2, etc. Starting from the Line 1, the lines with odd numbers, *i.e.*, Line 3, Line 5, etc., are those meant for only *ch*-coefficients, which are kept within [ ], and are all +ve. The same pattern is followed exclusively for *sh*-coefficients, for which the lines are marked with even numbers, *i.e.*, Line 2, Line 4, Line 6, etc., and all the numbers belonging to these lines are all *+ve* also.

For ***ch nx* coefficients** shown in **Figure 2.1** [refer to (2.7a)]:

Each term of the expanded form of *ch nx*, is composed of the product of the coefficient [C], $ch^p x$, and $sh^q x$. In the first term $p = n$ and $q = 0$, in the next term $p = n - 2$, and $q = 2$, and so on. In fact, once the first term is known, in the next term, $(ch^p x.sh^q x)$, p is reduced by 2, and q is increased by 2. This way the chain of terms proceed till the last term, either becomes $(ch^1 x.sh^{n-1} x)$, or simply $(sh^n\, x)$, alongwith the corresponding coefficient.

All the first terms for $(n = 0, 1, 2, 3, \ldots n)$, begin with coefficient corresponding to the Line 1, *i.e.*, [1] multiplied by $(ch^n x\ .sh^0 x)$.

That is, for $n = 1$, row 1, for chx, in row 1, the term is, [1]chx,
for $n = 2$, row 2, for $ch\ 2x$, in row 1, the term is, [1]$ch^2 x$
for $n = 3$, row 3, for $ch\ 3x$, in row 3, the term is, [1]$ch^3 x$
for $n = 8$, row 8, for $ch\ 8x$, in row 8, the term is, [1]$ch^8 x$

and so on.

All the second terms for $(n = 0, 1, 2, 3, \ldots n)$, begin with coefficient corresponding to the Line 3, *i.e.*, [coefficient] multiplied by $(ch^{n-2} x.sh^2 x)$.

That is, for $n = 1$, row 1, for chx, in row 1, the term is nil,
for $n = 2$, row 2, for $ch\ 2x$, in row 2, the term is, [1]$sh^2 x$
for $n = 3$, row 3, for $ch\ 3x$, in row 3, the term is, [3]$chx\ sh^2 x$
for $n = 8$, row 8, for $ch\ 8x$, in row 8, the term is, [28]$ch^6 x\ sh^2 x$

and so on.

For **sh nx coefficients in Figure 2.1** [refer to (2.7b)]:

After filling in the coefficients for, ch nx, the gaps (which are also in between the two ch nx coefficients), in each row, are left only for the coefficients for sh nx, to occupy. For each value of n, each expansion in this case of sh nx, starts with the coefficients D, written within ( ), multiplied by, (n $ch^{n-1}x$. shx). The next term that follows is, (D) (n $ch^{n-3}.sh^3x$), and so on. In fact, once the first term is known, in the next term, ($ch^px.sh^qx$), p is reduced by 2, and q is increased by 2. This chain process continues till the last term ($ch^px.sh^qx$), either becoms (ch^1x $.sh^{n-1}x$), or simply (sh^nx), alongwith the coefficient. The former or the later form corresponds to when n is even or odd, respectively.

It is thus evident that, the Pascal's triangle of numbers can also be effectively and gainfully applied for the expansion of ch nx and sh nx, with utmost ease, eliminating altogether straineous direct expansion. Note that, each row is completely filled up by the corresponding expanded expression of ch nx, as well as, sh nx, without any overlap and spill over. Further, the complete expansion obtained by summing up the expansion of ch nx, as well as, sh nx, in terms of ch^px and sh^qx, shows that, each term corresponding to, ( )sh nx in (ch^px. sh^qx), follows the gradual decrease of p, and gradual increase of q, than the previous term meant for [ ]ch nx, in steps of 1, as is in the case of binomial expansion of $(x + y)^n$.

The sh nx-coefficients **Figure 2.1** for the odd numbered lines along NE to SW direction, namely, Line 1, Line 3, Line 5, .., are shown with (±) sign. The (+) sign corresponds to the sh^qx terms in the expansion of, $sh\ nx + ch\ nx = (shx + chx)^n$, and (–) sign corresponds to the sh^qx terms in the expansion of, $sh\ nx - ch\ nx = (shx - chx)^n$.

Thus the statement in (2.6a) that, $(ch\ nx + sh\ nx) = (chx + shx)^n$, is verified and validated alternatively by the proposed use of Pascal's triangle of numbers.

2.1.8 Representation of Expansion for, cos *nx* and sin *nx*

Coefficients for expansion of, cos *nx* and sin *nx* through Pascal's triangle of numbers

Using the trigonometric formulas

$$\sin(x + y) = \sin x \cos y + \cos x \sin y$$
$$\sin(x - y) = \sin x \cos y - \cos x \sin y$$
$$\cos(x + y) = \cos x \cos y - \sin x \sin y$$
$$\cos(x - y) = \cos x \cos y + \sin x \sin y$$

For cos *nx* coefficients:

By adjusting y and x, in them on the actual expansion, we have:

for **cos *nx*, (*n* = 0, 1, 2, 3, ...*n*)**

[that is, for cos*x*, cos 2*x*, cos 3*x*, cos 4*x*, ..., cos *nx*]

Coefficients of cos nx are shown within [ ].

$$\cos 0x = [1]$$
$$\cos x = [1]\cos x$$
$$\cos 2x = [1]\cos^2 x - [1]\sin^2 x$$
$$\cos 3x = [1]\cos^3 x - [3]\cos x \sin^2 x$$
$$\cos 4x = [1]\cos^4 x - [6]\cos^2 x \sin^2 x + [1]\sin^4 x$$
$$\cos 5x = [1]\cos^5 x - [10]\cos^3 x \sin^2 x + [5]\cos x \sin^4 x$$

$$\cos 6x = [1]\cos^6 x - [15]\cos^4 x \sin^2 x + [15]\cos^2 x \sin^4 x - [1]\sin^6 x$$
$$\cos 7x = [1]\cos^7 x - [21]\cos^5 x \sin^2 x + [35]\cos^3 x \sin^4 x - [7]\cos x \sin^6 x$$
$$\cos 8x = [1]\cos^8 x - [28]\cos^6 x \sin^2 x + [70]\cos^4 x \sin^4 x - [28]\cos^2 x \sin^6 x + [1]\sin^8 x$$

and so on. (2.8)

For sin *nx* coefficients:

Similarly, on actual expansion, we have :

for **sin *nx* , (*n* = 0, 1, 2, 3, …*n*)**

[that is, for sin*x*, sin 2*x*, sin 3*x*, sin 4*x*, …., sin *nx*]

Coefficients of *sh nx* are shown within ( ).

$$\sin 0x = 0 = (\text{nil})$$
$$\sin x = (1)\sin x$$
$$\sin 2x = (2)\cos x \sin x$$
$$\sin 3x = (3)\cos^2 x \sin x - (1)\sin^3 x$$
$$\sin 4x = (4)\cos^3 x \sin x - (4)\cos x \sin^3 x$$
$$\sin 5x = (5)\cos^4 x \sin x - (10)\cos^2 x \sin^3 x + (1)\sin^5 x$$
$$\sin 6x = (6)\cos^5 x \sin x - (20)\cos^3 x \sin^3 x + (6)\cos x \sin^5 x$$
$$\sin 7x = (7)\cos^6 x \sin x - (35)\cos^4 x \sin^3 x + (21)\cos^2 x \sin^5 x - (1)\sin^7 x$$
$$\sin 8x = (8)\cos^7 x \sin x - (56)\cos^5 x \sin^3 x + (56)\cos^3 x \sin^5 x - (8)\cos x \sin^7 x$$

(2.9)

and so on.

The same Pascal's triangle of numbers can be seen to describe the expansion of cosine and sine of multiple angles, *viz.*, cos *nx*, and sin *nx*. In **Figure 2.2**, the coefficients are shown. Those marked within [ ] are for cos *nx*, whereas, those within ( ) are for sin *nx*.

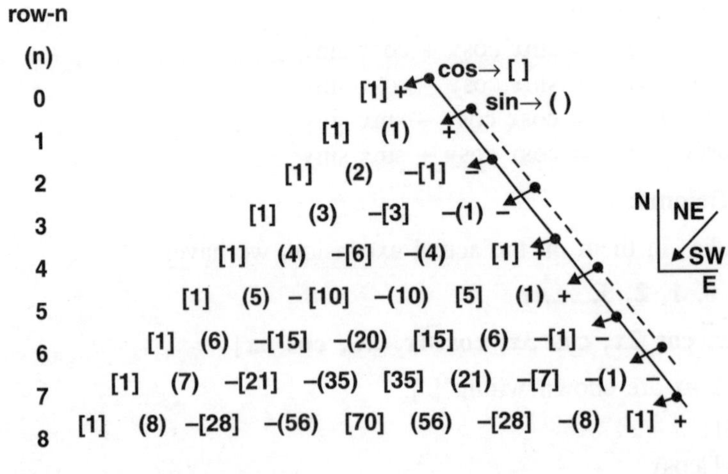

Fig. 2.2: Coefficients of circular functions of multiple angles through Pascal's triangle

Marked are the Line 1, Line 2, Line 3, etc., similar to the case of hyperbolic functions *ch nx* and *sh nx*, descending from the N–E corner to S–W corner. Line 1, Line 3, Line 5 etc., (odd line numbers), are for the coefficients for cos *nx*, and Line 2, Line 4, Line 6 etc., (even line numbers), are for the coefficients for sin *nx*.

The expansion of cos *nx*, as well as, sin *nx* for each value of *n*, (n = 0, 1, 2, 3, ...n), follows exactly the same pattern and same identical form of the corresponding *ch nx* and *sh nx* ones, except that:

(1) ***ch*** and ***sh*** operators are to be replacd by **cos** and **sin** respectively, and

(2) every alternate term beginning from the 1st one, *i.e.*, the 2nd, 4th, 6th, etc., are –ve, (these are +ve in the case of Hyperbolic functions).

The reason that Pascal's triangle of numbers, (which actually are the binomial coefficients, nCr), appear in the expansion of, cos *nx* and sin *nx*, will be imminent as follows:

Let us consider the expansion of, cos 5*x* and sin 5*x*, using Coci-function as

$$\cos 5x + i \sin 5x = e^{i5x} = (e^{ix})^5 = (\cos x + i \sin x)^5$$

Expanding binomially

$$\begin{aligned}(\cos x + i \sin x)^5 &= \cos^5 x + 5 \cos^4 x\,(i \sin x) + 10\cos^3 x\,(i \sin x)^2 \\ &\quad + 10 \cos^2 x\,(i \sin x)^3 + 5\cos x\,(i \sin x)^4 + (i \sin x)^5 \\ &= \cos^5 x + 5i \cos^4 x \sin x - 10\cos^3 x \sin^2 x - 10i \cos^2 x \sin^3 x \\ &\quad + 5 \cos x \sin^4 x + i \sin^5 x\end{aligned}$$

Thus

$$\begin{aligned}\cos 5x + i \sin 5x &= [\cos^5 x - 10 \cos^3 x \sin^2 x + 5 \cos x \sin^4 x] \\ &\quad + i\,[5 \cos^4 x \sin x - 10 \cos^2 x \sin^3 x + \sin^5 x]\end{aligned}$$

Separating the real and imaginary parts

$$\begin{aligned}\cos 5x &= [1] \cos^5 x - [10] \cos^3 x \sin^2 x + [5] \cos x \sin^4 x \\ \sin 5x &= (5) \cos^4 x \sin x - (10) \cos^2 x \sin^3 x + (1) \sin^5 x\end{aligned} \qquad (2.10)$$

Interestingly, it is to be noticed that, if it is required to express, [cos 5*x* + sin 5*x*], in terms of, $(\cos x + \sin x)^5$, it can be readily done without any difficulty by consulting the Pascal's triangle of numbers, which are exploited and proposed as presented in **Figure 2.2**.

Thus

$$\begin{aligned}&(\cos 5x + \sin 5x) \\ &= (\cos x + \sin x)^5 - 2\,[[10] \cos^3 x \sin^2 x + (10) \cos^2 x \sin^3 x]\end{aligned} \qquad (2.11)$$

2.1.9 Few Products of Two Cohy 1 Functions when Compiled

Generalized Form of the Product of Two Cohy-Functions

Expressions (2.5*h*) in generalized form, for *m* and *n* = 1, 2, 3, ... , are:

$$\begin{aligned}f^{mix} f^{nix} &= f^{(m+n)ix} \\ f^{-mix} f^{-nix} &= f^{-(m+n)ix} \\ f^{mix} f^{-mix} &= ch^m\, 2x, \text{ [it is } \neq 1] \\ f^{imx} f^{-imx} &= ch\, 2mx, \text{ [it is } \neq 1]\end{aligned} \qquad (2.12)$$

CHAPTER 2

Cohy 2 Function [with variable (x, y)]

Recollect the pattern (2.6) and on actual expansion

$$\begin{aligned}
f^{ix} f^{iy} &= (chx + i\ shx)\ (chy + i\ shy) = ch(x - y) + i\ sh(x + y)\\
f^{ix} f^{-iy} &= (chx + i\ shx)\ (chy - i\ shy) = ch(x + y) + i\ sh(x - y)\\
f^{-ix} f^{iy} &= (chx - i\ shx)\ (chy + i\ shy) = ch(x + y) - i\ sh(x - y)\\
f^{-ix} f^{-iy} &= (chx - i\ shx)\ (chy - i\ shy) = ch(x - y) - i\ sh(x + y)
\end{aligned} \tag{2.13}$$

from which

$$ch(x - y) = \frac{1}{2}(f^{ix} f^{iy} + f^{-ix} f^{-iy})$$

$$i\ sh(x - y) = \frac{1}{2}(f^{ix} f^{-iy} - f^{-ix} f^{iy}) \tag{2.14}$$

Complex f function

Similar to hyperbolic terms, [written here, cos*hy* as c*hy*, and, sin*hy* as s*hy*], with complex hyperbolic functions, let us define, Complex *f*-function, or in short, Cof-function, made up of '*cf y*'and '*sf y*', to be identified for 'cosine – *f* ', and 'sine – *f* ' of the function *y*, respectively,

Extending the **Cohy–functions** involving two variables, *x* and *y*, and selecting suitable functions for, *ch*(*x* + *y*), *ch*(*x* – *y*), *sh*(*x* + *y*) and, *sh*(*x* – *y*), and rearranging we have using (2.14)

$$\begin{aligned}
f^{i(x-y)} &= ch(x - y) + i\ sh(x - y)\\
&= f^{ix}\,[(f^{iy} + f^{-iy})]/2 - f^{-ix}\,[(f^{iy} - f^{-iy})]/2
\end{aligned}$$

Thus,

$$\begin{aligned}
f^{i(x-y)} &= f^{ix}\ cf\ y - f^{-ix}\ sf\ y\\
f^{-i\,(x-y)} &= f^{ix}\ cf\ y + f^{-ix}\ sf\ y\\
f^{i\,(x+y)} &= f^{ix}\ cf\ y + f^{-ix}\ sf\ y\\
f^{-i\,(x+y)} &= f^{-ix}\ cf\ y - f^{ix}\ sf\ y
\end{aligned} \tag{2.15}$$

where, *cf y*, and *sf y* are defined in (2.16).

Thus, the following Cohy 2 forms once again, are represented by adopting new terminology, namely, **Cof-functions** (Complex f-Functions), namely, cf *x*, sf *x* etc., which are defined as

$$\begin{aligned}
\mathbf{cf\ y} &= \mathbf{(f^{iy} + f^{-iy})/2 = chy}\\
\mathbf{sf\ y} &= \mathbf{(f^{iy} - f^{-iy})/2 = i\ shy}
\end{aligned} \tag{2.16}$$

so that

$$\begin{aligned}
cf\ y \pm sf\ y &= chy \pm i\ shy = f^{\pm iy}\\
cf\ y.sf\ y &= i\ sh\ 2y/2
\end{aligned} \tag{2.17}$$

$$\begin{aligned}
cf^{\,2}\ y + sf^{\,2}\ y &= ch^2y - sh^2y = 1\\
cf^{\,2}y - sf^{\,2}y &= ch^2y + sh^2y = ch\ 2y
\end{aligned} \tag{2.18}$$

from which

$$\begin{aligned}
cf^{\,2}y &= (1 + ch\ 2x)/2 \Rightarrow cf\ y = \text{sqrt}\ [(1 + ch\ 2x)/2]\\
sf^{\,2}y &= (1 - ch\ 2x)/2 \Rightarrow sf\ y = \text{sqrt}\ [(1 - ch\ 2x)/2]
\end{aligned} \tag{2.19}$$

Thus in one place we write the Cohy 2 forms in terms of Cohy 1 – , and Cof-functions.

Collecting formulas together [from (2.15)]

$$
\begin{aligned}
f^{i(x+y)} &= f^{ix} . \text{cf } y + f^{-ix} . \text{sf } y\\
f^{-i(x+y)} &= f^{-ix} . \text{cf } y - f^{ix} . \text{sf } y\\
f^{i(x-y)} &= f^{ix} . \text{cf } y - f^{-ix} . \text{sf } y\\
f^{-i(x-y)} &= f^{-ix} . \text{cf } y + f^{ix} . \text{sf } y
\end{aligned} \qquad (2.20)
$$

2.1.10 Complex Single Nbic Functions (Cosn Function)

Extending the reasoning of the of Coci, and Cohy-functions, let us further define complex functions made up of the combination of circular and hyperbolic functions. They are: Cosn1 and Cosn 2 functions. The former type is the function of single variable (x, x), while the later type is of two variables (x, y).

Cosn 1 function [with variable (x, x)]

$$
\begin{aligned}
&e^{-iN/(x,\, x)}\\
&= \cos N(x, x) + i \sin N(x, x) = C_1 + i\, S_1\\
&= \cos x\, chx - \sin x\, shx + i \sin x\, chx + i \cos x\, shx\\
&= (\cos x + i \sin x)\,(chx + i\, shx)\\
&= e^{ix}.\, f^{ix}
\end{aligned} \qquad (2.21a)
$$

$$
\begin{aligned}
&e^{-iN/(x,\, x)}\\
&= \cos N(x, x) - i \sin N(x, x) = C_1 - i\, S_1\\
&= \cos x\, chx - \sin x\, shx - i \sin x\, chx - i \cos x\, shx\\
&= (\cos x - i \sin x)\,(ch\, x - i\, shx)\\
&= e^{-ix}.\, f^{-ix}
\end{aligned} \qquad (2.21b)
$$

Similarly

$$
\begin{aligned}
&e^{\,iN/(x,\, -x)}\\
&= \cos N(x, -x) + i \sin N(x, -x) = C_2 + i\, S_2\\
&= \cos x\, chx + \sin x\, shx + i \sin x\, chx - i \cos x\, shx\\
&= (\cos x + i \sin x)\,(chx - i\, shx)\\
&= e^{\,ix}.\, f^{-ix}
\end{aligned} \qquad (2.21c)
$$

$$
\begin{aligned}
&e^{-iN/(x,\, x)}\\
&= \cos N(x, -x) - i \sin N(x, -x) = C_2 - i\, S_2\\
&= \cos x\, chx + \sin x\, shx - i \sin x\, chx + i \cos x\, shx\\
&= (\cos x - i \sin x)\,(chx + i\, shx)\\
&= e^{-ix}.\, f^{ix}
\end{aligned} \qquad (2.21d)
$$

Collecting formulas together

$$
\begin{aligned}
e^{iN/(x,\, x)} &= e^{ix}.\, f^{ix}\\
e^{-iN/(x,\, x)} &= e^{-ix}.\, f^{-ix}\\
e^{iN/\,(x,\, -x)} &= e^{ix}.\, f^{-ix}\\
e^{-iN/\,(x,\, x)} &= e^{\,-ix}.\, f^{ix}
\end{aligned} \qquad (2.22a)
$$

Cosn 1 function [with variable (ix, ix)]

On replacing x, by (ix), (2.22a) when recast, appear as

$$e^{iN/(ix,\, ix)} = \cos N(ix, ix) + i \sin N(ix, ix)$$

$$
\begin{aligned}
&= \cos x\, chx + \sin x\, shx - \sin x\, chx - \cos x\, shx \\
&= (\cos x - \sin x)\,(chx - shx) \\
&= \cos N(x, -x) - \sin N(x, x) \\
&= f^{-x}\, e^{-x} && (2.22b)\\
e^{-iN/(ix,\ ix)} &= \cos N(ix, ix) - i \sin N(ix, ix) \\
&= \cos x\, chx + \sin x\, shx + \sin x\, chx + \cos x\, shx \\
&= (\cos x + \sin x)\,(chx + shx) \\
&= \cos N(x, -x) + \sin N(x, x) \\
&= f^{x}\, e^{x} && (2.22c)\\
e^{iN/(ix,\ -ix)} &= \cos N(ix, -ix) + i \sin N(ix, -ix) \\
&= \cos x\, chx - \sin x\, shx + \cos x\, shx + \sin x\, chx \\
&= (\cos x + \sin x)\,(chx - shx) \\
&= \cos N(x, x) + \sin N(x, -x) \\
&= f^{x}\, e^{-x} && (2.22d)\\
e^{-iN/(ix,\ -ix)} &= \cos N(ix, -ix) - i \sin N(ix, -ix) \\
&= \cos x\, chx - \sin x\, shx + \cos x\, shx - \sin x\, chx \\
&= (\cos x - \sin x)\,(chx + shx) \\
&= \cos N(x, x) - \sin N(x, -x) \\
&= f^{-x}\, e^{x} && (2.22e)
\end{aligned}
$$

Collecting formulas together

$$
\begin{aligned}
e^{iN/(ix,\ ix)} &= \cos N(x, -x) + \sin N(x, x) \\
&= (\cos x - \sin x)\,(chx - shx) \\
&= f^{-x}\, e^{-x} \\
e^{-iN/(ix,\ ix)} &= \cos N(x, -x) - \sin N(x, x) \\
&= (\cos x + \sin x)\,(chx + shx) \\
&= f^{x}\, e^{x} \\
e^{iN/(ix,\ -ix)} &= \cos N(x, x) + \sin N(x, -x) \\
&= (\cos x + \sin x)\,(chx - shx) \\
&= f^{x}\, e^{-x} \\
e^{-iN/(ix,\ -ix)} &= \cos N(x, x) - \sin N(x, -x) \\
&= (\cos x - \sin x)\,(chx + shx) \\
&= f^{-x}\, e^{x} && (2.22f)
\end{aligned}
$$

Note

As described later, the following two expressions can be described through triple Nbic functions, $N_{3e}(x, x)$, as

$$
\begin{aligned}
\cos N_{3e}(x, x) &= \cos 2x - \sin 2x\, sh\, 2x \\
\sin N_{3e}(x, x) &= \sin 2x + \cos 2x\, sh\, 2x && (2.23a)
\end{aligned}
$$

so that,

$$e^{\pm iN_{3e}/(x,\ x)} = \cos N_{3e}(x, x) \pm i \sin N_{3e}(x, x)$$

is understood. (2.23b)

Then on the strength of (2.23)

$$
\begin{aligned}
[e^{iN/(x,\,x)}]^2 &= [\cos N(x, x) + i \sin N(x, x)]^2 \\
&= e^{2ix} (f^{ix}.\, f^{ix}) = (\cos 2x + i \sin 2x)(1 + i\, sh\, 2x) \\
&= (\cos 2x - \sin 2x\, sh\, 2x) + i(\sin 2x + \cos 2x\, sh\, 2x) \\
&= \cos N_{3e}(x, x) + i \sin N_{3e}(x, x) \\
&= e^{iN_{3e}/(x,\,x)} && (2.24a) \\
[e^{-iN/(x,\,x)}]^2 &= [\cos N(x, x) - i \sin N(x, x)]^2 \\
&= e^{-2ix} (f^{-ix}.\, f^{-ix}) = (\cos 2x - i \sin 2x)(1 - i\, sh\, 2x) \\
&= (\cos 2x - \sin 2x\, sh\, 2x) - i(\sin 2x + \cos 2x\, sh\, 2x) \\
&= \cos N_{3e}(x, x) - i \sin N_{3e}(x, x) \\
&= e^{-iN_{3e}/(x,\,x)} && (2.24b)
\end{aligned}
$$

Complex Single Nbic Function Ratio, R_{nsn}

To start with, let us investigate the form of the ratio, R_{nsn}, for complex Single Nbic functions. It is expected that, R_{nsn} will be a certain function of x, as seen for the case of Complex hyperbolic functions, R_{nhy} (2.5j).

For a particular case, ($n = 2$), let us observe that

$$R_{2sn} = \frac{\cos N(2x) + i \sin N(2x)}{[\cos N(x) + i \sin N(x)]^2} \tag{2.24c}$$

Remembering that

$$
\begin{aligned}
(f^{ix})^2 &= (chx + i\, shx)^2 = f^{ix} f^{ix} = f^{2ix} = 1 + i\, sh\, 2x \\
f^{i2x} &= ch\, 2x + i\, sh\, 2x
\end{aligned}
$$

[Note that $(f^{ix})^2 = f^{2ix}\ (\neq f^{i2x})$] (2.24d)

Expand the numerator of (2.24c)

$$
\begin{aligned}
e^{iN/2x} &= e^{iN/(2x,\,2x)} = \cos N\, 2x + i \sin N\, 2x \\
&= \cos N(2x, 2x) + i \sin N(2x, 2x) \\
&= [\cos 2x\, ch\, 2x - \sin 2x\, sh\, 2x] + i[\sin 2x\, ch\, 2x + \cos 2x\, sh\, 2x] \\
&= (\cos 2x + i \sin 2x)[ch\, 2x + i\, sh\, 2x] \\
&= e^{2ix} f^{i.2x} && (2.24d')
\end{aligned}
$$

In view of (2.22), expand

$$e^{iN/x} = \cos Nx + i \sin Nx = e^{ix} f^{ix}$$

Squaring both sides.

$$(\cos Nx + i \sin Nx)^2 = (e^{ix})^2 (f^{ix})^2 = e^{2ix} (f^{ix} f^{ix}) = e^{2ix} f^{2ix} \tag{2.24e}$$

Then simplify (2.24c) with (2.24d and 2.24e)

$$
\begin{aligned}
R_{2sn} &= \frac{\cos N(2x) + i \sin N(2x)}{[\cos N(x) + i \sin N(x)]^2} \\
&= e^{iN/2x} / (e^{iN/x})^2 = (e^{2ix} f^{i2x}) / (e^{2ix} f^{2ix}) = f^{i2x} / f^{2ix} && (2.24f)
\end{aligned}
$$

Since, $f^{2ix} \neq f^{i2x}$, the ratio of the two functions, for Nbic function, yields

$$R_{2sn} = f^{i2x} / f^{2ix},\ (\neq 1) \tag{2.24g}$$

CHAPTER 2

General case

As a general derivation of R_{nsn}, we follow an alternative procedure of direct derivation.

From the very definition of single Nbic functions, (1.31), we write

$$\begin{aligned}
\cos N\, nx &= \cos N(nx, nx) = \cos nx\; ch\, nx - \sin nx\; sh\, nx \\
\sin N\, nx &= \sin N(nx, nx) = \sin nx\; ch\, nx + \cos nx\; sh\, nx \\
\cos N\, nx + i \sin N\, nx &= ch\, nx\,(\cos nx + i \sin nx) \\
&\quad + i\, sh\, nx\,(\cos nx + i \sin nx) \\
&= (ch\, nx + i\, sh\, nx)(\cos nx + i \sin nx) \\
&= f^{\,inx} \cdot e^{nix}
\end{aligned}$$

Similarly

$$\begin{aligned}
\cos Nx &= \cos N(x, x) = \cos x\; chx - \sin x\; shx \\
\sin Nx &= \sin N(x, x) = \sin x\; chx + \cos x\; shx \\
\cos N\, x + i \sin Nx &= chx\,(\cos x + i \sin x) + i\, shx(\cos x + i \sin x) \\
&= (chx + i\, shx)(\cos x + i \sin x) \\
&= f^{\,ix} \cdot e^{ix} \\
(\cos Nx + i \sin Nx)^n &= [f^{\,ix}\, e^{ix}]^n = e^{nix}\,(f^{\,ix})^n
\end{aligned}$$

$$\begin{aligned}
R_{nsn} &= \frac{\cos N(nx) + i \sin N(nx)}{[\cos N(x) + i \sin N(x)]^n} \\
&= [f^{\,inx}\, e^{nix}] / [e^{nix}\,(f^{\,ix})^n] \\
&= f^{\,inx} / f^{\,nix} \\
&= [f^{\,inx}\, f^{\,-nix}] / ch^n\, 2x \qquad (2.24h)
\end{aligned}$$

This expression is the same one as that for the Complex hyperbolic function ratio, R_{nhy} (2.5j). Thus, $R_{nsn} = R_{nhy}$. (2.24i)

Differentiation and Integration of Cosn1 Function

Now we have in hand, that

$$e^{\,ix} = \cos x + i \sin x\,,\; e^{-ix} = \cos x - i \sin x$$

$$\frac{d}{dx} e^{\,ix} = -\sin x + i \cos x = i\,(\cos x + i \sin x) = i\, e^{\,ix}$$

$$\frac{d}{dx} e^{-ix} = -\sin x - i \cos x = -i\,(\cos x - i \sin x) = -i\, e^{-ix} \qquad (2.25a)$$

So that in view of (2.25a)

$$\begin{aligned}
\frac{d}{dx}(e^{ix} + e^{-ix}) &= \frac{d}{dx} e^{\,ix} + \frac{d}{dx}\, e^{-ix} \\
&= i\,(e^{\,ix} - e^{-ix}) = i\,(2\, i \sin x) = -2 \sin x
\end{aligned}$$

also noting that, $e^{ix} + e^{-ix} = 2 \cos x$

then

$$\frac{d}{dx}(e^{\,ix} + e^{\,-ix}) = \frac{d}{dx}(2 \cos x) = -2 \sin x \qquad (2.25b)$$

Similarly

$$\frac{d}{dx}(e^{ix} - e^{-ix}) = \frac{d}{dx}e^{ix} - \frac{d}{dx}e^{-ix}$$
$$= i\,(e^{ix} + e^{-ix}) = i\,(2\cos x) = 2\,i\cos x$$

also on subtracting

$$\frac{d}{dx}(e^{ix} - e^{-ix}) = \frac{d}{dx}(2\,i\sin x) = 2\,i\cos x \qquad (2.25c)$$

Collecting formulas together

$$\frac{d}{dx}\boldsymbol{e^{ix} = i\,e^{ix}}$$
$$\frac{d}{dx}\boldsymbol{e^{-ix} = -\,i\,e^{-ix}} \qquad (2.25d)$$

Similarly for these two ***f*** functions

$$f^{ix} = (chx + i\,shx)$$
$$f^{-ix} = (chx + i\,shx)$$

the differential rule is:

$$\frac{d}{dx}\boldsymbol{f^{ix}} = shx + i\,chx = i\,(chx - i\,shx) = \boldsymbol{i\,f^{-ix}}$$
$$\frac{d}{dx}\boldsymbol{f^{-ix}} = shx - i\,chx = -\,i\,(chx + i\,shx) = \boldsymbol{-\,i\,f^{ix}} \qquad (2.26a)$$

Further

$$\frac{d}{dx}(f^{ix} + f^{-ix}) = i\,f^{-ix} - i\,f^{ix} = -\,i\,(f^{ix} - f^{-ix}) = 2\,shx$$
$$\frac{d}{dx}(f^{ix} - f^{-ix}) = i\,f^{-ix} + i\,f^{ix} = i\,(f^{ix} + f^{-ix}) = 2\,i\,chx \qquad (2.26b)$$

where

$$f^{ix} + f^{-ix} = 2\,ch\boldsymbol{x}$$
$$f^{ix} - f^{-ix} = 2\,i\,sh\boldsymbol{x}$$

Again, we have from (2.16)

$$\text{cf}\,x = \frac{1}{2}\,(f^{ix} + f^{-ix}) = \mathbf{chx}$$
$$\text{sf}\,x = \frac{1}{2}\,(f^{ix} - f^{-ix}) = \boldsymbol{i}\ \mathbf{shx} \qquad (2.26c)$$

from which we have

$$\text{cf}^{\,2}x + \text{sf}^{\,2}x = [chx]^2 + [i\,shx]^2 = ch^2x - sh^2x = 1 \qquad (2.26d)$$

Thus, remembering the differentiation rule of plane trigonometric functions, *viz.*, $\frac{d}{dx}(\sin x) = \cos x$, and, $\frac{d}{dx}(\cos x) = -\sin x$, we get similar form for Cof-functions, but multiplied by, i, as

$$\frac{d}{dx} \text{sf } x = i.\text{cf } x$$

$$\frac{d}{dx} \text{cf } x = -\, i.\text{sf } x \tag{2.26e}$$

Now call (2.22), and operate $\frac{d}{dx}(\;)$, using $(\;)'$ symbol for it, to obtain the differentiation expression as

$$\begin{aligned}(e^{iN/(x,\,x)})' &= (e^{ix} f^{\,ix})' \\ &= e^{ix} (f^{\,ix})' + (e^{ix})' f^{\,ix} = i\, e^{ix} (f^{\,ix} + f^{\,-ix}) = 2\, i\, e^{ix}\, chx\end{aligned}$$

On similar procedure, thus we have the differentiation expressions

$$\begin{aligned}(e^{iN/(x,\,x)})' &= 2\, i\, e^{ix}\, chx \\ (e^{-iN/(x,\,x)})' &= -2\, i\, e^{-ix}\, chx \\ (e^{iN/(x,\,-x)})' &= 2\, e^{ix}\, shx \\ (e^{-iN/(x,\,-x)})' &= 2\, e^{-ix}\, shx\end{aligned} \tag{2.27}$$

Integration

We have

$$\frac{d}{dx} f^{\,ix} = i\, f^{\,-ix} \quad \text{so, } I(f^{\,-ix})\, dx = \int f^{\,-ix}\, dx = -\, i\, f^{\,ix}$$

$$\frac{d}{dx} f^{\,-ix} = -\, i\, f^{\,ix} \quad \text{so, } I(f^{\,ix})\, dx = \int f^{\,ix}\, dx = i\, f^{\,-ix} \tag{2.28a}$$

From (2.26e)

$$\frac{d}{dx}[\text{sf } x] = i.\text{cf } x \Rightarrow \int \text{cf } x\, dx = -\, i\, \text{sf } x$$

$$\frac{d}{dx}[\text{cf } x] = -\, i.\text{sf } x \Rightarrow \int \text{sf } x\, dx = i\, \text{cf } x \tag{2.28b}$$

In (2.28b), subtract one from the other

$$\frac{d}{dx}(\text{cf } x - \text{sf } x) = -\, i\, (\text{cf } x + \text{sf } x) \tag{2.28c}$$

Then

$$\int (\text{cf } x + \text{sf } x)\, dx = i\, (\text{cf } x - \text{sf } x) \tag{2.28d}$$

Cosn 2 Function [with variable (*x*, *y*)]

Extending the derivation with variables (*x*, *y*), using the various Nbic functions defined earlier, we can write each Cosn2-function as a product of Coci-function in (*x*, *x*), and Cohy1-function in (*y*, *y*), as follows:

$$\begin{aligned} e^{iN/(x,\,y)} &= \cos N(x,\ y) + i\ \sin N(x,\ y) = C1^* + i\ S1^* \\ &= e^{ix}.\ f^{\,iy} \\ e^{-iN/(x,\,y)} &= \cos N(x,\ y) - i\ \sin N(x,\ y) = C1^* - i\ S1^* \\ &= e^{-ix}.\ f^{\,-iy} \\ e^{iN/(x,\,-y)} &= \cos N(x,\ -y) + i\ \sin N(x,\ -y) = C2^* + i\ S2^* \\ &= e^{\,ix}.\ f^{\,-iy} \\ e^{-iN/(x,-y)} &= \cos N(x,\ -y) - i\ \sin N(x,\ -y) = C2^* - i\ S2^* \\ &= e^{\,-ix}.\ f^{\,iy} \end{aligned} \qquad (2.29)$$

Thus we get from (2.29)

$$\frac{e^{iN/(x,y)} + e^{-iN/(x,y)}}{2} = \cos N(x,\ y)$$

$$\frac{e^{iN/(x,y)} - e^{-iN/(x,y)}}{2} = i\ \sin N(x,\ y) \qquad (2.30)$$

2.1.11 Checking of Expressions

To Check 1.97 using (2.29)

$\cos N^2(x,\ y) + \sin N^2(x,\ y) = ch\ 2y$

$$\begin{aligned} \text{LHS} &= [\cos N(x,\ y) + i\ \sin N(x,\ y)]\ [\cos N(x,\ y) - i\ \sin N(x,\ y)] \\ &= (e^{ix}\ f^{\,iy})\ (e^{-ix}\ f^{\,-iy}) \\ &= (e^{\,ix}\ e^{\,-ix})\ (f^{\,iy}\ f^{\,-iy}) \\ &= (\cos^2 x + \sin^2 x)\ (ch\ 2y) = ch\ 2y = \text{RHS} \end{aligned}$$

Now from the first and second of (2.29)

$$e^{\,ix} = \frac{e^{iN/(x,y)}}{f^{iy}},\quad e^{-ix} = \frac{e^{-iN/(x,y)}}{f^{-iy}}$$

$$e^{ix}\ e^{-ix} = 1 = \frac{e^{iN/(x,y)}}{f^{iy}} \cdot \frac{e^{-iN/(x,y)}}{f^{-iy}}$$

Since $f^{\,iy} f^{\,-iy} = ch\ 2y$, and $e^{ix}\ e^{-ix} = 1$, we get

$$\mathbf{e^{iN/(x,\,y)}.\ e^{-iN/(x,\,y)} = ch\ 2y} \qquad (2.31a)$$

Now from the third and fourth of (2.29)

$$e^{ix} = \frac{e^{iN/(x,-y)}}{f^{-iy}},\quad e^{-ix} = \frac{e^{-iN/(x,-y)}}{f^{iy}}$$

$$e^{ix}.\ e^{-ix} = 1 = \frac{e^{iN/(x,-y)}}{f^{-iy}} \cdot \frac{e^{-iN/(x,-y)}}{f^{iy}}$$

CHAPTER 2

Since $f^{iy}.f^{-iy} = ch\ 2y$, and $e^{ix}.\ e^{-ix} = 1$, we get

$$\mathbf{e^{iN/(x,-y)}.\ e^{-iN/(x,-y)} = ch\ 2y} \tag{2.31b}$$

Now from the first and fourth of (2.29)

$$e^{ix} = \frac{e^{iN/(x,y)}}{f^{iy}},\quad e^{-ix} = \frac{e^{-iN/(x,-y)}}{f^{iy}}$$

$$e^{ix}.\ e^{-ix} = 1 = \frac{e^{iN/(x,y)}}{f^{iy}} \frac{e^{-iN/(x,-y)}}{f^{iy}}$$

Since $f^{iy}.f^{iy} = 1 + i\ sh\ 2y$, and $e^{ix}.\ e^{-ix} = 1$, we get

$$\mathbf{e^{iN/(x,y)}.\ e^{-iN/(x,-y)} = 1 + i\ sh\ 2y} \tag{2.31c}$$

Now from second and third of (2.29)

$$e^{-ix} = \frac{e^{-iN/(x,y)}}{f^{-iy}},\quad e^{ix} = \frac{e^{iN/(x,-y)}}{f^{-iy}}$$

$$e^{ix}.\ e^{-ix} = 1 = \frac{e^{iN/(x,-y)}}{f^{-iy}} \cdot \frac{e^{iN/(x,-y)}}{f^{-iy}}$$

Since $f^{-iy}.f^{-iy} = 1 - i\ sh\ 2y$, and $e^{ix}.\ e^{-ix} = 1$, we get

$$\mathbf{e^{iN/(x,-y)}.\ e^{-iN/(x,y)} = 1 - i\ sh\ 2y} \tag{2.31d}$$

Collecting formulas together

$$\begin{aligned}
&\mathbf{e^{iN/(x,y)}.\ e^{-iN/(x,y)} = ch\ 2y}\\
&\mathbf{e^{iN/(x,-y)}.\ e^{-iN/(x,-y)} = ch\ 2y}\\
&\mathbf{e^{iN/(x,y)}.\ e^{-iN/(x,-y)} = 1 + i\ sh\ 2y}\\
&\mathbf{e^{iN/(x,-y)}.\ e^{-iN/(x,y)} = 1 - i\ sh\ 2y}
\end{aligned} \tag{2.32}$$

General Case

The general form of (2.13) can be written as

$$f^{inx}.\ f^{imy} = ch(nx - my) + i\ sh(nx + my) \tag{2.33}$$

Note $\quad f^{ix} f^{-ix} = ch\ 2x,\ f^{ix} f^{ix} = 1 + i\ sh\ 2x$, then

$$ch(nx - my) = f^{i\frac{nx-my}{2}} . f^{-i\frac{nx-my}{2}}$$

$$i\ sh(nx + my) = f^{i\frac{nx-my}{2}} . f^{i\frac{nx-my}{2}} - 1$$

Adding,

$$f^{i\frac{nx-my}{2}} .\ [2\ cf(\frac{nx-my}{2}) - 1]$$

Thus

$$\mathbf{f^{inx}.\ f^{imy} = ch(nx - my) + i\ sh(nx + my)}$$

$$= f^{\,i\frac{nx-my}{2}} \;.\; [2\ cf(\frac{nx-my}{2})-1] \qquad (2.34)$$

where, the functions $cf\ (y)$ $sf\ (y)$, are defined in (2.16).

2.1.12 Complex *n* Function (Con function)

Let us define once again new functions with Cof-functions, on similar lines with Cohy-functions (which are made of hyperbolic terms, here in this case with, sechx, and thx), after rearranging. These functions are to be termed as **Complex *n*-function**, or in short, **Con-function**, as follows:

Define

$$\begin{aligned} n^{ix} &= \text{sech}x + i\ \text{th}x \\ n^{-ix} &= \text{sech}x - i\ \text{th}x \end{aligned} \qquad (2.35)$$

With the help of (2.4 c–d)

$$\begin{aligned} f^{\,ix} f^{\,ix} &= 1 + i\ sh\ 2x \qquad (\neq f^{\,2ix}) \\ f^{\,-ix} f^{\,-ix} &= 1 - i\ sh\ 2x \\ f^{\,ix} f^{\,-ix} &= ch\ 2x \end{aligned} \qquad (2.36a)$$

the Con-functions (2.35), on further simplification can be rewritten as

$$n^{ix} = \text{sech}\ x + i\ \text{th}x \quad = \frac{1}{chx} + i.\frac{shx}{chx} = \frac{1+i.shx}{chx}$$

$$= \frac{f^{ix/2}.f^{ix/2}}{f^{-ix/2}.f^{ix/2}} = \frac{f^{ix/2}}{f^{-ix/2}}$$

$$n^{-ix} = \text{sech}\ x - i\ \text{th}x \quad = \frac{1}{chx} - i.\frac{shx}{chx} = \frac{1-i.shx}{chx}$$

$$= \frac{f^{-ix/2}.f^{-ix/2}}{f^{-ix/2}.f^{ix/2}} = \frac{f^{-ix/2}}{f^{ix/2}} \qquad (2.36b)$$

Also

$$n^{\,i2x} = \text{sech}\ 2x + i\ \text{th}\ 2x = \frac{1}{ch\ 2x} + i.\frac{sh\ 2x}{ch\ 2x} = \frac{1+i.sh\ 2x}{ch\ 2x} = \frac{(chx+i.shx)^2}{ch\ 2x}$$

$$= \frac{f^{ix}.f^{ix}}{f^{-ix}.f^{ix}} = \frac{f^{ix}}{f^{-ix}}$$

Similarly

$$n^{-i2x} \quad = \text{sech}\ 2x - i\ \text{th}\ 2x = \frac{f^{-ix}}{f^{ix}} \qquad (2.36c)$$

Multiplying, n^{ix} and n^{-ix}, and n^{i2x} and n^{-i2x}

$$n^{ix}\, n^{-ix} = (\text{sech}x + i\ \text{th}x)\ (\text{sech}x - i\ \text{th}x) = \text{sech}^2x + \text{th}^2x = 1$$

$$n^{i\,2x}\ .\ n^{-i\,2x} = 1$$

CHAPTER 2

Following this way

$$n^{ikx} = \text{sech}(kx) + i\ \text{th}(kx) = \frac{f^{ikx/2}}{f^{-ikx/2}}$$

$$n^{-ikx} = \text{sech}(kx) - i\ \text{th}(kx) = \frac{f^{-ikx/2}}{f^{ikx/2}}$$

so that

$$n^{ikx}\, n^{-ikx} = 1 \tag{2.36d}$$

where k is any positive integer.

2.2 COMPLEX NUMBER SYSTEMS

2.2.1 Complex Circular Number (Coci number)

The complex number $a + i\,b$ when expressed through Coci-functions, be renamed as Coci number.

$$a + i\,b = r\,(\cos u + i\,\sin u) = r\,e^{iu}$$
$$a - i\,b = r\,(\cos u - i\,\sin u) = r\,e^{-iu}$$
$$a = r\,\cos u\ ,\ b = r\,\sin u$$

[where, u is expressed in degree]

$$r = \sqrt{a^2 + b^2}\ ,\ \tan u = b/a \tag{2.37}$$

r is always +ve irrespective of a is greater than or less than b.

2.2.2 Complex Hyperbolic Number (Cohy number)

The complex number $a + i\,b$ when expressed through Cohy-functions, is termed as Cohy-number.

$$a + i\,b = \text{R}\,(chx + i\,shx) = \text{R}\,f^{\,ix}$$
$$a - i\,b = \text{R}\,(chx - i\,shx) = \text{R}\,f^{\,-ix}$$
$$a = \text{R}\,chx,\ b = \text{R}\,shx$$

[where, x is expressed in radian]

$$\text{R} = \sqrt{a^2 - b^2}\,,\ thx = b/a \tag{2.38}$$

Case 1 (when $a > b$)

R is +ve, if a is greater than b.

To transfer e^{ix} in terms of $f^{\,ix}$ and *vice versa*

$$\text{R}\,(chx + i\,shx) = \text{r}\,(\cos u + i\,\sin u)$$

so

$$\text{R}\,chx = r\,\cos u\ ,\ \text{R}\,sh\,x = r\,\sin u$$

when $a > b$

$$r = \sqrt{a^2 + b^2}\,,\ R = \sqrt{a^2 - b^2}\,,\ \text{(both are +ve)} \tag{2.39a}$$

We have

$$r^2(\cos^2 u + \sin^2 u) = r^2 = R^2(ch^2x + sh^2x) = R^2\ ch^2x$$

$$r^2(\cos^2 u - \sin^2 u) = r^2 \cos 2u = R^2(ch^2x - sh^2x) = R^2$$

$$\Rightarrow \boldsymbol{r = R\sqrt{ch2x} \quad \text{or, } R = r\sqrt{\cos 2u}} \tag{2.39b}$$

$$R\ chx = r\cos u \Rightarrow \cos u = \frac{chx}{\sqrt{\cos 2u}}$$

$$R\ shx = r\sin u \Rightarrow \sin u = \frac{shx}{\sqrt{\cos 2u}}$$

$$R\,(chx + i\ shx) = r\,(\cos u + i\ \sin u)$$

$$\Rightarrow \boldsymbol{r\ e^{iu} = R\ f^{ix}} \tag{2.39c}$$

$$\Rightarrow \boldsymbol{e^{iu} = \frac{1}{\sqrt{ch2x}} \cdot f^{ix}, \text{ or, } f^{ix} = \frac{1}{\sqrt{ch2u}} \cdot e^{iu}}$$

$$e^{iu} \cdot f^{ix} = \frac{1}{\sqrt{ch2x}} \cdot \frac{1}{\sqrt{ch2u}} \cdot f^{ix} \cdot e^{iu}$$

$$\Rightarrow \boldsymbol{\cos 2u \ .\ ch\ 2x = 1} \tag{2.39d}$$

$$u = \frac{1}{2}\cos^{-1}\frac{1}{ch2x}$$

$$x = \frac{1}{2}ch^{-1}\frac{1}{\cos 2u} \tag{2.39e}$$

Also, $thx = \tan u$ (2.39f)

Case 2 (when $a < b$)

when $a < b$

R is complex, if a is less than b.

$r = \sqrt{a^2 + b^2}$, it is always +ve,

$R = \sqrt{a^2 - b^2}$, (it is complex)

Since in the Cohy-number, R is complex, when evaluated, it becomes (iR), where R is the mod value, *i.e.*, $R = \left|a^2 - b^2\right|$. So, multiply both sides by ($-i$).

$$(-i).(a + i\ b) = (-i).(iR)\ (chx + i\ shx),$$

or, $$b - i\ a = R\ (chx + i\ shx) \tag{2.39g}$$

then

$$b = R\ chx, \text{ and, } a = R(-\ shx)$$

Since, $ch(-ve) = +$ ve, and $sh(-$ ve$) = -$ve, we can write

$$b = R\ ch(-x), \text{ and, } a = R\ (-shx) \tag{2.39h}$$

Thus, in effect, it will be considered that, the given number is $(b - i\ a)$, where $b > a$, and proceed as usual for the case as described for $a > b$, with R as positive, with the number $(a - i\ b) = \mathrm{R}\ (chx + i\ shx)$.

The required value of x, for the present case $(a < b)$, will then be negative to that one obtained for the case, $a > b$.

The evaluation of R for two cases, namely when $a > b$, and $a < b$, has been demonstrated with numerical examples later.

2.2.3 Problems of Various Types of Complex Numbers in Terms of Various Types of Functions and Interrelations

Numerical Examples (Examples 2.2a–2.2b)

Example 2.2a (Equation 2.39a–f)

when $a > b$

$$a = 8,\ b = 6, \text{ so that the complex number is, } a + i\ b = 8 + 6\ i$$
$$8 + 6\ i = a + i\ b = \mathrm{R}\ (chx + i\ shx) = r\ (\cos u + i\ \sin u)$$
$$r^2 = 64 + 36 = 100 \Rightarrow r = 10\ ,\ \cos u = 0.8\ ,\ \sin u = 0.6 \qquad (2.40a)$$
$$\Rightarrow r\ (\cos u + i\ \sin u) = 10\ (0.8 + 0.6i) = 0.8 + 0.6i$$

Again

$$\mathrm{R}\ chx = 8\ ,\ \mathrm{R}\ shx = 6 \Rightarrow \mathrm{R}^2\ (ch^2x - sh^2x) = 64 - 36 = 28$$
$$\Rightarrow \mathrm{R} = 2\sqrt{7} = 5.292$$
$$\Rightarrow f^{\,ix} = 5.292\ (1.51186 + 1.133389i) \qquad (2.40b)$$

where

$$chx = 8/2\sqrt{7} = 4/\sqrt{7} = 1.51186 \Rightarrow x = 0.972957$$
$$shx = 6/2\sqrt{7} = 3/\sqrt{7} = 1.13389 \Rightarrow x = 0.972957 \quad \textbf{same} \qquad (2.40c)$$

Recheck

$$\text{Given } \cos u = 0.8,\ u = 36.87 \text{ deg} = 0.6435 \text{ rad} \Rightarrow \cos 2u = 0.28$$
$$\text{Given } x = 0.972957,\ 2x = 1.945914 \Rightarrow ch\ 2x = 3.57144$$
$$\Rightarrow \cos 2u\ .\ ch\ 2x = 0.28 \times 3.57144 = 1 \quad \textbf{(Checked)} \qquad (2.40d)$$

Recheck

$$\text{Given } x = 0.972957,\ 2x = 1.945914 \Rightarrow ch2x = 3.57144$$
$$\Rightarrow \mathrm{sech}\ 2x = 0.2799989 = 0.28,\ \cos^{-1}(0.28) = 73.73986 \text{ deg}$$
$$\Rightarrow u = \frac{1}{2}\ (73.73986 \text{ deg}) = 36.87 \text{ deg} \quad \textbf{(Checked)} \qquad (2.40e)$$

Example 2.2b (Equation 2.39g–h)

when $a < b$

$$r = \sqrt{a^2 + b^2}\ ,\ \mathrm{R} = \sqrt{a^2 - b^2}\ ,\ \textbf{(Complex)}$$

For example, in $6 + 8i$, $a = 6$, $b = 8$, $a < b$

$$6 + 8i = \sqrt{a^2 - b^2} \,.\, (chx + i\, shx) = 5.292\, i\, (chx + i\, shx) \qquad (2.41a)$$

multiply both sides by $(-i)$

$$(-i)\,(6 + 8i) = (-i)\, 5.292\, i\, (chx + i\, shx)$$
$$\Rightarrow (8 - 6i) = 5.292\, (chx + i\, shx)$$

Remember that $ch(-) = (+)$, $sh(-) = (-)$,

$$R\, chx = 8,\ x = -\, 0.9728297,$$
$$R\, shx = -\, 6,\ x = -\, 0.9728297$$
$$R = \sqrt{|a^2 - b^2|} = 5.292$$
$$x = -\, 0.9728297 \qquad (2.41b)$$

Thus

when $a > b$, $R = \sqrt{a^2 - b^2}$, (+ve)

$$a + i\, b = 8 + 6i = R(chx + i\, shx),\ a = 8,\ b = 6$$
$$\boldsymbol{x = +\ ve} \qquad (2.41c)$$

when $a < b$, $a = 6$, $b = 8$, say $6 + 8i$

$$R \Rightarrow \sqrt{a^2 - b^2} \ \ , \textbf{(Complex)}$$

Take then

$$b - i\, a = 8 - 6i = R\,(chx + i\, shx)\ ,\ a = 8,\ b = 6$$
$$R = \sqrt{a^2 - b^2}\ ,\ (+ve)$$

$\boldsymbol{x = -ve}$ (same value as for the previous case) (2.41d)

Checking of problem

Check that

$$f^{\,ix} f^{\,-ix} = \text{ch } 2x$$

$$f^{\,ix} = \frac{1}{\sqrt{\text{ch } 2u}} \,.\, e^{iu},\ f^{\,-ix} = \frac{1}{\sqrt{\text{ch } 2u}} \,.\, e^{-iu},$$ **and *vice versa***

Now $$f^{\,ix} f^{\,-ix} = \frac{1}{\sqrt{ch\ 2u}} e^{iu} \frac{1}{\sqrt{ch\ 2u}} e^{-iu} = \frac{e^0 = 1}{ch\ 2u} = ch\ 2x \quad \textbf{(Checked)}$$

Check that

$$f^{\,ix} f^{\,ix} = 1 + i\ sh\ 2x$$

$$\text{LHS} = (e^{iu})^2 \cdot \frac{1}{\cos\ 2u} = e^{2iu} \frac{1}{\cos\ 2u} = \frac{\cos\ 2u + i \cdot \sin\ 2u}{\cos\ 2u} = 1 + i \,.\, \tan\ 2u$$

Now

$$\mathbf{\cos\ 2u = sech\ 2x}$$

$$\cos^2 2u = 1 - \sin^2 2u = \frac{1}{ch^2 2x} \cdot \sin^2 2u = 1 - \text{sech}^2\, 2x = \tanh^2 2x$$

$$\Rightarrow \sin^2 2u = \tanh^2 2x \Rightarrow \mathbf{\sin 2u = \tanh 2x}$$
$$\Rightarrow \mathbf{\tan 2u} = \sin 2u / \cos 2u = \tanh 2x \,.\, \cosh 2x = \mathbf{sh\ 2x}$$

with this

$$\text{LHS} = 1 + i \,.\, \tan 2u = 1 + i \,.\, \sin 2x \ \textbf{(Checked)} \tag{2.41e}$$

If we recollect that

$$ch^2x - sh^2x = 1, \text{ and also, } \sec^2 t - \tan^2 t = 1,$$

Replace **ch**x and **sh**x by **sec** t and **tan** t, respectively. Then

$$8 + 6i = \text{R}(chx + i\ shx) = \text{R}(\sec t + i \tan t) = r\ (\cos u + i \sin u)$$
$$\sec t = 1.51186 \Rightarrow t = 48.59 \text{ deg}$$
$$\tan t = 1.133389 \Rightarrow t = 48.59 \text{ deg} \ \ \textbf{(same)}$$
$$[\sin t = 0.75, \text{ and } \cos t = 0.6614]$$

$$\text{R} \sec t = r \cos u$$
$$\text{R} \tan t = r \sin u$$
$$\text{R}^2 (\sec^2 t - \tan^2 t) = \text{R}^2 = r^2 (\cos^2 u - \sin^2 u) = r^2 \cos 2u$$
$$\text{R}^2 (\sec^2 t + \tan^2 t) = \text{R}^2(1 + 2 \tan^2 t) = r^2 (\cos^2 u + \sin^2 u) = r^2$$

$$\Rightarrow \mathbf{R} = \boldsymbol{r}\ \sqrt{\cos 2u}\,,\ \boldsymbol{r} = \mathbf{R}\ \sqrt{1+2\tan^2 t}$$
$$\Rightarrow \mathbf{(\cos 2u)\ (1 + 2 \tan^2 t) = 1}$$

Again, $\tan u = \tan t / \sec t = \sin t$

$$\Rightarrow \mathbf{\tan u = \sin t} \tag{2.41f}$$
$$\tan u = 6/8 = 0.75 = \sin t \Rightarrow t = 48.59 \text{ deg}$$

Similarly if

$$8 + 6i = \text{R}(chx + i\ shx) = \text{R}(\sec t + i \tan t) = r\ (\cos u + i \sin u)$$
$$= \text{R}\ (\text{cosec} y + i \cot y)$$
$$\Rightarrow \mathbf{\tan u = \cos y} \tag{2.41g}$$

2.2.4 Relation between Cosn Function, $e^{iN/(x,y)} = \cos N(x, y) + i \sin N(x, y)$, and the Product of Two Complex Numbers, $z_1 = a + i\,b$ and $z_2 = c + i\,d$

This is illustrated through the following numerical problem.

Numerical Examples

Example 2.3

Given

z_1 (x-function part) $= a + i\ b = 8 + 6i$, and z_2 (y-function part) $= c + i\ d = 12 + 5i$. To verify the direct product of $\mathbf{z_1\, z_2}$ and that using (2.13).

First transfer the given complex numbers in Coci and Cohy-form. This is done as follows. The extra result will be utilized as and when required.

Expressions of, $8 + 6i$, $12 + 5i$ and $12 - 5i$, in Coci, and Cohy-form.

$$8 + 6i = \text{R}_1\, f^{\,ip} = \text{R}_1(chp + i\ shp) = 5.292(1.5117 + 1.1339i)$$
$$\textbf{In } f^{ip} = \boldsymbol{chp + i\ shp}$$
$$\Rightarrow \text{R}_1 = 5.292,\ chp = 1.5119,\ shp = 1.1339,\ p = 0.97296 \text{ rad } (55.7468 \text{ deg}),$$
$$\sin p = 0.8266,\ \cos p = 0.5628 \tag{2.42a}$$

$8 + 6i = r_1\, e^{im} = r_1\,(\cos m + i \sin m) = 10\ (0.8+0.6\ i)$

In $\boldsymbol{e^{im} = \cos m + i \sin m}$

$\Rightarrow r_1 = 10$, $ch\ m = 1.5119$, $sh\ m = 1.1339$, $m = 0.97296$ rad (55.7468 deg), $\sin m = 0.8266$, $\cos m = 0.5628$ (2.42b)

$12 + 5i = R_2\, f^{\,iy} = R_2(chy + shy) = 10.9087\ (1.1 + 0.4583i)$

In $\boldsymbol{f^{\,iy} = chy + i\ shy}$

$\Rightarrow R_2 = 10.9087$, $chy = 1.1$, $shy = 0.4583$, $y = 0.4436$ rad (25.4164 deg),

$\sin y = 0.4292$, $\cos y = 0.9032$

$12 + 5i = R_2\, f^{\,iy} = 10.9087\ (1.1 + 1.4583i)$ (2.42c)

$12 + 5i = r_2\, e^{in} = r_2(\cos n + i \sin n) = 13\ (0.923 + 0.385i)$

In $\boldsymbol{e^{in} = \cos n + i \sin n}$

$\Rightarrow r_2 = 13$, $\cos n = 0.923$, $\sin n = 0.385$, $n = 22.62$ deg (= 0.3948 rad) (2.42d)

Thus the given numbers in Coci and Cohy form as

(a) $8 + 6i = 5.292\, f^{\,0.97298i} = 10\, e^{\,0.6435i}$

(b) $12 + 5i = 10.9087\, f^{\,0.4436i} = 13\, e^{\,0.3948i}$

(c) $12 - 5i = 10.9087\, f^{\,-0.4436i} = 13\, e^{\,-0.3948}$ (2.42e)

The direct product of $(a + i\ b)$, and $(c + i\ d)$, and the corresponding one from (2.13) in reference to Example 2.2 using the results (2.42e) are done as:

Using (2.42e), the product of two corresponding complex numbers are:

(a) × (b) gives

$$\Rightarrow (8 + 6i)(12 + 5i) = \mathbf{66 + 112}\boldsymbol{i}$$

Again, $(8 + 6i)(12 + 5i) = 5.292 \times 10.9087\, f^{\,0.97298i}.\, f^{\,0.4436i}$

$= 57.7288\, f^{\,0.97298i}.\, f^{\,0.4436i}$

use $\boldsymbol{f^{\,ix}\,.\,f^{\,iy} = ch(x - y) + i\ sh(x + y)}$, to get

$= 57.7288\ [ch(0.97298 - 0.44360) + i\ sh(0.97298 + 0.44360)]$

$= 57.7288\ (1.143 + 1.9402i)) = \mathbf{66 + 112}\boldsymbol{i}$

Again, $(8 + 6i)(12 + 5i) = 10 \times 13\ e^{0.6435i}.\ e^{0.3948i} = 130\ e^{1.0383i}$

Noting, 1.0383 rad = 59.4902 deg, cos(59.4902) = 0.5077, sin(59.4902) = 0.8615,

$$(8 + 6i)(12 + 5i) = 130(0.5077 + 0.8615\ i) = \mathbf{66 + 112}\boldsymbol{i}$$

(a) × (c) gives

$$\Rightarrow (8 + 6i)(12 - 5i) = \mathbf{126 + 32}\boldsymbol{i}$$

Again, $(8 + 6i)(12 - 5i) = 5.292 \times 10.9087\, f^{\,0.97298i}.\, f^{\,-0.4436i}$

$= 57.7288\, f^{\,0.97298i}.\, f^{\,-0.4436i}$

use $\boldsymbol{f^{\,ix}\,.\,f^{\,-iy} = ch(x + y) + i\ sh(x - y)}$, to get

$= 57.7288\ ch(0.97298 + 0.44360) + i\ sh(0.97298 - 0.44360)$

$= 57.7288(2.18769 + 0.5545i) = \mathbf{126 + 32}\boldsymbol{i}$

Again, $(8 + 6i)(12 - 5i) = 10 \times 13\ e^{0.6435i}.\ e^{\,-0.3948i} = 130\ e^{0.2487i}$

CHAPTER 2

Noting that, 0.2487 rad = 14.25 deg, cos(14.25) = 0.9692, sin(14.25) = 0.2461,

$$(8 + 6i)(12 - 5i) = 130(0.9692 + 0.2461i) = \mathbf{66 + 112i}$$

(b) × (c) gives

$$\Rightarrow (12 + 5i)(12 - 5i) = \mathbf{169}$$
$$\Rightarrow (12 + 5i)(12 - 5i) = 169\, e^{0.3948i}\, e^{-0.3948i} = 169\, e^{0} = \mathbf{169}$$

use $x = 0.4436$, $ch\ 2x = ch(0.8872) = 1.4204$, in $\boldsymbol{f^{ix} . f^{-ix} = ch\ 2x}$, to get

$$\Rightarrow (12 + 5i)(12 - 5i) = 10.987^2 f^{0.4436i} f^{-0.4436i} = 119 \times 1.4204 = \mathbf{169}$$

(a) × (a) gives

$$\Rightarrow (8 + 6i)(8 + 6i) = \mathbf{28 + 96i}$$

Again, $(8 + 6i)(8 + 6i) = 5.292 \times 5.292\, f^{0.97298i} . f^{0.97298i}$

$$= 28\, f^{0.97298i} . f^{0.97298i}$$

use $\boldsymbol{f^{ix} . f^{ix} = 1 + i\ sh\ 2x}$, to get

$$= 28\, [1 + 3.4287i] = \mathbf{28 + 96i}$$

Again, $(8 + 6i)(8 + 6i) = 10 \times 10\, e^{0.6435i} . e^{0.6435i} = 100\, e^{1.287i}$

Noting, 1.287 rad = 73.7397 deg, cos(73.7397) = 0.2800, sin(73.7397) = 0.9600,

$$(8 + 6i)(8 + 6i) = 100(0.2800 + 0.9600\, i) = \mathbf{28 + 96i}$$

(Verified)

Note: Powers are added in products of $e^{ix} . e^{iy} = e^{i(x+y)}$, but that for $f^{ix} . f^{iy} \neq f^{i(x+y)}$

Example 2.4

With the help of (2.13), *viz,*

$$\begin{aligned}
f^{ix} f^{iy} &= (chx + i\ shx)\ (chy + i\ shy) = ch(x - y) + i\ sh(x + y) \\
f^{ix} f^{-iy} &= (chx + i\ shx)\ (chy - i\ shy) = ch(x + y) + i\ sh(x - y) \\
f^{-ix} f^{iy} &= (chx - i\ shx)\ (chy + i\ shy) = ch(x + y) - i\ sh(x - y) \\
f^{-ix} f^{-iy} &= (chx - i\ shx)\ (chy - i\ shy) = ch(x - y) - i\ sh(x + y)
\end{aligned} \tag{2.43}$$

check whether $e^{iN/(x,\,y)} = 1 / e^{-iN/(x,\,y)}$ or not

Solution

$$\frac{1}{e^{iN/(x,y)}} = \frac{\cos N(x,y) - i \sin N(x,y)}{[\cos N(x,y) + i \sin N(x,y)][\cos N(x,y) - i \sin N(x,y)]}$$

$$= \frac{\cos N(x,y) - i \sin N(x,y)}{\cos N^2(x,y) + \sin N^2(x,y)}$$

$$= \frac{\cos N(x,y) - i \sin N(x,y)}{ch\ 2y} = e^{-iN/(x,\,y)} / ch\ 2y$$

$$\Rightarrow 1 / e^{iN/(x,\,y)} = e^{-iN/(x,\,y)} / ch\ 2y$$

$$\Rightarrow \boldsymbol{e^{iN/(x,\,y)} . e^{-iN/(x,\,y)} = ch\ 2y}, \textbf{ (it is not equal to 1)} \tag{2.44a}$$

Alternatively

$$e^{i\,N/(x,\,y)} = \cos N(x,\ y) + i\ \sin N(x,\ y) = e^{ix}.\ f^{\,iy}$$
$$e^{-iN/(x,\,y)} = \cos N(x,\ y) - i\ \sin N(x,\ y) = e^{-ix}.\ f^{\,-iy}$$

$$\frac{1}{e^{iN/(x,y)}} = \frac{1}{e^{ix} f^{iy}} = e^{-ix} f^{-iy} \frac{1}{ch\ 2y} = e^{-iN/(x,y)} \frac{1}{ch\ 2y} \quad \textbf{Checked}$$

$$e^{iN/(x,\,y)}.\ e^{iN/(x,\,-y)} = e^{2\,ix}.\ ch\ 2y$$
$$e^{iN/(x,\,y)} / e^{iN/(x,\,-y)} = f^{\,iy}.\ (1/f^{\,-iy})$$
$$= (f^{\,iy})\ (f^{\,iy}) / ch\ 2y = [1 + i\ sh\ 2y] / ch\ 2y \tag{2.44b}$$

Example 2.5

Find the number a + *ib* when the other number *c* + *id* is given, using the formula: $e^{iN/(x,\,y)} = \cos N(x,\ y) + i\ \sin N(x,\ y) = e^{ix}.\ f^{\,iy}$

and also

$$a + ib = R_1 f^{\,ix} = \mathbf{R_1(chx + i\ shx)} = r_1\ (\cos\ u_1 + i\ \sin\ u_1)$$
$$c + id = R_2 f^{\,iy} = \mathbf{R_2(chy + i\ shy)} = r_2\ (\cos\ u_2 + i\ \sin\ u_2)$$
$$(a + ib)\ (c + id) = R_1 f^{\,ix}.\ R_2 f^{\,iy}$$

where

$$f^{\,ix} = (chx + i\ shx)$$
$$f^{\,iy} = (chy + i\ shy)$$

With

a = 8, b = 6, R_1 = 5.292, chx = 1.5117, shx = 1.1339,

x = 0.97296 rad (55.7468 deg), sin$x$ = 0.8266, cos$x$ = 0.5628

c = 12, d = 5, R_2 = 10.9087, chy = 1.1, shy = 0.4583,

y = 0.4436 rad (25.4164 deg), sin$y$ = 0.4292, cos$y$ = 0.9032

$(a + i\ b)\ (c + i\ d) = R_1 f^{\,ix}.\ R_2 f^{\,iy} = 66 + 112i$, (multiplying directly)

Solution

Relation between $e^{iN/(x,\,y)}$ and $(a + i\ b)\ (c + i\ d)$

$R_1\ (chx + i\ shx) = a + i\ b \Rightarrow chx = a/R_1,\ shx = b/R_1$

$\Rightarrow$ $\mathbf{x = ch^{-1}(a/R_1) = 0.97288\ (55.7422\ deg)}$, or $= sh^{-1}(b/R_1)$

$R_2\ (chy + i\ shy) = c + i\ d \Rightarrow chy = c/R_2$

$= 1.1,\ shy = d/R_2 = 0.4582$

NOTE : $e^{iN/(x,\,y)} = \cos N(x,\ y) + i\ \sin N(x,\ y) = e^{ix}.\ f^{\,iy}$

$e^{ix} = \cos x + i\ \sin x = \cos[ch^{-1}(a/R_1)] + i\ \sin[ch^{-1}(a/R_1)] = \mathbf{g + i\ h}$

$\Rightarrow 0.563 + 0.827i$

$f^{\,iy} = chy + i\ shy = c/R_2 + i\ d/R_2 = \mathbf{m + i\ n}$

$\Rightarrow 1.1 + 0.4583i$

So

$$e^{ix}.\ f^{\,iy} = e^{i\,N/(x,\,y)} = \cos N(x,\ y) + i\ \sin N(x,\ y)$$
$$= (c_1\ e_1 - f_1\ d_1) + i\ (d_1\ e_1 - f_1\ c_1) = \mathbf{p + i\ q}$$
$$= 0.2403 + 1.1672i$$

This gives p = **cosN**(x, y) = 0.2403, q = **sinN**(x, y) = 1.1672

Now Again

$$e^{ix} = \cos[ch^{-1}(a/R_1)] + i \sin[ch^{-1}(a/R_1)] = g + i\,h$$
$$f^{iy} = c/R_2 + i\,d/R_2 = m + i\,n$$
$$e^{ix}.f^{iy} = (g + i\,h)(m + i\,n) = (g\,m - h\,n) + i\,(g\,n + h\,m) = u + i\,v$$

Separating Real and Imaginary parts

$$\cos[ch^{-1}(a/R_1)]\,.\,c/R_2 - \sin[ch^{-1}(a/R_1)]\,.\,d/R_2 = p = g\,m - h\,n$$
$$\cos[ch^{-1}(a/R_1)]\,.\,d/R_2 + \sin[ch^{-1}(a/R_1)]\,.\,c/R_2 = q = g\,n + h\,m$$

Put $\boldsymbol{k} = \boldsymbol{ch^{-1}(a/R_1)}$, and thus solve for $\boldsymbol{k}$

$$c\,.\,\cos k - d\,.\,\sin k = p\,.\,R_2$$
$$d\,.\,\cos k + c\,.\,\sin k = q\,.\,R_2$$
$$\sin k = R_2\,(qc - pd)\,/(c^2 + d^2)$$
$$\cos k = R_2\,(pc + qd)\,/(c^2 + d^2)$$
$$\Rightarrow \tan k = (qc - pd)/(pc + qd)$$
$$\Rightarrow k = ch^{-1}(a/R_1) = \tan^{-1}[(qc - pd)/(pc + qd)]$$

Now $(qc - pd)/(pc + qd)$

$$= (1.1672 \times 12 - 0.2403 \times 5)/(0.2403 \times 12 + 1.1672 \times 5)$$
$$= 12.8049/8.7196 = 1.46852$$
$$\Rightarrow k = 55.7468 \text{ deg} = 0.97296 \text{ rad} \Rightarrow chx = a/R_1 = 1.5119$$
$$\Rightarrow \textbf{if } R_1 = 5.292\textbf{, then } a = 8$$

2.2.5 Alternative Definition of Nbic Function (in terms of sin and cos)

Using these

$$\sin(ix) = i\,shx$$
$$\cos(ix) = chx$$
$$(1 - i)^2 = -2i$$
$$(1 + i)^2 = 2i$$
$$(1 + i)(1 - i) = 2 \tag{2.45}$$

we have

$$\sin(x + iy) = \sin x \cos iy + \cos x \sin iy = \sin x\,chy + i\cos x\,shy$$
$$\sin(x - iy) = \sin x \cos iy - \cos x \sin iy = \sin x\,chy - i\cos x\,shy$$

adding : $\quad$ **sin$x$ ch$y$** $= [\sin(x + iy) + \sin(x - iy)]/2$

subtracting: $\quad$ **cos$x$ sh$y$** $= -i\,[\sin(x + iy) - \sin(x - iy)]/2$ $\qquad$ (2.46a)

from (2.46a)

adding : $\quad$ **sinN(x, y)** $= [(1 - i)\sin(x + iy) + (1 + i)\sin(x - i\,y)]/2$

subtracting : $\quad$ **sinN(x, $-y$)** $= [(1 + i)\sin(x + iy) + (1 - i)\sin(x - i\,y)]/2$

$$\cos(x + iy) = \cos x \cos iy - \sin x \sin iy = \cos x\,chy - i\sin x\,shy$$
$$\cos(x - iy) = \cos x \cos iy + \sin x \sin iy = \cos x\,chy + i\sin x\,shy$$

Again

adding : $\quad$ **cos$x$ ch$y$** $= [\cos(x + iy) + \cos(x - iy)]/2$

subtracting : $\quad$ **sin$x$ sh$y$** $= i\,[\cos(x + iy) - \cos(x - iy)]/2$ $\qquad$ (2.46b)

from (2.46b)

subtracting : $\quad$ **cosN(x, y)** $= [(1 - i)\cos(x + iy) + (1 + i)\cos(x - iy)]/2$

adding : $\quad$ **cosN(x, $-y$)** $= [(1 + i)\cos(x + iy) + (1 - i)\cos(x - iy)]/2$

Collecting formulas together

Collecting from 2.46a and 2.46b

$$\begin{aligned}
\mathbf{sinN}(x, y) &= [(1 - i) \sin(x + iy) + (1 + i) \sin(x - iy)]/2\\
\mathbf{sinN}(x, -y) &= [(1 + i) \sin(x + iy) + (1 - i) \sin(x - iy)]/2\\
\mathbf{cosN}(x, y) &= [(1 - i) \cos(x + iy) + (1 + i) \cos(x - iy)]/2\\
\mathbf{cosN}(x, -y) &= [(1 + i) \cos(x + iy) + (1 - i) \cos(x - iy)]/2
\end{aligned} \qquad (2.47a)$$

The four equations (2.47a) when arranged in matrix form, appear as

[S] = [K] [T]

$$\begin{bmatrix} \sin N(x,y) \\ \sin N(x,-y) \\ \cos N(x,y) \\ \cos N(x,-y) \end{bmatrix} = \begin{bmatrix} (1-i)/2 & (1+i)/2 & 0 & 0 \\ (1+i)/2 & (1-i)/2 & 0 & 0 \\ 0 & 0 & (1-i)/2 & (1+i)/2 \\ 0 & 0 & (1+i)/2 & (1-i)/2 \end{bmatrix} \begin{bmatrix} \sin(x+iy) \\ \sin(x-iy) \\ \cos(x+iy) \\ \cos(x-iy) \end{bmatrix} \qquad (2.47b)$$

where,

$[S]_{4\times1}$ = Nbic function matrix

where the elements of [S] matrix are:

$$s_{11} = \text{sinN}(x, y),\ s_{21} = \text{sinN}(x, -y),\ s_{31} = \text{cosN}(x, y),\ s_{41} = \text{cosN}(x, -y)$$

$[K]_{4\times4}$ = coefficient matrix, which can be partitioned into two symmetric sub-matrices each of $[2 \times 2]$ size, the elements of which are:

$$k_{11} = [B]_{2\times2},\ k_{12} = [0]_{2\times2},\ k_{21} = [0]_{2\times2},\ k_{22} = [B]_{2\times2}$$

The elements of the submatrice $[B]_{2\times2}$ are:

$$b_{11} = (1 - i)/2,\ b_{12} = (1 + i)/2,\ b_{21} = (1 + i)/2,\ b_{22} = (1 - i)/2$$

$[T]_{4\times1}$ = complex circular function matrix

where the the elements of [T] matrix are:

$$\begin{aligned} t_{11} &= \sin(x + iy),\ t_{21} = \sin(x - iy),\\ t_{31} &= \cos(x + iy),\ t_{41} = \cos(x - iy) \end{aligned} \qquad (2.47c)$$

Numerical Example

Example 2.6

Using (2.47a), verify (1.97) that : $\mathbf{cosN^2}(x, y) + \mathbf{sinN^2}(x, y) = ch\ 2y$

Solution

We have, $(1- i)^2 = - 2i$, $(1 + i)^2 = 2i$, $(1 + i)(1 - i) = 2$

With this the LHS when simplified, becomes

$$\begin{aligned}
\textbf{LHS} &= [- 2i \cos^2(x + iy) + 2i \cos^2(x - iy) + 4 \cos(x + iy) \cos(x - iy)\\
&\quad - 2i \sin^2(x + iy) + 2i \sin^2(x - iy) + 4 \sin (x + iy) \sin (x - iy)]/4\\
&= [- 2i + 2i + 4 \cos (2iy)]/4 = \cos(2iy) = ch\ 2y = \textbf{RHS} \quad \textbf{(Verified)}
\end{aligned}$$

2.2.6 Alternative Definition of Nbic Function (in terms of sinh and cosh)

Using these

$$\begin{aligned}
sh(ix) &= i \sin x\\
ch(ix) &= \cos x\\
sh(y + ix) &= sh\ [i\ (x - iy)] = i \sin (x - iy)\\
sh(y - ix) &= sh\ [- i\ (x - iy)] = -i \sin(x + iy)\ i \sin (x + iy)
\end{aligned}$$

CHAPTER 2

$$ch(y + ix) = ch[i(x - iy)] = \cos(x - iy)$$
$$ch(y - ix) = ch[-i(x + iy)] = \cos(x + iy) \qquad (2.48)$$

we have

$$sh(x + iy) = shx\ ch\ iy + chx\ sh\ iy = i[\sin y\ chx - i \cos y\ shx]$$
$$sh(x - iy) = shx\ ch\ iy - chx\ sh\ iy = -i\ [\sin y\ chx + i \cos y\ shx]$$
$$sh(x + iy) - sh(x - iy) = 2i \sin y\ chx$$
$$sh(x + iy) + sh(x - iy) = 2 \cos y\ shx \qquad (2.49a)$$

$$\sin y\ chx = i\ [sh(x - iy) - sh(x + iy)]/2$$
$$\cos y\ shx = [sh(x - iy) + sh(x + iy)]/2 \qquad (2.49b)$$

Thus adding and subtracting, finally

$$\mathbf{sinN(y, x)} = [(1 + i)\ sh(x - iy) + (1 - i)\ sh(x + iy)]/2$$
$$\mathbf{sinN(y, -x)} = -[(1 + i)\ sh(x + iy) + (1 - i)\ sh(x - iy)]/2$$
$$\mathbf{cosN(y, x)} = [(1 + i)\ ch(x + iy) + (1 - i)\ ch(x + iy)]/2$$
$$\mathbf{sinN(y, -x)} = [(1 - i)\ ch(x + iy) + (1 + i)\ ch(x - iy)]/2 \qquad (2.50a)$$

Collecting formulas together

Interchanging x and y, and collecting formulas at one place

$$\mathbf{sinN(x, y) = [(1 + i)\ sh(y - ix) + (1 - i)\ sh(y + ix)]/2}$$
$$\mathbf{sinN(x, -y) = -[(1 - i)\ sh(y - ix) + (1 + i)\ sh(y + ix)]/2}$$
$$\mathbf{cosN(x, y) = [(1 + i)\ ch(y + ix) + (1 - i)\ ch(y - ix)]/2}$$
$$\mathbf{cosN(x, -y) = [(1 - i)\ ch(y + ix) + (1 + i)\ ch(y - ix)]/2} \qquad (2.50b)$$

Numerical Example

Example 2.7

Using (2.48), verify (1.97) that:

$$\cos N^2(x, y) + \sin N^2(x, y) = ch\ 2y$$

Solution

Recall (2.48), namely

$$sh(y + ix) = sh\ [i\ (x - iy)] = i \sin\ (x - iy)$$
$$sh(y - ix) = sh\ [-i\ (x - iy)] = -i \sin\ (x + iy)$$
$$ch(y + ix) = ch\ [i\ (x - iy)] = \cos\ (x - iy)$$
$$ch(y - ix) = ch\ [-i\ (x - iy)] = \cos\ (x + iy)$$

With this the LHS of (1.14) when simplified, becomes

$$\mathbf{LHS} = [-2i \sin^2(x + iy) + 2i \sin^2(x - iy) + 4 \sin(x + iy) \sin(x - iy)$$
$$- 2i \cos^2(x + iy) + 2i \cos^2(x - iy) + 4 \cos\ (x + iy) \cos\ (x - iy)]/4$$
$$= [-2i + 2i + 2(\cos 2iy - \cos 2x) + 2(\cos 2iy + \cos 2x)]/4$$
$$= \cos(2iy) = ch\ 2y = \mathbf{RHS} \quad \textbf{(Verified)}$$

❑❑❑

3

DOUBLE NbIC FUNCTION, $N_2(x, y)$

3.1 CATEGORIES OF DOUBLE Nbic FUNCTION

How the double Nbic functions can be generated on separating the real and imaginary parts from the product of the complex single Nbic function and the complex circular functions (or, complex hyperbolic functions), have been discussed in details, in Chapter 1. The variations of the double Nbic functions thus generated are recorded in (1.27e).

Similar to the single Nbic Functions, the double Nbic Functions, $N_2(x, y)$, *viz.*, $N_{2e}(x, y)$, and $N_{2f}(x, y)$, under two categories can be written. The first category (Category 1), is derived from the product of complex single Nbic function, $N/(x, y)$, with $e^{\pm iy}$, and the other category (Category 2), is from the product of complex single Nbic function, $N/(x, y)$, with $f^{\pm iy}$. Each category (*viz.*, Category 1, and Category 2) again is identified under two types. They are: **E type** as $\mathbf{E_1}$ **and** $\mathbf{E_2}$ (1.25d, e), **and F type** as $\mathbf{F_1}$ **and** $\mathbf{F_2}$ (1.26b, c).

Category 1 : (**$\mathbf{E_1}$ type**), *i.e.*, $N_{2e}/(x, y) = N/(x, y).\ e^{\pm iy} = e^{\pm iN/(x,y)}.e^{\pm iy}$

(**$\mathbf{E_2}$ type**), *i.e.*, $N_{2e}/(x, y) = N/(x, x).\ e^{\pm iy} = e^{\pm iN/(x,x)}.e^{\pm iy}$

Category 2 : (**$\mathbf{F_1}$ type**), *i.e.*, $N_{2f}(x, y) = N/(x, y).\ f^{\pm iy} = e^{\pm iN/(x,y)}.f^{\pm iy}$

(**$\mathbf{F_2}$ type**), *i.e.*, $N_{2f}(x, y) = N/(x, x).\ f^{\pm iy} = e^{\pm iN/(x,x)}.f^{\pm iy}$ (3.1)

where, the **suffix**, *e* or *f*, indicates that, the corresponding Nbic function is generated on multiplying $F/(x, y)$ by, $e^{\pm iy}$, or $f^{\pm iy}$, as the case may be.

For all Single Nbic functions, these suffixes, *e* or *f*, are meaningless since, they are developed from the product of only, $e^{\pm iy}$, and, $f^{\pm iy}$.

However, as far as possible, these suffixes would be avoided to ease in the mathematical clumsiness, until badly needed for identification.

In Chapter 1, how the different groups of double Nbic functions can be generated from the Complex double Nbic functions, are shown in details in (1.25d–1.26d), and the relevant Double Nbic functions derived therefrom as the real and imaginary parts, are listed in (1.27e).

The definitions of double Nbic functions are as follows:

Recall here the relevant expressions from (1.27e) for $N_{2e}(x, y)$ and $N_{2f}(x, y)$ corresponding to: E_1 and F_1, and, E_2 and F_2 types [with variables indicated by suffixes, A indicating the Nbic function as, $N(x, y)$ or $N(x, x)$, and B indicating the Nbic function as, $N(x, -y)$ or $N(x, -x)$]. Thus, there are : (E_{1A}, E_{2A}, E_{1B}, E_{2B}) varieties for E type, and (F_{1A}, F_{2A}, F_{1B}, F_{2B}) varieties for F type double Nbic functions.

3.1.1 Category 1: (E type)

E_{1A} type $N_{2e}(x, y)$:

From (1.25d), we get on expanding

$$\begin{aligned}\sin N_{2e}(x, y) &= \sin N(x, y) \cos y + \cos N(x, y) \sin y \\ &= [\sin x\, ch y + \cos x\, sh y] \cos y \\ &\quad + [\cos x\, ch y - \sin x\, sh y] \sin y \\ &= \sin (x + y)\, ch y + \cos (x + y)\, sh y \\ &= \sin N[(x + y), y]\end{aligned}$$

$$\begin{aligned}\cos N_{2e}(x, y) &= \cos N(x, y) \cos y - \sin N(x, y) \sin y \\ &= [\cos x\, ch y - \sin x\, sh y] \cos y \\ &\quad - [\sin x\, ch y + \cos x\, sh y] \sin y \\ &= \cos(x + y)\, ch y - \sin(x + y)\, sh y \\ &= \cos N[(x + y), y]\end{aligned}$$

E_{2A} type $N_{2e}(x, y)$:

From (1.25e) we get on expanding

$$\begin{aligned}\sin N_{2e}(x, y) &= \sin N(x, x) \cos y + \cos N(x, x) \sin y \\ &= [\sin x\, ch x + \cos x\, sh x] \cos y \\ &\quad + [\cos x\, ch x - \sin x\, sh x] \sin y \\ &= \sin(x + y)\, ch x + \cos(x + y)\, sh x \\ &= \sin N[(x + y), x]\end{aligned}$$

$$\begin{aligned}\cos N_{2e}(x, y) &= \cos N(x, x) \cos y - \sin N(x, x) \sin y \\ &= [\cos x\, ch x - \sin x\, sh x] \cos y \\ &\quad - [\sin x\, ch x + \cos x\, sh x] \sin y \\ &= \cos(x + y)\, ch x - \sin(x + y)\, sh x \\ &= \cos N[(x + y), x]\end{aligned}$$

E_{1B} type $N_{2e}(x, -y)$:

$$\begin{aligned}\sin N_{2e}(x, -y) &= \sin N(x, y) \cos y - \cos N(x, y) \sin y \\ &= \sin N_{2e} [(x - y), y]\end{aligned}$$

$$\begin{aligned}\cos N_{2e}(x, -y) &= \cos N(x, y) \cos y + \sin N(x, y) \sin y \\ &= \cos N_{2e} [(x - y), y]\end{aligned}$$

E_{2B} type $N_{2e}(x, -y)$:

$$\begin{aligned} \sin N_{2e}(x, -y) &= \sin N(x, x) \cos y - \cos N(x, x) \sin y \\ &= -\sin N_{2e}[(x - y), x] \\ \cos N_{2e}(x, -y) &= \cos N(x, x) \cos y + \sin N(x, x) \sin y \\ &= \cos N_{2e}[(x - y), x] \end{aligned} \tag{3.2a}$$

3.1.2 Particular Case

when $y = x$

$N_{2e}(x, x)$ [$=N_{2e}(x) = N_{2e}\, x$]:

(E_{1A} and E_{2A} type)

$$\begin{aligned} \sin N_{2e}(x, x) &= \sin N(x, x) \cos x + \cos N(x, x) \sin x \\ &= \sin N(2x, x) \\ \cos N_{2e}(x, x) &= \cos N(x, x) \cos x - \sin N(x, x) \sin x \\ &= \cos N(2x, x) \end{aligned}$$

$N_{2e}(x, -x)$:

(E_{1B} and E_{2B} type)

$$\begin{aligned} \sin N_{2e}(x, -x) &= \sin N(x, x) \cos x - \cos N(x, x) \sin x \\ &= shx \\ \cos N_{2e}(x, -x) &= \cos N(x, x) \cos x + \sin N(x, x) \sin x \\ &= chx \end{aligned} \tag{3.2b}$$

3.1.3 Category 2: (F type)

F_{1A} type $N_{2f}(x, y)$:

From (1.26b) we get on expanding

$$\begin{aligned} \sin N_{2f}(x, y) &= \sin N(x, y)\, chy + \cos N(x, y)\, shy \\ &= [\sin x\, chy + \cos x\, shy]\, chy \\ &\quad + [\cos x\, chy - \sin x\, shy]\, shy \\ &= \sin x + \cos x\, sh\, 2y \\ \cos N_{2f}(x, y) &= \cos N(x, y)\, chy - \sin N(x, y)\, shy \\ &= [\cos x\, chy - \sin x\, shy]\, chy \\ &\quad - [\sin x\, chy + \cos x\, shy]\, shy \\ &= \cos x - \sin x\, sh\, 2y \end{aligned}$$

F_{2A} type $N_{2f}(x, y)$:

$$\begin{aligned} \sin N_{2f}(x, y) &= \sin N(x, x)\, chy + \cos N(x, x)\, shy \\ &= [\sin x\, chx + \cos x\, shx]\, chy \\ &\quad + [\cos x\, chx - \sin x\, shx]\, shy \\ &= \sin N(x, y)\, chx + \cos N(x, y)\, shx \\ &= \sin N_{2f}[(x, y), x] \\ \cos N_{2f}(x, y) &= \cos N(x, x)\, chy - \sin N(x, x)\, shy \\ &= [\cos x\, chx - \sin x\, shx]\, chy \\ &\quad - [\sin x\, chx + \cos x\, shx]\, shy \\ &= \cos N(x, y)\, chx - \sin N(x, y)\, shx \\ &= \cos N_{2f}[(x, y), x] \end{aligned}$$

F_{1B} type $N_{2f}(x, -y)$:

$$\begin{aligned}\sin N_{2f}(x, -y) &= \sin N(x, y)\, ch y - \cos N(x, y)\, sh y \\ &= [\sin x\, ch y + \cos x\, sh y]\, ch y \\ &\quad - [\cos x\, ch y - \sin x\, sh y]\, sh y \\ &= \sin x\,(ch^2 y + sh^2 y) \\ &= \sin x\, ch\, 2y \\ \cos N_{2f}(x, -y) &= \cos N(x, y)\, ch y + \sin N(x, y)\, sh y \\ &= [\cos x\, ch y - \sin x\, sh y]\, ch y \\ &\quad + [\sin x\, ch y + \cos x\, sh y]\, sh y \\ &= \cos x\,(ch^2 y + sh^2 y) \\ &= \cos x\, ch\, 2y\end{aligned}$$

F_{2B} type $N_{2f}(x, -y)$:

$$\begin{aligned}\sin N_{2f}(x, -y) &= \sin N(x, x)\, ch y - \cos N(x, x)\, sh y \\ &= [\sin x\, ch x + \cos x\, sh x]\, ch y \\ &\quad - [\cos x\, ch x - \sin x\, sh x]\, sh y \\ &= \sin N(x, -y)\, ch x - \cos N(x, -x)\, sh x \\ &= \sin N_{2f}[(x, -y), x] \\ \cos N_{2f}(x, -y) &= \cos N(x, x)\, ch y + \sin N(x, x)\, sh x \\ &= [\cos x\, ch x - \sin x\, sh x\,]\, ch y \\ &\quad + [\sin x\, ch x + \cos x\, sh x]\, sh y \\ &= \cos N(x, -y)\, ch x - \sin N(x, -y)\, sh x \\ &= \cos N_{2f}[(x, -y), x]\end{aligned} \qquad (3.3a)$$

3.1.4 Particular Case

when $y = x$

$N_{2f}(x, x)$ $[N_{2f}(x) = N_{2f}\, x]$:

(F_{1A} and F_{2A} type)

$$\begin{aligned}\sin N_{2f}(x, x) &= \sin N(x, x)\, ch x + \cos N(x, x)\, sh x \\ &= \sin x + \cos x\, sh\, 2x \\ \cos N_{2f}(x, x) &= \cos N(x, x)\, ch x - \sin N(x, x)\, sh x \\ &= \cos x - \sin x\, sh\, 2x\end{aligned}$$

$N_{2f}(x, -x)$:

(F_{1B} and F_{2B} type)

$$\begin{aligned}\sin N_{2f}(x, -x) &= \sin N(x, x)\, ch x - \cos N(x, x)\, sh x \\ &= \sin x\, ch\, 2x \\ \cos N_{2f}(x, -x) &= \cos N(x, x)\, ch x + \sin N(x, x)\, sh x \\ &= \cos x\, ch\, 2x\end{aligned} \qquad (3.3b)$$

We discuss here few properties of the Double Nbic functions, $N_{2e}(x, y)$, and $N_{2f}(x, y)$, belonging to both the Categories along with the corresponding Complex functions $N_{2e}/(x, y)$, and $N_{2f}/(x, y)$.

3.1.5 Complex Double Nbic Function: $N_{2e}/(x, y)$

The double Nbic functions can be derived from the corresponding Complex double Nbic functions which have been shown as the product of single Nbic functions and either of the complex circular function or complex hyperbolic function, as described in (3.1). In (1.25d–1.26d), these products are described, particularly when one of the multiplicant complex Nbic function is either of (x, y) or (x, x). The other formulations with (x, –y) and (x, –x) are discussed here along with the previously formulated expressions (which are recalled here as ready reckoner).

Category 1: $N_{2e}/(x, y)$:

This category 1 belongs to E type.

$$\begin{aligned}
e^{iN/(x,\,y)}\, e^{iy} &= [\cos N(x, y) + i \sin N(x, y)]\,(\cos y + i \sin y)\\
&= [\cos N(x, y)\cos y - \sin N(x, y)\sin y]\\
&\quad + i\,[\sin N(x, y)\cos y + \cos N(x, y)\sin y]\\
&= \cos N_{2e}(x, y) + i \sin N_{2e}(x, y)\\
&= e^{iN_{2e}/[(x,\,y),\,y]}\\
&= e^{iN/[(x+y),\,y]}\\
e^{iN/x}\, e^{iy} &= (\cos Nx + i \sin Nx)(\cos y + i \sin y)\\
&= [\cos Nx \cos y - \sin Nx \sin y]\\
&\quad + i\,[\sin Nx \cos y + \cos Nx \sin y]\\
&= \cos N_{2e}(x, y) + i \sin N_{2e}(x, y)\\
&= e^{iN_{2e}/[(x,\,x),\,y]} = e^{iN/[(x+y),\,x]} \qquad (3.4a)\\
e^{iN/(x,\,y)}\, e^{-iy} &= (\cos N(x, y) + i \sin N(x, y)]\,(\cos y - i \sin y)\\
&= [\cos N(x, y)\cos y + \sin N(x, y)\sin y]\\
&\quad + i\,[\sin N(x, y)\cos y - \cos N(x, y)\sin y]\\
&= \cos N_{2e}[(x, y), -y] + i \sin N_{2e}[(x, y), -y]\\
&= e^{iN_{2e}/[(x,\,y),\,-y]}\\
&= e^{iN/[(x-y),\,y]}\\
e^{iN/x}\, e^{-iy} &= (\cos Nx + i \sin Nx)(\cos y - i \sin y)\\
&= [\cos Nx \cos y + \sin Nx \sin y]\\
&\quad + i\,[\sin Nx \cos y - \cos Nx \sin y]\\
&= \cos N_{2e}(x, -y) + i \sin N_{2e}(x, -y)\\
&= e^{iN2e/[(x,\,x)\,-y]} \qquad (3.4b)\\
&= e^{iN/[(x-y),\,x]}
\end{aligned}$$

Double Nbic function $N_{2e}(x, y)$:

Note that, the expansions are different but identifications are same for (E_{1A} and E_{2A}) and also for (E_{1B} and E_{2B}) types. To avoid ambiguity, they are also identified, side by side, in alternative way.

Thus separating the real and imaginary parts, we arrive at double Nbic function $N_{2e}(x, y)$ of E type as

E_{1A} type:

$$\cos N_{2e}(x, y) = \cos N(x, y)\cos y - \sin N(x, y)\sin y = \cos N_{2e}[(x, y), y]$$
$$\sin N_{2e}(x, y) = \sin N(x, y)\cos y + \cos N(x, y)\sin y = \sin N_{2e}[(x, y), y]$$

E_{1B} type:

$$\cos N_{2e}(x, -y) = \cos N(x, y)\cos y + \sin N(x, y)\sin y = \cos N_{2e}[(x, y), -y]$$
$$\sin N_{2e}(x, -y) = \sin N(x, y)\cos y - \cos N(x, y)\sin y = \sin N_{2e}[(x, y), -y]$$

E_{2A} type:

$$\cos N_{2e}(x, y) = \cos Nx \cos y - \sin Nx \sin y = \cos N_{2e}[(x, x), y]$$
$$\sin N_{2e}(x, y) = \sin Nx \cos y + \cos Nx \sin y = \sin N_{2e}[(x, x), y]$$

E_{2B} type:

$$\cos N_{2e}(x, -y) = \cos Nx \cos y + \sin Nx \sin y = \cos N_{2e}[(x, x), -y]$$
$$\sin N_{2e}(x, -y) = \sin Nx \cos y - \cos Nx \sin y = \sin N_{2e}[(x, x), -y] \quad (3.4c)$$

Complex double Nbic function, $N_{2f}/(x, y)$:

Category 2: $N_{2f}/(x, y)$:

This category 2 belongs to F type.

$$\begin{aligned} e^{iN/(x, y)} f^{iy} &= [\cos N(x, y) + i \sin N(x, y)] (chy + i\, shy) \\ &= [\cos N(x, y)\, chy - \sin N(x, y)\, shy] \\ &\quad + i [\sin N(x, y)\, chy + \cos N(x, y)\, shy] \\ &= \cos N_{2f}(x, y) + i \sin N_{2f}(x, y) \\ &= e^{iN_{2f}/[(x, y), y]} \\ e^{iN/x} f^{iy} &= (\cos Nx + i \sin Nx)(chy + i\, shy) \\ &= [\cos Nx\, chy - \sin Nx\, shy] \\ &\quad + i [\sin Nx\, chy + \cos Nx\, shy] \\ &= \cos N_{2f}(x, y) + i \sin N_{2f}(x, y) \\ &= e^{iN_{2f}/[(x, x), y]} \end{aligned} \quad (3.4d)$$

$$\begin{aligned} e^{iN/(x, y)} f^{-iy} &= (\cos N(x, y) + i \sin N(x, y) (chy - i\, shy) \\ &= [\cos N(x, y)\, chy + \sin N(x, y)\, shy] \\ &\quad + i [\sin N(x, y)\, chy - \cos N(x, y)\, shy] \\ &= \cos N_{2f}[(x, y), -y] + i \sin N_{2f}[(x, y), -y] \\ &= e^{i N_{2f}/[(x,y), -y]} \\ e^{iN/x} e^{-iy} &= (\cos Nx + i \sin Nx)(chy - i\, shy) \\ &= [\cos Nx\, chy + \sin Nx\, shy] \\ &\quad + i [\sin Nx\, chy - \cos Nx\, shy] \\ &= \cos N_{2f}(x, y) + i \sin N_{2f}(x, -y) \\ &= e^{iN_{2f}/[(x, x), -y]} \end{aligned} \quad (3.4e)$$

Double Nbic function $N_{2f}(x, y)$:

Note that, the expansions are different but identifications are same for (F_{1A} and F_{2A}) and also for (F_{1B} and F_{2B}) types. To avoid ambiguity, they are also identified, side by side, in an alternative way. Thus separating the real and imaginary parts, we arrive at double Nbic function $N_{2f}(x, y)$ of F type as:

F_{1A} type:

$$\cos N_{2f}(x, y) = \cos N(x, y)\, chy - \sin N(x, y)\, shy = \cos N_{2f}[(x, y), y]$$
$$\sin N_{2f}(x, y) = \sin N(x, y)\, chy + \cos N(x, y)\, shy = \sin N_{2f}[(x, y), y]$$

F_{1B} type:

$$\cos N_{2f}(x, -y) = \cos N(x, y)\, chy + \sin N(x, y)\, shy = \cos N_{2f}[(x, y), -y]$$
$$\sin N_{2f}(x, -y) = \sin N(x, y)\, chy - \cos N(x, y)\, shy = \sin N_{2f}[(x, y), -y]$$

F_{2A} type:

$$\cos N_{2f}(x, y) = \cos Nx\ chy - \sin Nx\ shy = \cos N_{2f}[(x, x), y]$$
$$\sin N_{2f}(x, y) = \sin Nx\ chy + \cos Nx\ shy = \sin N_{2f}[(x, x), y]$$

F_{2B} type:

$$\cos N_{2f}(x, -y) = \cos Nx\ chy + \sin Nx\ shy = \cos N_{2f}[(x, x), -y]$$
$$\sin N_{2f}(x, -y) = \sin Nx\ chy - \cos Nx\ shy = \sin N_{2f}[(x, x), -y] \quad (3.4f)$$

[**Note:** For F type functions $N_{2f}/(x, y)$ and $N_{2f}(x, y)$, all these exercise is not necessary. Simply change **sin** and **cos** of E type to **sh** and **ch** for F type, respectively.]

The **suffix**, *e* or *f*, indicates that, the corresponding Nbic function is generated on multiplying $N/(x, y)$ by, $e^{\pm iy}$, or $f^{\pm iy}$, as the case may be, to be grouped under Category 1 or Category 2, respectively, (which is synonymous with the E type or F type, respectively).

For all single Nbic functions, these suffixes, *e* or *f*, are meaningless since, they are developed from the product of, $e^{\pm iy}$, and, $e^{\pm iy}$, only.

However, as far as possible, these suffixes would be avoided to ease in the mathematical clumsiness, untill badly needed for identification.

3.2 TRANSFER OF DOUBLE Nbic FUNCTION IN TERMS OF ANOTHER DOUBLE Nbic FUNCTION OR SINGLE Nbic FUNCTION

Expressions obtained here are the illustration of double Nbic functions expressed in terms of single Nbic functions. Double Nbic functions with one type of variable which can also be expressed, in some cases, in another pattern of double Nbic functions with a different type of variable are also written side by side, where ever possible.

Recall the basic definition of the single Nbic functions (1.24b). Using these basic definition, the expansion of sin and cos of $N_2(x, y)$, and their particular forms $N_2(x, x)$, of both E and F types are, in most cases, found to yield functions in terms of single Nbic functions. However, those functions which do not result in the single Nbic functions, or yield double Nbic functions of other forms, are also included side by side, wherever possible. We consider here, A type $[N_{2e}(x, y), N_{2e}(x, x)]$, and B type $[N_{2e}(x, -y)$ and $N_{2e}(x, -x)]$.

3.2.1 For E_{1A} and E_{2A} type $N_{2e}(x, y)$

[taken from (1.25d, e)]

$$\mathbf{\cos N_{2e}(x, y)} = \cos N(x, y) \cos y - \sin N(x, y) \sin y$$
$$= \cos(x + y)\ chy - \sin(x + y)\ shy$$
$$= \mathbf{\cos N[(x + y), y]}$$

$$\mathbf{\sin N_{2e}(x, y)} = \sin N(x, y) \cos y + \cos N(x, y) \sin y$$
$$= \sin(x + y)\ chy + \cos(x + y)\ shy$$
$$= \mathbf{\sin N[(x + y), y]}$$

so that

$$\mathbf{\tan N_{2e}(x, y)} = \mathbf{\sin N_{2e}(x, y) / \cos N_{2e}(x, y)}$$
$$= \sin N[(x + y), y] / \cos N[(x + y), y] = \mathbf{\tan N[(x + y), y]}$$

$$\textbf{tanN}_{2e}(x, y) = \frac{\sin(x+y).chy+\cos(x+y).shy}{\cos(x+y).chy-\sin(x+y).shy} = \frac{\tan(x+y)+thy}{1-\tan(x+y).thy} \tag{3.5a}$$

For $\mathbf{E_{2A}}$ **type** $\mathbf{N_{2e}(x, y)}$:

$$\begin{aligned}
\textbf{cosN}_{2e}(x, y) &= \text{cosN}(x, x)\ \cos y - \text{sinN}(x, x)\ \sin y \\
&= \cos(x + y)\ chx - \sin(x + y)\ shx \\
&= \textbf{cosN}[(x + y), x]
\end{aligned}$$

Also alternatively

$$\begin{aligned}
\textbf{cosN}_{2e}(x, y) &= [\cos x\ chx - \sin x\ shx]\ \cos y \\
&\quad - [\sin x\ chx + \cos x\ shx]\ \sin y \\
&= [\cos y\ chx - \sin y\ shx]\ \cos x \\
&\quad - [\sin y\ chx + \cos y\ shx]\ \sin x \\
&= \text{cosN}(y, x)\ \cos x - \text{sinN}(y, x)\ \sin x \\
&= \textbf{cosN}_{2e}[(y, x), x]
\end{aligned}$$

$$\begin{aligned}
\textbf{sinN}_{2e}(x, y) &= \text{sinN}(x, x)\ \cos y + \text{cosN}(x, x)\ \sin y \\
&= \sin(x + y)\ chx + \cos(x + y)\ shx \\
&= \textbf{sinN}[(x + y), x]
\end{aligned}$$

Also alternatively

$$\begin{aligned}
\textbf{sinN}_{2e}(x, y) &= [\sin x\ chx + \cos x\ shx]\ \cos y \\
&\quad + [\cos x\ chx - \sin x\ shx]\ \sin y \\
&= [\sin y\ chx + \cos y\ shx]\ \cos x \\
&\quad + [\cos y\ chx - \sin y\ shx]\ \sin x \\
&= \text{sinN}(y, x)\ \cos x + \text{cosN}(y, x)\ \sin x \\
&= \textbf{sinN}_{2e}[(y, x), x]
\end{aligned}$$

so that

$$\begin{aligned}
\textbf{tanN}_{2e}(x, y) &= \textbf{sinN}_{2e}(x, y) / \textbf{cosN}_{2e}(x, y) \\
&= \text{sinN}[(x + y), x] / \text{cosN}[(x + y), x] = \textbf{tanN}[(x + y), x]
\end{aligned}$$

Also

$$\textbf{tanN}_{2e}(x, y) = \text{sinN}_{2e}[(y, x), x] / \text{cosN}_{2e}[(y, x), x] = \textbf{tanN}_{2e}[(y, x), x]$$

Also

$$\textbf{tanN}_{2e}(x, y) = \frac{\sin(x+y).chx+\cos(x+y).shx}{\cos(x+y).chx-\sin(x+y).shx} = \frac{\tan(x+y)+thx}{1-\tan(x+y).thx} \tag{3.5b}$$

For $\mathbf{E_{1B}}$ **type** $\mathbf{N_{2e}(x, y)}$:

$$\begin{aligned}
\textbf{cosN}_{2e}(x, -y) &= \text{cosN}(x, y)\ \cos y + \text{sinN}(x, y)\ \sin y \\
&= \cos(x - y)\ chy - \sin(x - y)\ shy \\
&= \textbf{cosN}[(x - y), y]
\end{aligned}$$

Also alternatively

$$\begin{aligned}
\textbf{cosN}_{2e}(x, -y) &= [\cos y\ chy + \sin y\ shy]\ \cos x \\
&\quad + [\sin y\ chy - \cos y\ shy]\ \sin x \\
&= \text{cosN}(y, -y)\ \cos x + \text{sinN}(y, -y)\ \sin x \\
&= \textbf{cosN}_{2e}\ [(y, -y), -x]
\end{aligned}$$

$$\mathbf{sinN_{2e}(x, -y)} = \text{sinN}(x, y)\ \cos y - \text{cosN}(x, y)\ \sin y$$
$$= \sin(x - y)\ chy + \cos(x - y)\ shy$$
$$= \mathbf{sinN[(x - y), y]}$$

Also alternatively

$$\mathbf{sinN_{2e}(x, -y)} = [\sin y\ chy - \cos y\ shy]\ \cos x$$
$$- [\cos y\ chy + \sin y\ shy]\ \sin x$$
$$= \text{sinN}(y, -y)\ \cos x - \text{cosN}(y, -y)\ \sin x$$
$$= \mathbf{sinN_{2e}\ [(y, -y), -x]}$$

so that

$$\mathbf{tanN_{2e}(x, -y)} = \text{sinN}_{2e}(x, -y)/\text{cosN}_{2e}(x, -y)$$
$$= \text{sinN}[(x - y), y]/\text{cosN}[(x - y), y] = \mathbf{tanN[(x - y), y]}$$

Also

$$\mathbf{tanN_{2e}(x, -y)} = -\ \text{sinN}_{2e}[(y, -y), -x]/\text{cosN}_{2e}[(y, -y), -x]$$
$$= \mathbf{-\ tanN_{2e}[(y, -y), -x]}$$

Also

$$\mathbf{tanN_{2e}(x, -y)}\ \frac{\sin(x-y).chx + \cos(x-y).shx}{\cos(x-y).chx - \sin(x-y).shx} = \frac{\tan(x-y) + thx}{1 - \tan(x-y).thx} \tag{3.5c}$$

For $\mathbf{E_{2B}}$ **type** $\mathbf{N_{2e}(x, y)}$:

$$\mathbf{cosN_{2e}(x, -y)} = \text{cosN}(x, x)\ \cos y + \text{sinN}(x, x)\ \sin y$$
$$= \cos(x - y)\ chx - \sin(x - y)\ chx$$
$$= \mathbf{cosN[(x - y), x]}$$

Also alternatively

$$\mathbf{cosN_{2e}(x, -y)} = [\cos x\ chx - \sin x\ shx]\ \cos y$$
$$+ [\sin x\ chx + \cos x\ shx]\ \sin y$$
$$= [\cos y\ chx + \sin y\ shx]\ \cos x$$
$$+ [\sin y\ chx - \cos y\ shx]\ \sin x$$
$$= \text{cosN}(y, -x)\ \cos x + \text{sinN}(y, -x)\ \sin x$$
$$= \mathbf{cosN_{2e}[(y, -x), -x]}$$

$$\mathbf{sinN_{2e}(x, -y)} = \text{sinN}(x, x)\ \cos y - \text{cosN}(x, x)\ \sin y$$
$$= \sin(x - y)\ chx + \cos(x - y)\ shx$$
$$= \mathbf{sinN[(x - y), x]}$$

Also alternatively

$$\mathbf{sinN_{2e}(x, -y)} = [\sin x\ chx + \cos x\ shx]\ \cos y$$
$$- [\cos x\ chx - \sin x\ shx]\ \sin y$$
$$= [\cos y\ chx + \sin y\ shx]\ \sin x$$
$$- [\sin y\ chx - \cos y\ shx]\ \cos x$$
$$= \text{cosN}(y, -x)\ \sin x - \text{sinN}(y, -x)\ \cos x$$
$$= \mathbf{-\ sinN_{2e}[(y, -x), -x]}$$

CHAPTER 3

so that

$$\mathbf{tanN_{2e}(x, -y)} = \sin N_{2e}(x, -y)/\cos N_{2e}(x, -y)$$
$$= \sin N[(x-y), x]/\cos N[(x-y), x] = \mathbf{tanN[(x-y), x]}$$

Also

$$\mathbf{tanN_{2e}(x, -y)} = -\sin N_{2e}[(y, -x), -x]/\cos N_{2e}[(y, -x), -x]$$
$$= -\mathbf{tanN_{2e}[(y, -x), -x]} \quad (3.5d)$$

3.2.2 Particular Case (E type)

When $y = x$, we arrive at these particular forms.

For $\mathbf{E_{1A}}$ **and** $\mathbf{E_{2A}}$ **type** $\mathbf{N_{2e}(x, x)}$**:**

[taken from (1.26a and 1.27a)]

$$\mathbf{cosN_{2e}(x, x)} = \cos N_{2e}(x) = \cos Nx \cos x - \sin Nx \sin x = \mathbf{cosN(2x, x)}$$

$$\mathbf{sinN_{2e}(x, x)} = \sin N_{2e}(x) = \sin Nx \cos x + \cos Nx \sin x = \mathbf{sinN(2x, x)}$$

so that

$$\mathbf{tanN_{2e}(x, x)} = \sin N_{2e}(x, x)/\cos N_{2e}(x, x)$$
$$= \sin N(2x, x)/\cos N(2x, x)$$
$$= \mathbf{tanN(2x, x)}$$

Also putting $y = x$ in (3.5b)

$$\mathbf{tanN_{2e}(x, x)} = \frac{\tan 2x + thx}{1 - \tan 2x.thx} \quad (3.5e)$$

In **Figure 3.1**, the variation of $N_{2e}(x)$ vs. x is shown.

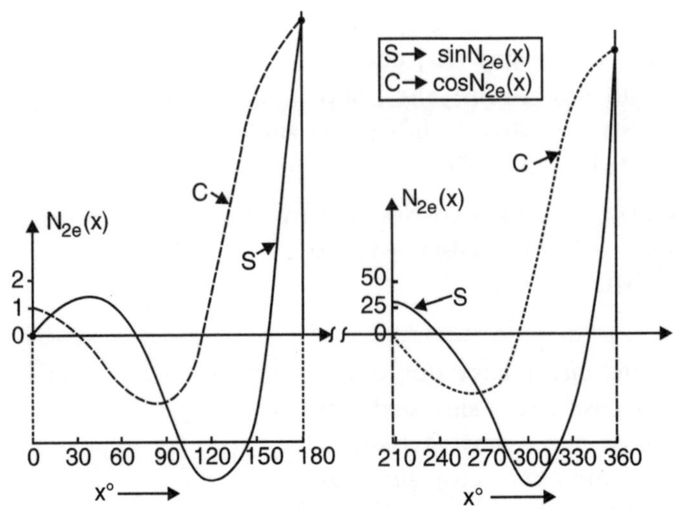

Fig. 3.1 : $N_{2e}(x)$ vs. x

For $\mathbf{E_{1B}}$ **and** $\mathbf{E_{2B}}$ **type** $\mathbf{N_{2e}(x, x)}$**:**

[taken from (3.2b)]

$$\begin{aligned}\textbf{cosN}_{2e}(x, -x) &= \text{cosN}(x, x)\cos x + \text{sinN}(x, x)\sin x = \textbf{ch}x \\ \textbf{sinN}_{2e}(x, -x) &= \text{sinN}(x, x)\cos x - \text{cosN}(x, x)\sin x = \textbf{sh}x\end{aligned}$$

so that

$$\begin{aligned}\textbf{tanN}_{2e}(x, -x) &= \text{sinN}_{2e}(x, -x)/\text{cosN}_{2e}(x, -x) \\ &= shx/chx = \textbf{th}x\end{aligned} \tag{3.5f}$$

Note: The derivation of E_{1A} and E_{2A} and, E_{1B} and E_{2B} type $N_{2e}(x, y)$ and, $N_{2e}(x, x)$ when screened, reveals the following:

All of them can be expressed in terms of Single Nbic functions.

Additionally, E_{1B} and E_{2B} type of $N_{2e}(x, y)$ functions can also be expressed in terms of double Nbic functions of different variables.

For all the four Nbic functions, *viz.*, E_{1A} and E_{2A} and, E_{1B} and E_{2B} type of $N_{2e}(x, y)$, the tan-formulas of (3.5a–3.5e) except (3.5f), do conform to the trigonometric tan(A + B) formula.

Expression (3.5f) of the tan-formula results solely in simple hyperbolic function ***thx***, which establishes the interchangeability of N_{2e} functions with the hyperbolic function.

Now consider, A type $[N_{2f}(x, y), N_{2f}(x, x)]$, and B type $[N_{2f}(x, -y)$ and $N_{2f}(x, -x)]$ which belong to F type.

3.2.3 For F_{1A} and F_{2A} type $N_{2f}(x, y)$

[taken from (3.3a)]

$$\begin{aligned}\textbf{cosN}_{2f}(x, y) &= \text{cosN}(x, y)\, chy - \text{sinN}(x, y)\, shy = \textbf{cos}x - \textbf{sin}x\ \textbf{sh}\ 2y \\ \textbf{sinN}_{2f}(x, y) &= \text{sinN}(x, y)\, chy + \text{cosN}(x, y)\, shy = \textbf{sin}x + \textbf{cos}x\ \textbf{sh}\ 2y\end{aligned}$$

so that

$$\textbf{tanN}_{2f}(x, y) = \textbf{sinN}_{2f}(x, y)/\textbf{cosN}_{2f}(x, y) \tag{3.5g}$$

$$\begin{aligned}&= [\sin x + \cos x\ sh\ 2y]/[\cos x - \sin x\ sh\ 2y] \\ &= \frac{\tan x + sh\ 2y}{1 - \tan x . sh\ 2y}\end{aligned} \tag{3.5h}$$

For $\mathbf{F_{2A}}$ **type** $\mathbf{N_{2f}(x, y)}$:

[taken from (3.3a)]

$$\begin{aligned}\textbf{cosN}_{2f}(x, y) &= \text{cosN}(x, x)\, chy - \text{sinN}(x, x)\, shy \\ &= [\cos x\ chx - \sin x\ shx]\, chy \\ &\quad - [\sin x\ chx + \cos x\ shx]\, shy \\ &= \cos x\ ch\,(x - y) - \sin x\ sh\,(x + y) \\ &= [\cos x\ chy - \sin x\ shy]\, chx \\ &\quad - [\sin x\ chy + \cos x\ shy]\, shx \\ &= \text{cosN}(x, y)\, chx - \text{sinN}(x, y)\, shx \\ &= \textbf{cosN}_{2f}[(x, y), x]\end{aligned}$$

$$\begin{aligned}\textbf{sinN}_{2f}(x, y) &= \text{sinN}(x)\, chy + \text{cosN}(x)\, shy \\ &= [\sin x\ chx + \cos x\ shx]\, chy \\ &\quad + [\cos x\ chx - \sin x\ shx]\, shy \\ &= \sin x\ ch(x - y) + \cos x\ sh(x + y) \\ &= [\cos x\ shy + \sin x\ chy]\, chx\end{aligned}$$

$$+ [\cos x\ chy - \sin x\ shy]\ shx$$
$$= \sin N(x, y)\ chx + \cos N(x, y)\ shx$$
$$= \mathbf{\sin N_{2f}\ [(x, y), x]}$$

so that

$$\mathbf{\tan N_{2f}\ (x, y)} = \sin N_{2f}\ [(x, y), x] / \cos N_{2f}\ [(x, y), x]$$
$$= \frac{\sin x.ch(x-y) + \cos x.sh(x+y)}{\cos x.ch(x-y) - \sin x.sh(x+y)}$$

Divide the numerator and denominator by [cos*x* *ch*(*x* – *y*)] and simplify, to get

$$\tan N_{2f}(x, y) = \frac{\tan x + \dfrac{sh(x+y)}{ch(x-y)}}{1 - \tan x.\dfrac{sh(x+y)}{ch(x-y)}}\ ;\ \frac{sh(x+y)}{ch(x-y)} = \frac{thx + thy}{1 - thx.thy} \qquad (3.5i)$$

For $\mathbf{F_{1B}}$ **type** $\mathbf{N_{2f}\ (x, y)}$: [taken from (3.3a)]

$$\mathbf{\cos N_{2f}\ (x, -y)} = \cos N(x, y)\ chy + \sin N(x, y)\ shy$$
$$= \mathbf{\cos x\ ch\ 2y}$$
$$\mathbf{\sin N_{2f}\ (x, -y)} = \sin N(x, y)\ chy - \cos N(x, y)\ shy$$
$$= \mathbf{\sin x\ ch\ 2y}$$

so that

$$\mathbf{\tan N_{2f}\ (x, -y)} = \sin x\ ch\ 2y / \cos x\ ch\ 2y = \mathbf{\tan x} \qquad (3.5j)$$

In **Figure 3.2**, the variation of $N_{2f}(x)$ vs. x is shown.

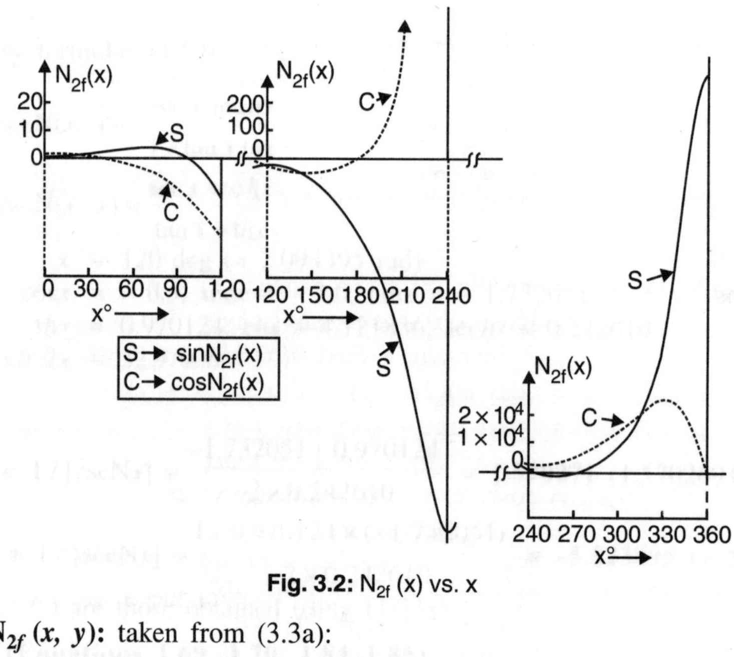

Fig. 3.2: N_{2f} (x) vs. x

For $\mathbf{F_{2B}}$ **type** $\mathbf{N_{2f}\ (x, y)}$: taken from (3.3a):

$$\begin{aligned}\textbf{cosN}_{2f}(x, -y) &= \cos N(x, x)\, chy + \sin N(x, x)\, shx\\ &= \cos x\, ch(x + y) - \sin x\, sh(x - y)\\ &= \textbf{cosN}_{2f}[(x, -y), x]\\ \textbf{sinN}_{2f}(x, -y) &= \sin N(x, x)\, chy - \cos N(x, x)\, shy\\ &= \sin x\, ch(x + y) + \cos x\, sh(x - y)\\ &= \textbf{sinN}_{2f}[(x, -y), x]\end{aligned}$$

so that

$$\begin{aligned}\textbf{tanN}_{2f}(x, -y) &= \sin N_{2f}[(x, -y), x] / \cos N_{2f}[(x, -y), x]\\ &= \frac{\sin x.ch(x+y) + \cos x.sh(x-y)}{\cos x.ch(x+y) - \sin x.sh(x-y)}\end{aligned}$$

Divide the numerator and denominator by [$\cos x\, ch(x + y)$] and simplify, to get

$$\tan N_{2f}(x, -y) = \frac{\tan x + \dfrac{sh(x-y)}{ch(x+y)}}{1 - \tan x.\dfrac{sh(x-y)}{ch(x+y)}}; \quad \frac{sh(x-y)}{ch(x+y)} = \frac{thx - thy}{1 + thx.thy} \tag{3.5k}$$

3.2.4 Particular Case

For F_{1A} **and** F_{2A} **type** $N_{2f}(x, x)$: [taken from (3.3b)]

$$\begin{aligned}\textbf{cosN}_{2f}(x, x) &= \cos N(x, x)\, chx - \sin N(x, x)\, shx\\ &= \textbf{cos}\boldsymbol{x} - \textbf{sin}\boldsymbol{x}\ \boldsymbol{sh}\ \textbf{2}\boldsymbol{x}\\ \textbf{sinN}_{2f}(x, x) &= \sin N(x, x)\, chx + \cos N(x, x)\, shx\\ &= \textbf{sin}\boldsymbol{x} + \textbf{cos}\boldsymbol{x}\ \boldsymbol{sh}\ \textbf{2}\boldsymbol{x}\end{aligned} \tag{3.5l}$$

so that

$$\textbf{tanN}_{2f}(x, x) = \sin N_{2f}(x, x) / \cos N_{2f}(x, x) = \frac{\tan x + sh\ 2x}{1 - \tan x.sh\ 2x} \tag{3.5m}$$

For F_{1B} **and** F_{2B} **type** $N_{2f}(x, x)$: [taken from (3.3b)]

$$\begin{aligned}\textbf{cosN}_{2f}(x, -x) &= \cos N(x, x)\, chx + \sin N(x, x)\, shx\\ &= \textbf{cos}\boldsymbol{x}\ \boldsymbol{ch}\ \textbf{2}\boldsymbol{x}\\ \textbf{sinN}_{2f}(x, -x) &= \sin N(x, x)\, chx - \cos N(x, x)\, shx\\ &= \textbf{sin}\boldsymbol{x}\ \boldsymbol{ch}\ \textbf{2}\boldsymbol{x}\end{aligned}$$

so that

$$\textbf{tanN}_{2f}(x, -x) = \sin N_{2f}(x, -x) / \cos N_{2f}(x, -x) = \textbf{tan}\boldsymbol{x} \tag{3.5n}$$

It can be noted from (3.5m) and (3.5h), and from (3.5j) and (3.5n), that:

$$(F_{1B} \text{ or } F_{2B} \text{ type})\ \textbf{tanN}_{2f}(x, -x) = (F_{1B} \text{ type})\ \textbf{tanN}_{2f}(x, -y) \tag{3.5o}$$

Note: The derivation of F_{1A} and F_{2A} and, F_{1B} and F_{2B} type $N_{2f}(x, y)$ and, $N_{2f}(x, x)$ when screened, reveal the following:

None of them can be expressed in terms of Single Nbic functions.

Except $N_{2f}(x, y)$ functions of the type F_{2A} and F_{2B}, none of these functions has its alternative form of N_{2f} function with different of variables.

The tan-formulas of (3.5h, 3.5i, 3.5k and 3.5m) do conform to the trigonometric tan (A + B) formula.

CHAPTER 3

Expressions (3.5j and 3.5n) of the tan-formulas result in simple trigonometric **tan$x$**, which establish the interchangeability of N_{2f} functions with the circular function.

Collecting formulas together

Double Nbic functions in terms of single Nbic functions

$$\begin{aligned} \cos N_{2e}(x, y) &= \cos N_{2e}[(y, x), x] = \cos N[(x + y), x] \\ \sin N_{2e}(x, y) &= \sin N_{2e}[(y, x), x] = \sin N[(x + y), x] \\ \cos N_{2e}(x, -y) &= \cos N_{2e}[(y, -x), -x] = \cos N[(x - y), x] \\ \sin N_{2e}(x, -y) &= -\sin N_{2e}[(y, -x), x] = \sin N[(x - y), x] \end{aligned} \tag{3.6}$$

$$\begin{aligned} \cos N_{2e}(x, x) &= \cos N(2x, x) \\ \sin N_{2e}(x, x) &= \sin N(2x, x) \end{aligned} \tag{3.7}$$

3.2.5 Complex Double Nbic Function $N_{2e}/(x, x)$ as a Particular Case of $N_{2e}/(x, y)$

when $y = x$, we have

$N_{2e}/(x, x)$:

$$\begin{aligned} e^{iN/x}\, e^{ix} &= [\cos N(x) + i \sin N(x)] (\cos x + i \sin x) \\ &= [\cos N(x) \cos x - \sin N(x) \sin x] \\ &\quad + i [\sin N(x) \cos x + \cos N(x) \sin x] \\ &= [(\cos x\, chx - \sin x\, shx) \cos x \\ &\quad - (\sin x\, chx + \cos x\, shx) \sin x)] \\ &\quad + i [(\sin x\, chx + \cos x\, shx) \cos x \\ &\quad + (\cos x\, chx - \sin x\, shx) \sin x)] \\ &= [\cos^2 x\, chx - 2\sin x \cos x\, shx - \sin^2 x\, chx] \\ &\quad + i [\cos^2 x\, shx + 2\sin x \cos x\, shx - \sin^2 x\, shx] \\ &= [\cos 2x\, chx - \sin 2x\, shx] + i [\sin 2x\, chx + \cos 2x\, shx] \\ &= \cos N(2x, x) + i \sin N(2x, x) \\ &= e^{iN/(2x,\, x)} \end{aligned} \tag{3.8a}$$

Single Nbic functions of $(2x, x)$

That the single Nbic functions $N(2x, x)$ can be direcly derived as the real and imaginary parts of the Complex Double Nbic function $N_{2e}(x, x)$, is clearly noticed in (3.8a).

The bracketed terms within [ ], can be easily identified to be single Nbic functions of $(2x, x)$, as given in (1.120) earlier.

Thus

$$\begin{aligned} e^{iN/x}\, e^{ix} &= \cos N(2x, x) + i \sin N(2x, x) \\ &= e^{iN/(2x,\, x)} = e^{2ix} f^{\;ix} \end{aligned}$$

Also

$$\begin{aligned} e^{iN/x}\, e^{ix} &= \cos N_{2e}(x, x) + i \sin N_{2e}(x, x) \\ &= e^{iN_{2e}/(x,\, x)} = e^{iN_{2e}/(x)} \end{aligned} \tag{3.8b}$$

Separating the real and imaginary parts, and in view of single Nbic function of $(2x, x)$, we get, $N_{2e}(x, x)$ [$=N_{2e}(x)$] as follows:

Double Nbic functions $N_{2e}(x, x)$ in Terms of $N(2x, x)$:

$$\begin{aligned} \cos N_{2e}(x, x) &= \cos Nx \cos x - \sin Nx \sin x = \cos N(2x, x) \\ \sin N_{2e}(x, x) &= \sin Nx \cos x + \cos Nx \sin x = \sin N(2x, x) \end{aligned} \tag{3.9a}$$

from which

$$\begin{aligned} \tan N_{2e}(x, x) &= \tan N(2x, x) \\ &= \sin N(2x, x) / \cos N(2x, x) \\ &= \sin N_{2e}(x, x) / \cos N_{2e}(x, x) \end{aligned}$$

Referring to (3.5e), we write

$$\tan N_{2e}(x, x) = \frac{\tan 2x + thx}{1 - \tan 2x.thx} \tag{3.9b}$$

Note: Relation (3.9a) is the interlink of the double Nbic function to be expressed in terms of the single Nbic function.

Similarly, expanding and simplifying

$$\begin{aligned} \cos N_{2e}(x, -x) &= \cos Nx \cos x + \sin Nx \sin x = chx \\ \sin N_{2e}(x, -x) &= \sin Nx \cos x - \cos Nx \sin x = shx \end{aligned} \tag{3.9c}$$

from which

$$\tan N_{2e}(x, -x) = \sin N_{2e}(x, -x) / \cos N_{2e}(x, -x) = shx / chx = thx \tag{3.9d}$$

Additionally, following similar procedure as adopted in getting (1.59), we have

$$\csc N_{2e}(x, x) = \frac{\sec x.\sec hx}{\tan x + thx}$$

$$\sec N_{2e}(x, x) = \frac{\sec x.\sec hx}{1 - \tan x.thx} \tag{3.9e}$$

Further

$$\begin{aligned} &2 \sin N_{2e}(x, x) \cos N_{2e}(x, x) = \sin 4x + \cos 4x\ sh\ 2x \\ &2 \sin N_{2e}(x, -x) \cos N_{2e}(x, -x) = sh\ 2x \\ &\cos N_{2e}{}^2(x, x) + \sin N_{2e}{}^2(x, x) = ch\ 2x \\ &\cos N_{2e}{}^2(x, x) - \sin N_{2e}{}^2(x, x) = \cos 4x - \sin 4x\ sh\ 2x \end{aligned} \tag{3.10a}$$

Now from (3.10a)

$$\begin{aligned} &[\sin 4x + \cos 4x\ sh\ 2x] / [\cos 4x - \sin 4x\ sh\ 2x] \\ = &[2 \sin N_{2e}(x, x) \cos N_{2e}(x, x)] / [\cos N_{2e}{}^2(x, x) - \sin N_{2e}{}^2(x, x)] \end{aligned}$$

Divide the numerator and denominator of the LHS by, cos $4x$, and divide the numerator and denominator of RHS by, cos $N_{2e}{}^2(x, x)$, to get

$$\begin{aligned} \text{LHS} &= [\tan 4x + sh\ 2x] / [1 - \tan 4x\ sh\ 2x] \\ \text{RHS} &= 2\,[\tan N_{2e}(x, x)] / [1 - \tan N_{2e}{}^2(x, x)] \end{aligned}$$

The LHS has the same form of the formula

$$\tan (A + B) = \frac{\tan A + \tan B}{1 - \tan A.\tan B}$$

CHAPTER 3

The RHS has the same form of the formula

$$\tan 2C = \frac{2\tan C}{1-\tan^2 C}$$

Hence, LHS = RHS, provides

$$\frac{2\tan N_{2e}(x,x)}{1-\tan N_{2e}{}^2(x,x)} = \frac{\tan 4x + sh2x}{1-\tan 4x.sh2x} \quad (3.10b)$$

Now

$$e^{iN/x}\, e^{-ix} = \cos N_{2e}(x,\ -x) + i\ \sin N_{2e}(x,\ -x) = e^{iN_{2e}/(x,-x)}$$
$$= chx + i\ shx = f^{ix}$$
$$\text{or, } e^{iN/x}\, e^{ix} = f^{\ ix}$$

Again, from (3.8b)

$$e^{iN/x}\, e^{ix} = e^{iN_{2e}/(x)}$$

so, multiplying

$$[e^{iN/x}\, e^{-ix}]\ [e^{iN/x}\, e^{ix}] = [e^{iN/x}]^2 = f^{ix}\, e^{iN_{2e}/(x)}$$

or, $$[e^{iN/x}]^2 / e^{iN_{2e}/(x)} = f^{\ ix} \quad (3.11)$$

Double Nbic function $N_{2f}(x,\ x)$, as viewed from Complex Nbic function ratio, R_{ndn}

Considering the case of, $N_{2f}(x,\ y) = N/(x,\ x).\ f^{\pm iy} = e^{\pm iN/(x,x)}.f^{\pm iy}$, (category 2), the definition and usefulness of double Nbic function will be imminent when we attempt to express Complex single Nbic Functions expressed through multiple angles, (say) of $(nx,\ nx)$.
Recollect the Complex single Nbic function ratio, R_{nsn} from (2.24h) here, once again.

$$R_{nsn} = \frac{\cos N(nx) + i\sin N(nx)}{[\cos N(x) + \sin N(x)]^2} = \frac{f^{inx}}{f^{nix}} \quad (3.12)$$

Then the numerator

$$[\cos N(nx) + i\ \sin N(nx)]$$
$$= [f^{\ inx}/f^{\ nix}]\ [chx + i\ shx]^n$$
$$= [e^{iN/x}/f^{\ ix}]^n\ [ch\ nx + i\ sh\ nx] \qquad \textbf{OK}$$
$$= [(\cos Nx + i\ \sin Nx)/(ch\ nx + i\ sh\ nx)]^n.\ [ch\ nx + i\ sh\ nx]$$
$$= [R_i]^n\ .(ch\ nx + i\ sh\ nx)$$

where

$$f^{\ nix} = [f^{\ ix}]^n = [chx + i\ shx]^n$$
$$[R_i]^n = [R_i(x)]^n = [(\cos Nx + i\ \sin Nx)/(chx + i\ shx)]^n$$

from which

$$R_i = (\cos Nx + i\ \sin Nx)/(chx + i\ shx) \quad (3.13a)$$

Multiply the numerator and denominator by, ($chx - i\ shx$), and simplify to get

$$R_i = [(\cos Nx\ chx + \sin Nx\ shx) + i\ (\sin Nx\ chx - \cos Nx\ shx)]\,/\,ch\ 2x \quad (3.13b)$$

A very close similarity is observed at the very form of the component parts (that is, the real and imaginary parts), of this equation (3.13b) representing R_i, with the corresponding ones of single Nbic functions, defined earlier, *viz.*,

$$(\cos x\ chx + \sin Nx\ shx) = \cos N(x, -x)$$
$$(\sin x\ chx - \cos x\ shx) = \sin N(x, -x) \quad (3.14)$$

In view of this similarity, it can be seen that, the expressions on the LHS of (3.14) fit in well with the double Nbic functions, F_{2B} type $N_{2f}(x, y)$, as already defined earlier in (3.3a).

Thus we define

$N_{2f}(x, y)$:

$$\cos N_{2f}(x, -x) = (\cos Nx\ chx + \sin Nx\ shx)$$
$$\sin N_{2f}(x, -x) = (\sin Nx\ chx - \cos Nx\ shx) \quad (3.15a)$$

and with these, (3.13b) becomes

$$R_i = [(\cos Nx\ chx + \sin Nx\ shx) + i\ (\sin Nx\ chx - \cos Nx\ shx)]\,/\,ch\ 2x$$
$$= [\cos N_{2f}(x, -x) + i\ \sin N_{2f}(x, -x)]\,/\,ch\ 2x \quad (3.16a)$$

which can be further condensed as

$$R_i = e^{iN_{2f}/(x,-x)}\,/\,ch\ 2x \quad (3.16b)$$

where

$$e^{iN_{2f}/(x,-x)} = \cos N_{2f}(x, -x) + i\ \sin N_{2f}(x, -x) \quad (3.17)$$

Double Nbic functions $N_{2f}(x, x)$, in terms of Single Nbic functions $N(x, x)$:

Thus define double Nbic functions $N_{2f}(x, x)$, in terms of single Nbic functions, after expanding, simplifying and rearranging as

$N(x, x)$:

$$\sin N_{2f}(x, x) = \sin N_{2f}(x)$$
$$= \sin Nx\ chx + \cos Nx\ shx = \sin x + \cos x\ sh\ 2x$$
$$\cos N_{2f}(x, x) = \cos N_{2f}(x)$$
$$= \cos Nx\ chx - \sin Nx\ shx = \cos x - \sin x\ sh\ 2x$$
$$\sin N_{2f}(x, -x)$$
$$= \sin Nx\ chx - \cos Nx\ shx = \sin x\ ch\ 2x$$
$$\cos N_{2f}(x, -x)$$
$$= \cos Nx\ chx + \sin Nx\ shx = \cos x\ ch\ 2x \quad (3.18)$$

Some Identities

Using (3.18), we arrive at these following identities, in terms of circular, and double Nbic functions of single variable, $N_{2f}(x, x)$:

[**Note**: The LHS of the following expressions, are actually the triple Nbic functions, which are discussed later elaborately.]

$$\mathbf{cosN}_{2f}^2\,(x, x) + \mathbf{sinN}_{2f}^2\,(x, x) = ch^2\,2x$$

$$\mathbf{cosN}_{2f}^2\,(x, x) - \mathbf{sinN}_{2f}^2\,(x, x)$$
$$= [\mathbf{cosN}^2x - \mathbf{sinN}^2x] - 2\,\mathbf{sinN}x\,\mathbf{cosN}x.\,sh\,2x$$

$$\cos x\,\mathbf{sinN}_{2f}\,(x, x) + \sin x\,\mathbf{cosN}_{2f}\,(x, x)$$
$$= \mathbf{sinN}x\,\mathbf{cosN}x + \mathbf{sinN}x\,\mathbf{cosN}x$$
$$= 2\,\mathbf{sinN}x\,\mathbf{cosN}x$$

$$\sin x\,\mathbf{sinN}_{2f}\,(x, x) + \cos x\,\mathbf{cosN}_{2f}\,(x, x) = 1$$

$$\sin x\,\mathbf{sinN}_{2f}\,(x, -x) + \cos x\,\mathbf{cosN}_{2f}\,(x, -x) = ch\,2x$$

$$\cos x\,\mathbf{cosN}_{2f}\,(x) - \sin x\,\mathbf{sinN}_{2f}\,(x) = \mathbf{cosN}x\,\mathbf{cosN}x - \mathbf{sinN}x\,\mathbf{sinN}x$$
$$= \mathbf{cosN}^2x - \mathbf{sinN}^2x \qquad (3.19a)$$

Further, in terms of single Nbic function N(x, x) and double Nbic function, $N_{2f}(x, x)$, we write:

$$\mathbf{cosN}x\,\mathbf{sinN}_{2f}(x, x) - \mathbf{sinN}x\,\mathbf{cosN}_{2f}(x, x) = shx\,ch\,2x$$
$$\mathbf{sinN}x\,\mathbf{sinN}_{2f}(x, x) + \mathbf{cosN}x\,\mathbf{cosN}_{2f}(x, x) = chx\,ch\,2x \qquad (3.19b)$$

3.2.6 Various tan-formulas of Circular, Hyperbolic, Nbic Functions, in the Form of, tan(A + B) Formula, and Interrelations

From (3.19b)

$$thx = shx / chx$$
$$= [\text{cosN}x\,.\text{sinN}_{2f}\,(x, x) - \text{sinN}x\,.\text{cosN}_{2f}\,(x, x)]$$
$$/ [\text{sinN}x\,.\text{sinN}_{2f}\,(x, x) + \text{cosN}x\,.\text{cosN}_{2f}\,(x, x)]$$

Divide the numerator and denominator by, cos $N_{2f}(x)$, using

$$\text{tanN}_{2f}\,(x, x) = [\text{sinN}_{2f}\,(x, x)] / [\text{cosN}_{2f}\,(x, x)], \text{ to get}$$

$$\mathbf{th}x = [\mathbf{tanN}_{2f}\,(x, x) - \mathbf{tanN}x] / [1 + \mathbf{tanN}_{2f}\,(x, x) \cdot \mathbf{tanN}x] \qquad (3.20)$$

[It resembles the expansion of tan (A – B) formula, where tan A = tan $N_{2f}(x, x)$, tan B = tanNx]

(3.20) on rearrangement, reduces to

$$\text{tanN}_{2f}\,(x, x) = [\text{tanN}x + thx] / [1 - \text{tanN}x.thx] \qquad (3.21a)$$

[It resembles expansion of tan(A + B) formula, where tan A = tan$x$, tan B = th$x$]

Also from (3.18)

$$\text{tanN}_{2f}\,(x, x) = \text{tanN}_{2f}\,(x) = (\sin x + \cos x\,sh\,2x) / (\cos x - \sin x\,sh\,2x)$$

Divide the numerator and denominator by cos x, to get

$$\text{tanN}_{2f}\,(x, x) = [\text{sinN}_{2f}\,(x, x)] / [\text{cosN}_{2f}\,(x, x)]$$
$$= [\tan x + sh\,2x] / [1 - \tan x\,sh\,2x] \qquad (3.21b)$$

[It resembles expansion of tan(A – B) formula, where tan A = tan$x$, tan B = sh 2x]

Again from (3.18)

$$\text{tanN}_{2f}\,(x, -x) = [\text{sinN}_{2f}\,(x, -x)] / [\text{cosN}_{2f}\,(x, -x)] = \tan x \qquad (3.21c)$$

[It resembles expansion of tan (A – B) formula, where tan A = tan $N_{2f}(x, x)$, tan B = tan$x$]

Collecting formulas together

$$\begin{aligned}
\mathbf{sinN_{2f}(x, x)} &= \mathbf{sinN_{2f}(x)} = \text{sinN}(x)\, chx + \text{cosN}(x)\, shx \\
&= \sin x + \cos x\, sh\, 2x \\
\mathbf{cosN_{2f}(x, x)} &= \mathbf{cosN_{2f}(x)} = \text{cosN}(x)\, chx - \text{sinN}(x)\, shx \\
&= \cos x - \sin x\, sh\, 2x \\
\mathbf{sinN_{2f}(x, -x)} &= \text{sinN}(x, x)\, chx - \text{cosN}(x, x)\, shx \\
&= \sin x\, ch\, 2x \\
\mathbf{cosN_{2f}(x, -x)} &= \text{cosN}(x, x)\, chx + \text{sinN}(x, x)\, shx \\
&= \cos x\, ch\, 2x \qquad (3.22a)
\end{aligned}$$

$$\boldsymbol{thx = [\tan N_{2f}(x, x) - \tan Nx]/[1 + \tan N_{2f}(x, x) + \tan Nx]}$$

$$\boldsymbol{\tan N_{2f}(x, x) = [\tan Nx + thx]/[1 - \tan Nx.thx]}$$

also

$$\boldsymbol{\tan N_{2f}(x, x) = [\tan x + sh\, 2x]/[1 - \tan x\, sh\, 2x]}$$

and

$$\boldsymbol{\tan N_{2f}(x, -x) = [\sin N_{2f}(x, -x)]/[\cos N_{2f}(x, -x)] = \tan x}$$

and

$$\boldsymbol{\tan x = [\tan N_{3e}(x) - \tan N_{2f}(x)]/[1 + \tan N_{3e}(x)\, \tan N_{2f}(x)]} \qquad (3.22b)$$

Additionally, following similar procedure as adopted in getting (1.59), we have

$$\text{cosecN}_{2f}(x, x) = \frac{\sec Nx.\sec hx}{\tan Nx + thx}$$

$$\text{secN}_{2f}(x, x) = \frac{\sec Nx.\sec hx}{1 - \tan Nx.thx} \qquad (3.22c)$$

Strangely enough, all these formulas (3.22a-c), conform to the mathematical structure of the ordinary trigonometric formula of

$$\tan(A + B) = [\tan A + \tan B]/[1 + \tan A \tan B].$$

Numerical Example

Example 3.1

Calculate $\tan N_{2f}(x, x)$ values using formula (3.21a), for $x = 60$ deg, and 240 deg, and verify the values using the alternative formula (3.21b).

Solution

For brevity, the suffix ***f*** in the following various double Nbic functions, $N_{2f}(x, x)$, particularly for this numerical check, has been omitted.

Given Data:

$$\begin{aligned}
x &= 1.0472 \text{ rad} = 60 \text{ deg} \\
&\sin x = 0.867,\ \cos x = 0.5,\ \tan x = 1.732, \\
&shx = 1.2494,\ chx = 1.6,\ thx = 0.7809, \\
&\text{sinN}x = 2.0106,\ \text{cosN}x = -0.2818,\ \text{tanN}x = -7.1348
\end{aligned}$$

CHAPTER 3

$$\begin{aligned}
2x &= 2.0944 \text{ rad} = 120 \text{ deg}\\
\sin 2x &= 0.867,\ \cos 2x = -0.5,\ \tan 2x = -1.732,\\
sh\ 2x &= 3.9987,\ ch\ 2x = 4.1218,\ th\ 2x = 0.9701,\\
\text{sinN } 2x &= 1.5703,\ \text{cosN } 2x = -5.5239,\ \text{tanN } 2x = -0.2844\\
3x &= 3.1416 \text{ rad} = 180 \text{ deg}\\
\sin 3x &= 0,\ \cos x = -1,\ \tan x = 0,\\
sh\ 3x &= 11.5920,\ ch\ 3x = 11.5920,\ thx = 1,\\
\text{sinN } 3x &= -11.5487,\ \text{cosN } 3x = -11.5920,\ \text{tanN } 3x = 1\\
4x &= 4.1888 \text{ rad} = 240 \text{ deg}\\
\sin 4x &= -0.867,\ \cos 4x = -0.5,\ \tan 4x = 1.732,\\
sh\ 4x &= 32.9639,\ ch\ 4x = 32.9791,\ th\ 4x = 0.9995,\\
\text{sinN } 4x &= -45.0427,\ \text{cosN } 4x = 12.0580,\ \text{tanN } 4x = -3.7355
\end{aligned}$$

Now

For $x = 4.1888$ (240 deg)

Using (3.21a):

$$\begin{aligned}
\text{tanN}_{2f}(x, x) &= [\text{tanN}x + thx]/[1 - \text{tanN}x\ thx]\\
&= [-3.7355 + 0.9995]/[1 + 3.7355 + 0.9995]\\
&= -2.736/4.7336 = -0.578
\end{aligned}$$

Using (3.21b):

$$\begin{aligned}
\text{tanN}_{2f}(x, x) &= [\tan x + sh\ 2x]/(1 - \tan x\ sh\ 2x)\\
&= [1.732 + (2 \times 32.9639 \times 32.9791)]/[1 - 1.732\ (2 \times 32.9639 \times 32.9791)]\\
&= -2175.9715/3764.7828 = -0.578
\end{aligned}$$

The two forms of $\text{tanN}_{2f}(x, x)$ are **same**.

For $x = 1.0472$ rad (= 60 deg)

Using (3.21a):

$$\begin{aligned}
\text{tanN}_{2f}(x, x) &= [\text{tanN}x + thx]/[1 - \text{tanN}x\ thx]\\
&= [-7.1348 + 0.7809]/[1 - 7.1348 \times 0.7809]\\
&= -6.3539/6.5716 = -0.9669
\end{aligned}$$

Using (3.21b):

$$\begin{aligned}
\text{tanN}_{2f}(x, x) &= [\tan x + sh\ 2x]/(1 - \tan x\ sh\ 2x)\\
&= [1.732 + 3.9981]/[1 - 1.732 \times 3.9981]\\
&= -5.7301/5.9247 = -0.9672
\end{aligned}$$

The two forms of $\text{tanN}_{2f}(x, x)$ are **same**.

3.2.7 Differentiation of Double Nbic Functions, $N'_2(x, x)$

[To represent $\frac{d}{dx}(\ )$, using ***dash*** symbol as $(\ )'$]

For $N'_{2e}(x, x)$:

Refer to (3.9a–3.9d), for the following double Nbic functions

$$\begin{aligned}
\text{sinN}_{2e}(x, x) &= \text{sinN}x \cos x + \text{cosN}x \sin x = \text{sinN}(2x, x)\\
\text{cosN}_{2e}(x, x) &= \text{cosN}x \cos x - \text{sinN}x \sin x = \text{cosN}(2x, x)
\end{aligned}$$

$$\sin N_{2e}(x, -x) = \sin Nx \cos x - \cos Nx \sin x = shx$$
$$\cos N_{2e}(x, -x) = \cos Nx \cos x + \sin Nx \sin x = chx \quad (3.23)$$

The first derivative, $\mathbf{N'_{2e}(x, x)}$, of these functions are:

$$\sin N'_{2e}(x, x) = \sin N'_{2e}(x) = \sin N'x \cos x - \sin Nx \sin x + \cos N'x \sin x + \cos Nx \cos x$$

Substituting,

$\sin N'x = 2 \cos x\, chx$, $\cos N'x = -2 \sin x\, chx$, in it, and simplifying

$$\sin N'_{2e}(x, x) = 3 \cos 2x\, chx - \sin 2x\, shx = 2 \cos 2x\, chx + \cos N(2x, x) \quad (3.24a)$$

Alternatively, by treating

$\sin N'_{2e}(x, x) = \sin N'(2x, x) = [\sin 2x\, chx + \cos 2x\, shx]'$

the required result is arrived at.

$$\cos N'_{2e}(x, x) = \cos N'x \cos x - \cos Nx \sin x - \sin N'x \sin x - \sin Nx \cos x$$

Substituting,

$\sin N'x = 2 \cos x\, chx$, $\cos N'x = -2 \sin x\, chx$, in it, and simplifying

$$\cos N'_{2e}(x, x) = -3 \sin 2x\, chx - \cos 2x\, shx = -2 \sin 2x\, chx - \sin N(2x, x) \quad (3.24b)$$

Alternatively, by treating

$\cos N'_{2e}(x, x) = \cos N'(2x, x) = [\cos 2x\, chx - \sin 2x\, shx]'$

the required result is arrived at.

Also

$$\sin N'_{2e}(x, -x) = chx$$
$$\cos N'_{2e}(x, -x) = shx \quad (3.24c)$$

For $N'_{2f}(x, x)$:

Refer to (3.18), for the following double Nbic functions.

$$\sin N_{2f}(x, x) = \sin Nx\, chx + \cos Nx\, shx = \sin x + \cos x\, sh\, 2x$$
$$\cos N_{2f}(x, x) = \cos Nx\, chx - \sin Nx\, shx = \cos x - \sin x\, sh\, 2x$$
$$\sin N_{2f}(x, -x) = \sin Nx\, chx - \cos Nx\, shx = \sin x\, ch\, 2x$$
$$\cos N_{2f}(x, -x) = \cos Nx\, chx + \sin Nx\, shx = \cos x\, ch\, 2x \quad (3.25)$$

The first derivative $\mathbf{N'_{2f}(x, x)}$, of these functions are:

$$\sin N'_{2f}(x, x) = \sin N'_{2f}(x) = \cos x + 2 \cos x\, ch\, 2x - \sin x\, sh\, 2x = \cos x + \cos x\, ch\, 2x + (\cos x\, ch\, 2x - \sin x\, sh\, 2x) = \cos x\,(1 + ch\, 2x) + \cos N(x, 2x)$$

Alternatively

since, $\cos N_{2f}(x, -x) = \cos x\, ch\, 2x$

$\sin N'_{2f}(x, x) = \sin N'_{2f}(x)$

CHAPTER 3

$$
\begin{aligned}
& = \cos x + \cos N_{2f}(x, -x) + \cos N(x, 2x) \\
\cos N'_{2f}(x, x) & = \cos N'_{2f}(x) \\
& = -\sin x - 2 \sin x\, ch\, 2x - \cos x\, sh\, 2x \\
& = -\sin x - \sin x\, ch\, 2x - (\sin x\, ch\, 2x + \cos x\, sh\, 2x) \qquad (3.26a) \\
& = -\sin x\, (1 + ch\, 2x) - \sin N(x, 2x)
\end{aligned}
$$

Alternatively

since, $\sin N_{2f}(x, -x) = \sin x\, ch\, 2x$

$$
\begin{aligned}
\cos N'_{2f}(x, x) & = \cos N'_{2f}(x) \\
& = -\sin x - \sin N_{2f}(x, -x) - \sin N(x, 2x)
\end{aligned}
$$

Similarly

$$
\begin{aligned}
\sin N'_{2f}(x, -x) & = 2 \sin x\, sh\, 2x + \cos x\, ch\, 2x \\
& = \sin x\, sh\, 2x + \cos N(x, -2x) \\
\cos N'_{2f}(x, -x) & = 2 \cos x\, sh\, 2x - \sin x\, ch\, 2x \\
& = \cos x\, sh\, 2x - \sin N(x, -2x) \qquad (3.26b)
\end{aligned}
$$

Collecting formulas together

For $N'_{2e}(x, x)$:

$$
\begin{aligned}
\sin N'_{2e}(x, x) & = 2 \cos 2x\, chx + \cos N(2x, x) \\
\cos N'_{2e}(x, x) & = -2 \sin 2x\, chx - \sin N(2x, x) \\
\sin N'_{2e}(x, -x) & = chx \\
\cos N'_{2e}(x, -x) & = shx
\end{aligned}
$$

For $N'_{2f}(x, x)$:

$$
\begin{aligned}
\sin N'_{2f}(x, x) & = \cos x\, (1 + ch\, 2x) + \cos N(x, 2x) \\
\cos N'_{2f}(x, x) & = -\sin x\, (1 + ch\, 2x) - \sin N(x, 2x) \\
\sin N'_{2f}(x, -x) & = \sin x\, sh\, 2x + \cos N(x, -2x) \\
\cos N'_{2f}(x, -x) & = \cos x\, sh\, 2x - \sin N(x, -2x) \qquad (3.27)
\end{aligned}
$$

3.2.8 Complex Double Nbic Function Ratio, R_{ndn}

In (2.3d), (2.5j)and (2.24h), the expressions of the Complex circular function ratio (R_{nci}), Complex hyperbolic function ratio (R_{nhy}), and Complex single Nbic function ratio (R_{nsn}), are given.

For the case of double Nbic function, $N_{2f}(x, x)$, this ratio, R_{ndn} is defined as

$$
\begin{aligned}
R_{ndn} = & [\cos N_{2f}(nx) + i \sin N_{2f}(nx)] \\
& / [\cos N_{2f}(x) + i \sin N_{2f}(x)]^n \qquad (3.28)
\end{aligned}
$$

To simplify the numerator and the denominator, use (3.25), that is

$$
\begin{aligned}
\cos N_{2f}(x, x) & = \cos N_{2f}(x) = \cos Nx\, chx - \sin Nx\, shx \\
& = \cos x - \sin x\, sh\, 2x \\
\sin N_{2f}(x, x) & = \sin N_{2f}(x) = \sin Nx\, chx + \cos Nx\, shx \\
& = \sin x + \cos x\, sh\, 2x \qquad (3.29)
\end{aligned}
$$

In view of (3.29), the different parts of (3.28) are

$$
\begin{aligned}
\cos N_{2f}(nx) & = \cos nx - \sin nx\, sh\, 2nx \\
i \sin N_{2f}(nx) & = i \sin nx + i \cos nx\, sh\, 2nx
\end{aligned}
$$

$$\cos N_{2f}(x) = \cos x - \sin x \; sh\; 2x$$
$$i \sin N_{2f}(x) = i \sin x + i \cos x \; sh\; 2x \tag{3.30a}$$

adding suitable terms

$$\cos N_{2f}(nx) + i \sin N_{2f}(nx) = (\cos nx + i \sin nx)(1 + i\; sh\; 2nx)$$
$$= e^{inx}[f^{\;inx}]^2$$
$$[\cos N_{2f}(x) + i \sin N_{2f}(x)]^n = (\cos x + i \sin x)^n (1 + i\; sh\; 2x)^n$$
$$= (e^{ix})^n [f^{\;2ix}]^n \tag{3.30b}$$

So

$$R_{ndn} = [\cos N_{2f}(nx) + i \sin N_{2f}(nx)] / [\cos N_{2f}(x) + i \sin N_{2f}(x)]^n$$
$$= (1 + i\; sh\; 2nx)/(1 + i\; sh\; 2x)^n$$
$$= [f^{\;inx}]^2 / [f^{\;2ix}]^n \tag{3.31}$$

❑❑❑

4

TRIPLE Nbic FUNCTION, $N_3(x, y)$

4.1 TRIPLE Nbic FUNCTION, $N_{3e}(x, y)$

4.1.1 Definition of Triple Nbic Function, $N_{3e}(x, y)$

Define afresh in a different fashion triple Nbic function, $[\mathbf{N_{3e}}(x, y)]$, as:

$$\begin{aligned}
\sin N_{3e}(x, y) &= \sin Nx \cos Ny + \cos Nx \sin Ny \\
\cos N_{3e}(x, y) &= \cos Nx \cos Ny - \sin Nx \sin Ny \\
\sin N_{3e}(x, -y) &= \sin Nx \cos Ny - \cos Nx \sin Ny \\
\cos N_{3e}(x, -y) &= \cos Nx \cos Ny + \sin Nx \sin Ny
\end{aligned} \tag{4.1a}$$

Now expand

$$\begin{aligned}
\sin N_{3e}(x, y) &= \sin Nx \cos Ny + \cos Nx \sin Ny \\
&= (\sin x\ ch x + \cos x\ sh x)\ (\cos y\ ch y - \sin y\ sh y) \\
&\quad + \cos x\ ch x - \sin x\ sh x)\ (\sin y\ ch y + \cos y\ sh y) \\
&= \sin\ (x + y)\ ch(x - y) + \cos(x + y)\ sh(x + y)
\end{aligned}$$

$$\begin{aligned}
\cos N_{3e}(x, y) &= \cos Nx \cos Ny - \sin Nx \cos Ny \\
&= (\cos x\ ch x - \sin x\ sh x)\ (\cos y\ ch y - \sin y\ sh y) \\
&\quad - (\sin x\ ch x + \cos x\ sh x)\ (\sin y\ ch y + \cos y\ sh y) \\
&= \cos(x + y)\ ch(x - y) - \sin(x + y)\ sh(x + y)
\end{aligned}$$

Similarly, we get

$$\begin{aligned}
\sin N_{3e}(x, -y) &= \sin(x - y)\ ch(x + y) + \cos(x - y)\ sh(x - y) \\
\cos N_{3e}(x, -y) &= \cos(x - y)\ ch(x + y) - \sin(x - y)\ sh(x - y)
\end{aligned} \tag{4.1b}$$

Collecting formulas together

For $N_{3e}(x, y)$:

$$\begin{aligned}
\sin N_{3e}(x, y) &= \sin(x + y)\ ch(x - y) + \cos(x + y)\ sh(x + y) \\
\cos N_{3e}(x, y) &= \cos(x + y)\ ch(x - y) - \sin(x + y)\ sh(x + y) \\
\sin N_{3e}(x, -y) &= \sin(x - y)\ ch(x + y) + \cos(x - y)\ sh(x - y) \\
\cos N_{3e}(x, -y) &= \cos(x - y)\ ch(x + y) - \sin(x - y)\ sh(x - y)
\end{aligned} \tag{4.1c}$$

4.1.2 Particular Case, $N_{3e}(x, x)$

When, $y = x$

In the light of (1.45) and (1.48), (4.1a) after simplification, produces the results as **$[N_{3e}(x, x)]$:**

$$\begin{aligned}\sin N_{3e}(x, x) &= \sin N_{3e}(x)\\ &= \sin Nx \cos Nx + \cos Nx \sin Nx\\ &= 2 \sin Nx \cos Nx\\ &= \sin 2x + \cos 2x \; sh \; 2x\\ \cos N_{3e}(x, x) &= \cos N_{3e}(x)\\ &= \cos Nx \cos Nx - \sin Nx \sin Nx\\ &= \cos N^2 x - \sin N^2 x\\ &= \cos 2x - \sin 2x \; sh \; 2x \end{aligned} \tag{4.1d}$$

[Form (4.1d) can be obtained from (1.29a) in the light of (1.27c) and for $y = x$]

In **Figure 4.1,** the variation of $N_{3e}(x)$ vs. x is shown.

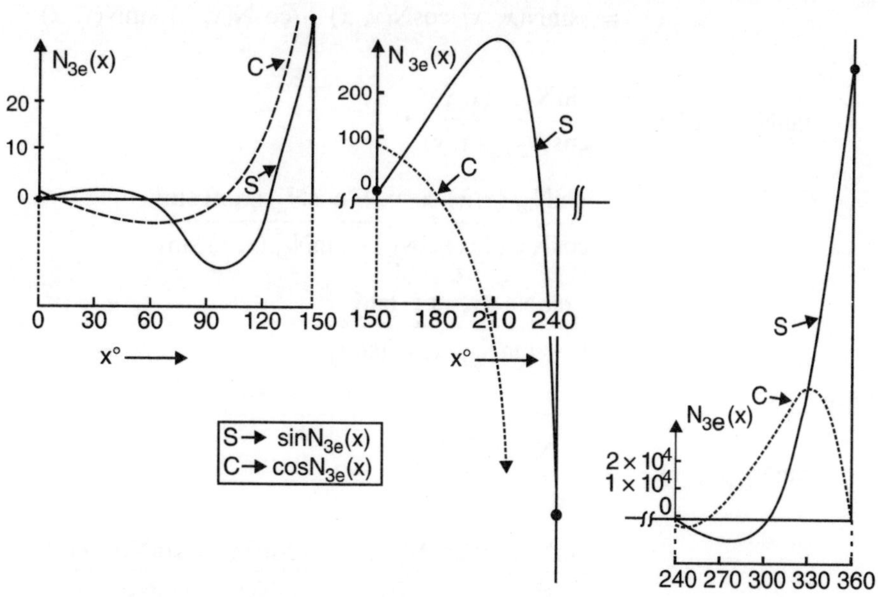

Fig. 4.1: N_{3e} (x) vs. x

[**Note:** (4.1d) shows that triple Nbic function $\sin N_{3e}(x, x)$ and $\cos N_{3e}(x, x)$ formed out of variables of single arguments (x, x), can be expressed in terms of relations written in terms of double argument $2x$.]

4.1.3 Interchangeability of Third Order Nbic Functions

It has been shown in (1.28h) and (1.28i) that, some form of the third order Nbic functions follow the interchangeability, *e.g.*, $N_{2f,e}(x, x) = N_{2e,f}(x, x)$. These functions are identified with two suffixes (**e** as well as **f**), and not with single suffix (either **e** or **f**). In this connection, let us observe the expanded form of these functions. That is why they are categorized in a different group, namely **Category M (Mixed category)** involving EF type and FE type. In this connection, let us observe the expanded form of these functions. When recalled here:

FE type: $N_{2f,e}(x, y)$:

$$\begin{aligned}
\cos N_{2f,e}(x, y) &= \cos N_{2f}(x, x)\cos y - \sin N_{2f}(x, x)\sin y \\
&= \cos N(x, x)\cos N(y, x) - \sin N(x, x)\sin N(y, x) \\
\sin N_{2f,e}(x, y) &= \sin N_{2f}(x, x)\cos y + \cos N_{2f}(x, x)\sin y \\
&= \sin N(x, x)\cos N(y, x) + \cos N(x, x)\sin N(y, x) \\
\cos N_{2f,e}(x, -y) &= \cos N_{2f}(x, x)\cos y + \sin N_{2f}(x, x)\sin y \\
&= \cos N(x, x)\cos N(y, -x) + \sin N(x, x)\sin N(y, -x) \\
\sin N_{2f,e}(x, -y) &= \sin N_{2f}(x, x)\cos y - \cos N_{2f}(x, x)\sin y \\
&= \sin N(x, x)\cos N(y, x) - \cos N(x, x)\sin N(y, x)
\end{aligned} \tag{4.2a}$$

So that

$$\begin{aligned}
\tan N_{2f,e}(x, y) &= \frac{\sin N_{2f,e}(x, y)}{\cos N_{2f,e}(x, y)} \\
&= \frac{\sin N_{2f}(x, x)\cos y + \cos N_{2f}(x, x)\sin y}{\cos N_{2f}(x, x)\cos y - \sin N_{2f}(x, x)\sin y} \\
&= \frac{\tan N_{2f}(x, x) + \tan y}{1 - \tan N_{2f}(x, x).\tan y}
\end{aligned} \tag{4.2b}$$

Also otherwise

$$\begin{aligned}
\tan N_{2f,e}(x, y) &= \frac{\sin N_{2f,e}(x, y)}{\cos N_{2f,e}(x, y)} \\
&= \frac{\sin N(x, x)\cos N(y, x) + \cos N(x, x)\sin N(y, x)}{\cos N(x, x)\cos N(y, x) - \sin N(x, x)\sin N(y, x)} \\
&= \frac{\tan N(x, x) + \tan N(y, x)}{1 - \tan N(x, x).\tan N(y, x)}
\end{aligned} \tag{4.2c}$$

Particular Case, $[N_{2f,e}(x, x)]$:

For $y = x$

$$\begin{aligned}
\cos N_{2f,e}(x, x) &= \cos N_{2f}(x, x)\cos x - \sin N_{2f}(x, x)\sin x \\
&= \cos 2x - \sin 2x \, sh\, 2x \\
\sin N_{2f,e}(x, x) &= \sin N_{2f}(x, x)\cos x - \cos N_{2f}(x, x)\sin x
\end{aligned}$$

$$= \sin 2x + \cos 2x\ sh\ 2x$$

$$\cos N_{2f,e}(x, -x) = \cos N(x, x) \cos N(x, -x) + \sin N(x, x) \sin N(x, -x)$$

$$= 1$$

$$\sin N_{2f,e}(x, -x) = \sin N(x, x) \cos N(x, -x) - \cos N(x, x) \sin N(x, -x)$$

$$= sh\ 2x \quad (4.2d)$$

[**Note:** It is important to observe that, the generation of $\sin N_{2f,e}(x, x)$ and $\cos N_{2f,e}(x, x)$ of (4.2d), which belong to Mixed category FE, can also be expressed through another alternative form, as $\sin N_{3e}(x, x)$ and $\cos N_{3e}(x, x)$ of (4.1a), respectively. This fact once again is elaborated later in (4.2o).]

For $y = x$, (4.2b) and (4.2c), give identical result. It is shown as follows:

Expand expression (4.2b) substituting suitably for the numerator and denominator

$$\tan N_{2f,e}(x, x) = \frac{\tan N_{2f}(x, x) + \tan x}{1 - \tan N_{2f}(x, x).\tan x}$$

Now we have

$$\tan N_{2f}(x, x) = \frac{\sin N(x, x)\ ch(x) + \cos N(x, x)\ sh(x)}{\cos N(x, x)\ ch(x) - \sin N(x, x)\ sh(x)}$$

$$= \frac{\tan N(x, x) + th(x)}{1 - \tan N(x, x).th(x)} = \frac{\sin x + \cos x.sh\ 2x}{\cos x - \sin x.sh\ 2x}$$

and

$$\tan N(x, x) = \frac{\sin x\ ch\ x + \cos x\ sh\ x}{\cos x\ ch\ x - \sin x\ sh\ x} = \frac{\tan x + thx}{1 - \tan x.thx}$$

With these we get on simplification

$$\tan N_{2f,e}(x, x) = \frac{\tan N_{2f}(x, x) + \tan x}{1 - \tan N_{2f}(x, x).\tan x}$$

$$= \frac{\sin 2x + \cos 2x.sh2x}{\cos 2x - \sin 2x.sh2x} \quad (4.2e)$$

Expand expression (4.2c) substituting suitably for the the numerator and denominator

$$\tan N_{2f,e}(x, x) = \frac{2 \tan N(x,x)}{1 - \tan N^2(x,x)}$$

$$= \frac{\sin 2x + \cos 2x.sh2x}{\cos 2x - \sin 2x.sh2x} = \frac{\tan 2x + sh\ 2x}{1 - \tan 2x.sh\ 2x} \quad (4.2f)$$

EF type: $N_{2e,f}(x, y)$:

$$\cos N_{2e,f}(x, y) = \cos N_{2e}(x, x)\ chy - \sin N_{2e}(x, x)\ shy$$

$$= \cos N(x, x) \cos N(x, y) - \sin N(x, x) \sin N(x, y)$$

$$\sin N_{2e,f}(x, y) = \sin N_{2e}(x, x)\ chy + \cos N_{2e}(x, x)\ shy$$

$$= \sin N(x, x) \cos N(y, y) + \cos N(x, x) \sin N(y, y)$$

$$\cos N_{2e,f}(x, -y) = \cos N_{2e}(x, x)\ chy + \sin N_{2e}(x, x)\ shy$$

$$= \cos N(x, x) \cos N(x, y) + \sin N(x, x) \sin N(x, y)$$

$$\sin N_{2e,f}(x, -y) = \sin N_{2e}(x, x)\, chy - \cos N_{2e}(x, x)\, shy$$

$$= \sin N(x, x) \cos N(y, -y) + \cos N(x, x) \sin N(y, -y) \qquad (4.2g)$$

So that

$$\tan N_{2e,f}(x, y) = \frac{\sin N_{2e,f}(x, y)}{\cos N_{2e,f}(x, y)}$$

$$= \frac{\sin N(x, x) \cos N(y, y) + \cos N(x, x) \sin N(y, y)}{\cos N(x, x) \cos N(x, y) - \sin N(x, x) \sin N(x, y)} \qquad (4.2h)$$

Also otherwise

$$\tan N_{2e,f}(x, y) = \frac{\sin N_{2e,f}(x, y)}{\cos N_{2e,f}(x, y)}$$

$$= \frac{\sin N_{2e}(x, x)\, chy + \cos N_{2e}(x, x)\, shy}{\cos N_{2e}(x, x)\, chy - \sin N_{2e}(x, x)\, shy}$$

$$= \frac{\tan N_{2e}(x, x) + th(y)}{1 - \tan N_{2e}(x, x).th(y)} \qquad (4.2i)$$

Particular case [$N_{2e,f}(x, x)$]:

For $y = x$

$$\cos N_{2e,f}(x, x) = \cos N_{2e}(x, x)\, chx - \sin N_{2e}(x, x)\, shx$$

$$= \cos 2x - \sin 2x\, sh\, 2x$$

$$\sin N_{2e,f}(x, x) = \sin N_{2e}(x, x)\, chx - \cos N_{2e}(x, x)\, shx$$

$$= \sin 2x + \cos 2x\, sh\, 2x$$

$$\cos N_{2e,f}(x, -x) = \cos N^2(x, x) + \sin N^2(x, x)$$

$$= ch\, 2x$$

$$\sin N_{2e,f}(x, -x) = 2 \sin N(x, x) \cos N(x, x) \qquad (4.2j)$$

$$= \sin 2x\, ch\, 2x$$

For $y = x$, (4.2h) and (4.2i), gives identical result.

Expand expression (4.2h) substituting suitably for the numerator and denominator

$$\tan N_{2e,f}(x, x) = \frac{2 \tan N(x,x)}{1 - \tan N^2(x,x)}$$

$$= \frac{\sin 2x + \cos 2x.sh\, 2x}{\cos 2x - \sin 2x.sh\, 2x} \qquad (4.2k)$$

Expand expression (4.2i) substituting suitably for the numerator and denominator

$$\tan N_{2e,f}(x, x) = \frac{\tan N_{2e}(x, x) + th\,(x)}{1 - \tan_{2e}(x,x).th(x)}$$

Note that

$$\tan N_{2e}(x,\ x) = \frac{\sin N_{2e}(x,x)}{\cos N_{2e}(x,x)} = \frac{\sin 2x.chx + \cos 2x.shx}{\cos 2x.chx - \sin 2x.shx} = \frac{\tan 2x + thx}{1 - \tan 2x.thx}$$

Thus substitute this expression of $\tan N_{2e}(x,\ x)$, we simplify to get

$$\tan N_{2e,f}(x,\ x) = \left[\frac{\sin 2x.chx + \cos 2x.shx}{\cos 2x.chx - \sin 2x.shx} + \frac{sh(x)}{ch(x)}\right]$$

$$\div \left[1 - \frac{\sin 2x.chx + \cos 2x.shx}{\cos 2x.chx - \sin 2x.shx}.\frac{sh(x)}{ch(x)}\right]$$

$$= \frac{\sin 2x + \cos 2x.sh\ 2x}{\cos 2x - \sin 2x.sh\ 2x} = \frac{\tan 2x + sh\ 2x}{1 - \tan 2x.\ sh\ 2x} \qquad (4.2l)$$

It can be easily shown that $N_{2f,e}(x,\ x) = N_{2e,f}(x,\ x)$, and further in view of (4.1d), we finally have:

$$\cos N_{2f,e}(x,\ x) = \cos N_{2e,f}(x,\ x) = \cos N_{3e}(x,\ x) = \cos 2x - \sin 2x\ sh\ 2x$$
$$\sin N_{2f,e}(x,\ x) = \sin N_{2e,f}(x,\ x) = \sin N_{3e}(x,\ x) = \sin 2x + \cos 2x\ sh\ 2x$$

and obviously

$$\tan N_{2f,\ e}(x,\ x) = \tan N_{2e,\ f}(x,\ x)$$
$$= \tan N_{3e}(x,\ x) = \frac{\sin 2x + \cos 2x.sh\ 2x}{\cos 2x - \sin 2x.sh\ 2x} = \frac{\tan 2x + sh\ 2x}{1 - \tan 2x.sh\ 2x} \qquad (4.2m)$$

This fact (4.2m) is evident also from equality of the expressions (4.2k) and (4.2l). Also in view of (4.1d), (4.2m) is equal to $\tan N_{3e}(x,\ x)$.

But it can be noticed that, $N_{2f,e}(x,\ -x) \neq N_{2e,f}(x,\ -x)$.

We have

$$\cos N_{2f,e}(x,\ -x) = 1$$
$$\cos N_{2e,f}(x,\ -x) = ch\ 2x$$

So, $\cos N_{2f,e}(x,\ -x) \neq \cos N_{2e,f}(x,\ -x)$

$$\sin N_{2f,e}(x,\ -x) = sh\ 2x$$
$$\sin N_{2e,f}(x,\ -x) = \sin 2x\ ch\ 2x$$

So, $\sin N_{2f,e}(x,\ -x) \neq \sin N_{2e,f}(x,\ -x)$ (4.2n)

Also using (3.18), for $\cos N_{2f}(x)$ and $\sin N_{2f}(x)$, we can alternatively arrive at the same result (4.2d), as:

Refer to the note following (4.2d).

$$\begin{aligned}\sin N_{3e}(x,\ x) &= \sin N_{3e}(x)\\ &= \sin N_{2f}(x)\cos x + \cos N_{2f}(x)\sin x\\ &= (\sin x + \cos x\ sh\ 2x)\cos x + (\cos x - \sin x\ sh\ 2x)\sin x\\ &= \sin 2x + \cos 2x\ sh\ 2x\end{aligned}$$

$$\begin{aligned}\cos N_{3e}(x,\ x) &= \cos N_{3e}(x)\\ &= \cos N_{2f}(x)\cos x - \sin N_{2f}(x)\sin x\end{aligned}$$

$$= (\cos x - \sin x\ sh\ 2x) \cos x - (\sin x + \cos x\ sh\ 2x) \sin x$$
$$= \cos 2x - \sin 2x\ sh\ 2x \qquad (4.2o)$$

Further

$$\sin N_{3e}(x, -x) = 0$$
$$\cos N_{3e}(x, -x) = ch\ 2x \qquad (4.2p)$$

From (4.2d and 4.2j), we get the following identity

$$\cos N^2_{2e,f}(x, -x) - \sin N^2_{2f,e}(x, -x) = ch^2\ 2x - sh^2\ 2x = 1$$

Also, we get the following identity [viewed according to (1.83)]

$$\cos N_{2f}(x, -x) \cos x + \sin N_{2f}(x, -x) \sin x$$
$$= \cos N(x, x) \cos N(x, -x) + \sin N(x, x) \sin N(x, -x)$$
$$= ch^2\ x - sh^2\ x = 1 \qquad (4.2q)$$

Now from (4.1c) on simplifying

$$\cos N_{3e}(x, y) + \sin N_{3e}(x, y)$$
$$= [\cos(x + y) + \sin(x + y)].ch(x - y)$$
$$+ [\cos(x + y) - \sin(x + y)].sh\ (x + y)$$

$$\cos N_{3e}(x, -y) + \sin N_{3e}(x, -y)$$
$$= [\cos\ (x - y) + \sin\ (x - y)].ch\ (x + y)$$
$$+ [\cos\ (x - y) - \sin\ (x - y)].sh\ (x - y) \qquad (4.3)$$

$$\cos N_{3e}(x, x) + \sin N_{3e}(x, x)$$
$$= (\cos 2x + \sin 2x) + (\cos 2x - \sin 2x)\ sh\ 2x$$

$$\cos N_{3e}(x, -x) + \sin N_{3e}(x, -x) = ch\ 2x \qquad (4.4a)$$

Thus on the strength of (4.2o)

$$\tan N_{3e}(x, x) = \tan N_{3e}(x)$$
$$= [\sin N_{3e}(x)] / [\cos N_{3e}(x)]$$
$$= [\sin 2x + \cos 2x\ sh\ 2x] / [\cos 2x - \sin 2x\ sh\ 2x]$$
$$= [\tan 2x + sh\ 2x] / [1 - \tan 2x\ sh\ 2x] \qquad (4.4b)$$

[It resembles expansion of tan(A + B) formula, where, tan A = tan $2x$, tan B = $sh\ 2x$]

Also, alternatively utilizing (4.2o), and dividing the numerator and denominator by, $\cos N^2 x$

$$\tan N_{3e}(x) = [\sin N_{3e}(x)] / [\cos N_{3e}(x)]$$
$$= [2 \sin Nx \cos Nx] / [\cos N^2 x - \sin N^2 x]$$
$$= 2(\tan Nx) / [1 - \tan N^2 x] \qquad (4.4c)$$

(4.4c) resembles the expansion of the trigonometric formula of tan 2A, where, tan A = tanNx

Also

$$\tan N_{3e}(x, -x) = 0 \qquad (4.4d)$$

Further, call (4.4b) and (3.21b). That is

$$\tan N_{3e}(x) = [\tan 2x + sh\ 2x] / [1 - \tan 2x\ sh\ 2x]$$
$$\tan N_{2f}(x) = [\tan x + sh\ 2x] / [1 - \tan x\ sh\ 2x] \qquad (4.4e)$$

from which

$$\tan 2x + sh\ 2x = \tan N_{3e}(x) - \tan N_{3e}(x).\tan 2x\ sh\ 2x$$
$$\tan x + sh\ 2x = \tan N_{2f}(x) - \tan N_{2f}(x).\tan x\ sh\ 2x$$

rearranging

$$sh\ 2x = [(\tan N_{3e}(x) - \tan 2x) + (\tan x - \tan N_{2f}(x))] / [(\tan N_{3e}(x) \tan 2x) - (\tan x \tan N_{2f}(x))] \quad (4.4f)$$

Additionally, following similar procedure as adopted in getting (1.59), we have

$$\text{cosec}N_{3e}(x) = \frac{\sec N_{2f}(x).\sec x}{\tan N_{2f}(x) + \tan x}$$

$$\sec N_{3e}(x) = \frac{\sec N_{2f}(x).\sec x}{1 - \tan N_{2f}(x).\tan x} \quad (4.5)$$

4.1.4 tan*x* in Terms of $\tan N_{3e}(x)$, and $\tan N_{2f}(x)$, through tan (A – B) Formula

By now, it has been made clear that, the triple Nbic functions $N_{3e}(x, y)$ and $N_{3f}(x, y)$ can be generated in several ways.

Call the very definition of triple Nbic Function of $N_{3e}(x, x)$ from (1.29a). **We will consider this triple Nbic Function $N_{3e}(x, x)$ for further discussion and reference**, particularly, for the **triple Nbic circle** [Chapter 6].

We have

$$\sin N_{3e}(x) = \sin N_{2f}(x) \cos x + \cos N_{2f}(x) \sin x \quad (4.6a)$$
$$\cos N_{3e}(x) = \cos N_{2f}(x) \cos x - \sin N_{2f}(x) \sin x \quad (4.6b)$$

Multiply (4.6b) by $\cos N_{2f}(x)$, and (4.6a) by $\sin N_{2f}(x)$, and add to yield

$$[\cos N_{2f}^2(x) + \sin N_{2f}^2(x)] \cos x = \cos N_{3e}(x) \cos N_{2f}(x) + \sin N_{3e}(x) \sin N_{2f}(x)$$

or,

$$[ch^2\ 2x] \cos x = \cos N_{3e}(x) \cos N_{2f}(x) + \sin N_{3e}(x) \sin N_{2f}(x) \quad (4.6c)$$

Multiply (4.6b) by $\sin N_{2f}(x)$ and (4.6a) by $\cos N_{2f}(x)$, and subtract to yield

$$[\cos N_{2f}^2(x) + \sin N_{2f}^2(x)] \sin x = \sin N_{3e}(x) \cos N_{2f}(x) - \cos N_{3e}(x) \sin N_{2f}(x)$$

or,

$$[ch^2\ 2x] \sin x = \sin N_{3e}(x) \cos N_{2f}(x) - \cos N_{3e}(x) \sin N_{2f}(x) \quad (4.6d)$$

where, according to (3.19a), $\cos N_{2f}^2(x) + \sin N_{2f}^2(x) = ch^2\ 2x$

Divide (4.6d) by (4.6c)

$$\tan x = [\sin N_{3e}(x) \cos N_{2f}(x) - \cos N_{3e}(x) \sin N_{2f}(x)] / [\cos N_{3e}(x) \cos N_{2f}(x) + \sin N_{3e}(x) \sin N_{2f}(x)]$$

Divide numerator and denominator by, $[\cos N_{3e}(x) \cos N_{2f}(x)]$, to get

$$\tan x = [\tan N_{3e}(x) - \tan N_{2f}(x)] / [1 + \tan N_{3e}(x) \tan N_{2f}(x)] \quad (4.7)$$

It can be easily identified that, (4.7) is of the same form as that of, tan (A – B) formula of plane trigonometry.

Now

$$\begin{aligned}
&\cos N_{3e}^2(x) + \sin N_{3e}^2(x) \\
&\quad = (\cos^2 2x + \sin^2 2x\, sh^2 2x - 2\cos 2x \sin 2x\, ch\, 2x) \\
&\quad\quad + (\sin^2 2x + \cos^2 2x\, sh^2 2x + 2\cos 2x \sin 2x\, ch\, 2x) \qquad (4.8a) \\
&\quad = (1 + sh^2 2x) = ch^2 2x \\
&\cos N_{3e}^2(x) - \sin N_{3e}^2(x) \\
&\quad = (\cos^2 2x + \sin^2 2x\, sh^2 2x - 2\cos 2x \sin 2x\, ch\, 2x) \\
&\quad\quad - (\sin^2 2x + \cos^2 2x\, sh^2 2x + 2\cos 2x \sin 2x\, ch\, 2x) \\
&\quad = (1 - sh^2 2x)\cos 4x - 2\sin 4x\, sh\, 2x \qquad (4.8b)
\end{aligned}$$

4.1.5 Single Nbic Function N(*x*, *x*), in terms Triple Nbic Function $N_{3e}(x, x)$

Let us calculate the values of sinN*x* and cosN*x*, and hence, tanN*x*, when the values of $\sin N_{3e}(x)$ and $\cos N_{3e}(x)$, are given.

Let, $\sin N_{3e}(x) = a$, and $\cos N_{3e}(x) = b$ (given).

In the light of (4.1d), the two criteria when viewed:

The first criterion is

$$2 \sin Nx \cos Nx = \sin N_{3e}(x) = b \qquad (4.9a)$$

$$\Rightarrow \sin N^2 x = b^2/(4 \cos N^2 x)$$

The second criterion is

$$\cos N^2 x - \sin N^2 x = \cos N_{3e}(x) = a \qquad (4.9b)$$

Substitute the value of $\sin N^2 x$, from the first criterion (4.9a), to the second criterion (4.9b), to get

$$\begin{aligned}
&\cos N^2 x - \sin N^2 x = \cos N_{3e}(x) = a \\
\Rightarrow\ &\cos N^2 x - b^2/(4 \cos N^2 x) = a \\
\Rightarrow\ &4 \cos N^4 x - 4a \cos N^2 x - b^2 = 0
\end{aligned}$$

Solving for $\cos N^2 x$,

$$\cos N^2 x = \frac{1}{2}\left[a \pm \sqrt{a^2 + b^2}\right] = \frac{1}{2}[a \pm ch2x], \text{ (take the +ve value)} \qquad (4.9c)$$

$$\Rightarrow \cos Nx = \pm\sqrt{\frac{a \pm ch\, 2x}{2}} \qquad (4.9d)$$

Since

$$\cos N^2 x + \sin N^2 x = ch\, 2x \qquad (4.9e)$$

$$\begin{aligned}
\sin N^2 x &= ch\, 2x - \cos N^2 x \\
&= ch\, 2x - \frac{1}{2}a \mp \frac{ch\, 2x}{2} \\
&= \frac{1}{2}[3ch\, 2x - a], \text{ or, } \frac{1}{2}[ch\, 2x - a]
\end{aligned}$$

From which

$$\sin\text{N}x = \pm\sqrt{\frac{3ch2x-a}{2}}, \text{ or, } \pm\sqrt{\frac{ch2x-a}{2}}, \qquad (4.9f)$$

Thus with sinNx and cosNx, get tanNx as

$$\tan\text{N}x = (\sin\text{N}x)/(\cos\text{N}x) \qquad (4.9g)$$

[Used here, sinNx from (4.9f), and cosNx, from (4.9d)].

Alternatively, we have

$$\tan\text{N}_{3e}(x) = [\sin\text{N}_{3e}(x)]/[\cos\text{N}_{3e}(x)] = b/a = c \qquad (4.10a)$$

Also we have from (4.4c)

$$\tan\text{N}_{3e}(x) = [2\tan\text{N}x]/[1 - \tan\text{N}^2 x] = c \qquad (4.10b)$$

$$\Rightarrow c\tan\text{N}^2 x + 2\tan\text{N}x - c = 0$$

Solving the quadratic equation for, tanNx

$$\tan\text{N}x = [-1 \pm (\sqrt{1+c^2})] / c \qquad (4.10c)$$

4.1.6 Similarities and Dissimilarities between $N_{2f}(x)$ and $N_{3e}(x)$ Functions

Similarities

Refer to the equation (4.2o) for $\sin N_{3e}(x)$ and $\cos N_{3e}(x)$, and (4.4b) for $\tan N_{3e}(x)$. The corresponding functions, $\sin N_{2f}(x)$ and $\cos N_{2f}(x)$, are given in (3.18), and $\tan N_{2f}(x)$ in (3.21b). There exist striking similarities in the mathematical structure between the corresponding pairs of, [$\sin N_{2f}(x)$ and $\sin N_{3e}(x)$], [$\cos N_{2f}(x)$ and $\cos N_{3e}(x)$], and, [$\tan N_{2f}(x)$ and $\tan N_{3e}(x)$]. The formulas are recalled here for comparison, as

$$\begin{aligned}
\sin\text{N}_{2f}(x) &= \sin x + \cos x\; sh\, 2x \\
\sin\text{N}_{3e}(x) &= \sin 2x + \cos 2x\; sh\, 2x \\
\cos\text{N}_{2f}(x) &= \cos x - \sin x\; sh\, 2x \\
\cos\text{N}_{3e}(x) &= \cos 2x - \sin 2x\; sh\, 2x \\
\tan\text{N}_{2f}(x) &= [\tan x + sh\, 2x]/[1 - \tan x\; sh\, 2x] \\
\tan\text{N}_{3e}(x) &= [\tan 2x + sh\, 2x]/[1 - \tan 2x\; sh\, 2x]
\end{aligned} \qquad (4.11a)$$

The comparison in (4.11a) reveals that: [$\sin x$, $\cos x$, $\tan x$] of double Nbic function [$N_{2f}(x)$], when replaced simply by [$\sin 2x$, $\cos 2x$, $\tan 2x$], respectively, the corresponding triple Nbic function [$N_{3e}(x)$], results. That is

$$[\sin x, \cos x, \tan x] \text{ of } N_{2f}(x) \Leftrightarrow [\sin 2x, \cos 2x, \tan 2x] \text{ of } N_{3e}(x) \qquad (4.11b)$$

Dissimilarities

The follwing do not maintain any similarities:

$$\begin{aligned}
&\sin\text{N}_{2f}(x, -x) = \sin x\; ch\, 2x, \text{ but, } \sin\text{N}_{3e}(x, -x) = 0 \\
&\cos\text{N}_{2f}(x, -x) = \cos x\; ch\, 2x, \text{ but, } \cos\text{N}_{3e}(x, -x) = ch\, 2x
\end{aligned} \qquad (4.12)$$

Some identities

We have these identities

For Hyperbolic function : $ch^2 x + sh^2 x = ch\, 2x$

CHAPTER 4

For single Nbic function : $\cos N^2 x + \sin N^2 x = ch\ 2x$

For triple Nbic function : $\cos N_{3e}{}^2(x) + \sin N_{3e}{}^2(x) = ch^2\ 2x$ (4.13)

So that

$$\cos N_{3e}(x, x) + \sin N_{3e}(x, x) = (\cos 2x + \sin 2x) + (\cos 2x - \sin 2x)\ sh\ 2x$$
$$\cos N_{3e}(x, -x) + \sin N_{3e}(x, -x) = ch\ 2x$$
$$\cos N_{3e}^2(x) + \sin N_{3e}^2(x) = (\cos N^2 x + \sin N^2 x)^2$$
$$= (ch^2 x + sh^2 x)^2 = ch^2 2x$$
$$= (\cos N^2 x + \sin N^2 x)(ch^2 x + sh^2 x)$$
$$\text{sqrt}\ [\cos N_{3e}^2(x) + \sin N_{3e}^2(x)]$$
$$= [\cos N_{2f}(x, -x) + \sin N_{2f}(x, -x)]/[\cos x + \sin x] \quad (4.14)$$

Alternatively

$$\cos x + \sin x = [\cos N_{2f}(x, -x) + \sin N_{2f}(x, -x)] \, / \, \text{sqrt}\ [\cos N_{3e}^2(x) + \sin N_{3e}^2(x)] \quad (4.15)$$

[read: sqrt( ) ⇒ square root of ( )]

This expression (4.15), is the relation of plane circular function written in terms of triple Nbic $[N_{3e}(x)]$, and double Nbic functions $[N_{2f}(x)]$.

Collecting formulas together

$$\sin N_{3e}(x) = \sin 2x + \cos 2x\ sh\ 2x$$
$$= \sin Nx \cos Nx + \cos Nx \sin Nx = 2 \sin Nx \cos Nx$$
$$= \sin N_{2f}(x) \cos x + \cos N_{2f}(x) \sin x$$
$$\cos N_{3e}(x) = \cos 2x - \sin 2x\ sh\ 2x$$
$$= \cos Nx \cos Nx - \sin Nx \sin Nx = \cos N^2 x - \sin N^2 x$$
$$= \cos N_{2f}(x) \cos x - \sin N_{2f}(x) \sin x$$
$$\tan N_{3e}(x) = [\tan 2x + sh\ 2x]/[1 - \tan 2x\ sh\ 2x]$$
$$= [2 \tan Nx]/[1 - \tan N^2 x]$$
$$\text{cosec} N_{3e}(x, x) = \frac{\sec N_{2f}(x).\sec x}{\tan N_{2f}(x) + \tan x}$$
$$\sec N_{3e}(x) = \frac{\sec N_{2f}(x).\sec x}{1 - \tan N_{2f}(x).\tan x}$$
$$\tan N_{3e}(x, -x) = 0$$
$$sh\ 2x = [(\tan N_{3e}(x) - \tan 2x) + (\tan x - \tan N_{2f}(x))] \, / \, [(\tan N_{3e}(x) \tan 2x) - (\tan x \tan N_{2f}(x)]$$
$$\cos N_{3e}^2(x) + \sin N_{3e}{}^2(x) = ch^2\ 2x \quad (4.16)$$

[**Note:** Three sets of formulas, each for $\sin N_{3e}(x)$ and $\cos N_{3e}(x)$ of (4.16), have been verified through numerical calculation in **Example 6.4** of Chapter 6.]

Numerical Examples

In the following examples, some of the derived mathematical expressions are verified with numerical data.

Example 4.1 (Equation 4.13)

Verify that, $\cos N_{3e}^2(x) + \sin N_{3e}^2(x) = ch^2\, 2x$
with, $x = 1.0472$ rad = 60 deg

Solution

Given that, $x = 1.0472$ rad (= 60 deg)

$$\sin x = 0.867,\ \cos x = 0.5,\ \tan x = 1.732$$
$$\sin 2x = 0.866,\ \cos 2x = 0.5$$
$$shx = 1.2494,\ chx = 1.6,\ thx = 0.7809$$
$$sh\, 2x = 3.9981,\ ch\, 2x = 4.1212$$
$$\sin Nx = 2.0106,\ \cos Nx = -\ 0.2818,\ \tan Nx = -7.1348$$

Using the relevant data in (4.11a)

$$\sin N_{3e}(x) = \sin 2x + \cos 2x\ sh\, 2x$$
$$= 0.866 + 0.5 \times 3.9981 = 2.8651$$
$$\cos N_{3e}(x) = \cos 2x - \sin 2x\ sh\, 2x$$
$$= 0.5 - 0.866 \times 3.9981 = -\ 2.9624$$

LHS of (4.13) $= \cos N_{3e}^2(x) + \sin N_{3e}^2(x)$
$= 8.2085 + 8.7755 = \mathbf{16.9840}$

RHS of (4.13) $= ch^2\, 2x = 4.1212 \times 4.1212 = \mathbf{16.9843}$

So, $\cos N_{3e}^2(x) + \sin N_{3e}^2(x) = ch^2\, 2x$ **(Checked)**

Example 4.2 (Equation 4.4f)

Verify (4.4f) that,

$$sh\, 2x = [(\tan N_{3e}(x) - \tan 2x) + (\tan x - \tan N_{2f}(x))] \,/\, [(\tan N_{3e}(x) \tan 2x) - (\tan x \tan N_{2f}(x))]$$

with, $x = 2.0944$, rad = 120 deg.

Solution

Given that

$$x = 2.0944,\ \text{rad} = 120\ \text{deg}$$
$$\tan x = -\ 1.73205,\ \tan 2x = 1.7321,\ sh\, 2x = \mathbf{32.9639}$$
$$\tan Nx = -0.28427,\ \tan N_{2f}(x) = 0.5376,\ \tan N_{3e}(x) = -\ 0.61852$$

Numerator of the RHS of (4.4f)

$$= [(\tan N_{3e}(x) - \tan 2x) + (\tan x - \tan N_{2f}(x))]$$
$$= (-\ 0.61852 - 1.7321 - 1.7321 - 0.5376) = -\ 4.62027$$

Denominator of the RHS of (4.4f)

$$= [(\tan N_{3e}(x) \tan 2x) - (\tan x \tan N_{2f}(x))]$$
$$= (-\ 0.61852)(1.7321) - (-\ 1.7321)(0.5376) = -\ 0.14019$$

Then the RHS of (4.4f)

$= (-\ 4.62027)/(-\ 0.14019) = \mathbf{32.9572} = sh\, 2x$ **(Checked)**

Example 4.3 (Equation 4.1d, 4.16)

Verify (4.1d, 4.16) that

$$\sin N_{3e}(x) = 2 \sin Nx . \cos Nx$$

$$\sin N_{3e}(x) = \sin N_{2f}(x)\cos x + \cos N_{2f}(x)\sin x$$
$$\cos N_{3e}(x) = \cos N^2 x - \sin N^2 x$$
$$\cos N_{3e}(x) = \cos N_{2f}(x)\cos x - \sin N_{2f}(x)\sin x$$

with, $x = -2.618$ rad $= -150$ deg

Solution

Given that

$$x = -2.618 \text{ rad} = -150 \text{ deg}$$
$$\sin x = -0.5, \cos x = -0.866$$
$$\sin Nx = 2.4589, \cos Nx = -9.3762,$$
$$\sin N_{2f}(x) = 80.8671, \cos N_{2f}(x) = -47.8434$$
$$\sin N_{3e}(x) = \mathbf{-46.1113}, \cos N_{3e}(x) = \mathbf{81.6669}$$

So

$$2\sin Nx .\cos Nx$$
$$= 2(2.4589)(-9.3762) = \mathbf{-46.1103} = \sin N_{3e}(x)$$
$$\sin N_{2f}(x)\cos x + \cos N_{2f}(x)\sin x$$
$$= (80.8671)(-0.866) + (-47.8434)(-0.5) = -46.1092 = \sin N_{3e}(x)$$
$$\cos N^2 x - \sin N^2 x$$
$$= (-9.3762)^2 - (2.4589)^2 = 81.6669 = \cos N_{3e}(x)$$
$$\cos N_{2f}(x)\cos x - \sin N_{2f}(x)\sin x$$
$$= (-47.8434)(-0.866) - (-80.8671)(-0.5) = \mathbf{81.8670}$$
$$= \cos N_{3e}(x)$$

(Checked)

Example 4.4 (Equation 4.10c)

Given the values of, $\sin N_{3e}(x)$, and $\cos N_{3e}(x)$, find the value of $\tan Nx$, and hence, the values of, $\sin Nx$ and $\cos Nx$

Using the given data:

$$x = 150 \text{ deg } (= 2.618 \text{ rad})$$
$$\cos N_{3e}(x) = 81.8671, \sin N_{3e}(x) = 46.1113$$

Solution

We have

$$a = 81.8671,\ b = 46.1113,\ c = b/a = 46.1113/81.8671 = 0.56325$$

Call (4.10c), and substitute the value of c in it

$$\tan Nx = [-1 \pm (\sqrt{1+c^2})]/c$$
$$= [-1 \pm (\sqrt{1+0.56325^2})]/0.56325$$
$$= [-1 \pm 1.1477]/0.56325$$
$$= -3.8181, +0.2625$$

Now take, $\tan Nx = +0.2625$ (say), and $ch\ 2x = 93.96$

It is now required to find the values of $\sin Nx$, and $\cos Nx$, from that of $\tan Nx$. [Refer to the derivation in this regard provived earlier in (1.53e) and (1.53f).]

We have

the first criterion is

$$\tan Nx = (\sin Nx) / (\cos Nx) = 0.2625$$
$$\Rightarrow \sin Nx = 0.2625 \cos Nx$$

or, $$\sin N^2 x = [0.2625 \cos Nx]^2$$

This expression of $\sin N^2 x$, is substituted in the second criterion, namely

$$\cos N^2 x + \sin N^2 x = ch\ 2x$$

to give

$$\Rightarrow \cos N^2 x\, [1 + 0.2625^2] = 93.96$$
$$\Rightarrow \cos Nx = \pm 9.3757, \text{ (actual value } = -9.3762)$$

and, $$\sin Nx = 0.2625 \cos Nx$$
$$= \pm 2.4611, \text{ (actual value is } = -2.4589)$$

Example 4.5 (Equations 1.59, 3.21a-b, 4.4b-c, 4.7)

Check the following Problems of tan-formulas through the tan-equivalent trigonometric formula

$$\tan\,(A + B) = \frac{\tan A + \tan B}{1 - \tan A.\tan B}, \text{ for } x = 150 \text{ deg } (= 2.618 \text{ rad})$$

where, A and B are the equivalent angles, (obtained from tan A and tan B, respectively), if the RHS of the following Problems are:

Problem 1: (1.59) : $\tan Nx = [\tan x + thx] / [1 - \tan x.thx]$

Problem 2a: (3.21a) : $\tan N_{2f}(x) = [\tan Nx + thx] / [1 - \tan Nx.thx]$

Problem 2b: (3.21b) : $\tan N_{2f}(x) = [\tan x + sh\ 2x] / [1 - \tan x.sh\ 2x]$

Problem 3a: (4.4c) : $\tan N_{3e}(x) = [2 \tan Nx] / [1 - \tan N^2 x]$

Problem 3b: (4.4b) : $\tan N_{3e}(x) = [\tan 2x + sh\ 2x] / [1 - \tan 2x + sh\ 2x]$

Problem 4: (4.7) : $\tan x = [\tan N_{3e}(x) - \tan N_{2f}(x)] / [1 + \tan N_{3e}(x).\tan N_{2f}(x)]$

Solution

Since the given various tan-formulas are written with: $\tan x$, $\sin 2x$, $\tan 2x$, thx, $\tan Nx$, $\tan N_{2f}(x)$, and $\tan N_{3e}(x)$, they need to be calculated for the given values of, $x = 150$ deg (2.618 rad), beforehand, and will be utilized whenever called for.

The RHS and LHS of the equivalent trigonometric formula are, tan (A + B), and $\frac{\tan A + \tan B}{1 - \tan A.\tan B}$, respectively,

where, A and B are the equivalent angles derived from, tan A and tan B, respectively.

Given data:

$$x = 2.618 \text{ rad} = 150 \text{ deg}$$

use (p, q, r : as the equivalent angles)

$$\sin x = 0.5$$
$$\cos x = -0.866$$

CHAPTER 4

$\tan x = -0.577$, [x = 150 deg = 2.618 rad]
$\sin 2x = -0.866$
$\cos 2x = 0.5$
$\tan 2x = -1.732$, [= tan p : p = – 60 deg = – 1.0472 rad]
$shx = 6.8136$
$chx = 6.8906$
$thx = 0.9894$, [= tan p : p = 44.6947 deg = 0.7801 rad]
$sh\ 2x = 93.9547$, [= tan q : q = 89.39 deg = 1.5602 rad]
$ch\ 2x = 93.96$
$th\ 2x = 0.9999$, [= tan p : p = 44.997 deg = 0.7853 rad]
$\sin Nx = -2.4589$
$\cos Nx = -9.3762$
$\tan Nx = 0.2623$, [= tan p : p = 14.7083 deg = 0.2567 rad]
$\sin N_{2f}(x) = -80.8671$
$\cos N_{2f}(x) = -47.8434$
$\tan N_{2f}(x) = 1.6902$, [= tan p : p = 59.1991 deg = 1.0366 rad]
$\sin N_{3e}(x) = 46.1113$
$\cos N_{3e}(x) = 81.8670$
$\tan N_{3e}(x) = 0.56325$, [= tan p : p = 29.39 deg = 0.5139 rad]

Thus

for Problem 1:

$\tan Nx = [\tan x + thx]/[1 - \tan x\ thx]$
tan A = $\tan x = -0.577$, A = 150 deg
tan B = $thx = 0.9894$, B = 44.6947 deg
A + B = 194.6947 deg, tan (A + B) = **0.2622**
LHS of Problem 1: $\tan Nx$ = tan (A + B) = **0.2622**
RHS of Problem 1: $[\tan x + thx]/[1 - \tan x.thx]$
= [– 0.577 + 0.9894]/[1 – (– 0.577) (0.9894)] = **0.2622**
Hence **LHS = RHS.**

for Problem 2a:

$\tan N_{2f}(x) = [\tan Nx + thx]/[1 - \tan Nx.thx]$
tan A = $\tan Nx$ = 0.2623, A = 14.5044 deg
tan B = thx = 0.9894, B = 44.6947 deg
A = 14.5044 deg, B = 44.6947 deg
A + B = 59.1991 deg, tan (A + B) = **1.6902**
LHS of Problem 2a: $\tan N_{2f}(x)$ = tan (A + B) = **1.69026**
RHS of Problem 2a: $[\tan Nx + thx]/[1 - \tan Nx\ .thx]$
= [0.2623 + 0.9894]/[1 – (0.2623) (0.9894)] = **1.69028**
Hence **LHS = RHS.**

for Problem 2b:

$\tan N_{2f}(x) = [\tan x + sh\ 2x]/[1 - \tan x.sh\ 2x]$
tan A = $\tan x$ = – 0.577, A = 150 deg

tan B = $sh\ 2x$ = 93.9558, B = 89.39 deg

A = 150 deg, B = 89.3902 deg

A + B = 239.3902 deg, tan (A + B) = **1.69025**

LHS of Problem 2b: $\tan N_{2f}(x)$ = tan (A + B) = **1.69025**

RHS of Problem 2b: [tanNx + thx] / [1 – tanNx.thx]

= [– 0.577 + 93.9547] / [1 – (– 0.577) (93.9547)] = **1.69126**

Hence **LHS = RHS.**

for Problem 3a:

$\tan N_{3e}(x)$ = [2 tanNx]/[1 – $\tan N^2 x$]

tan A = tan B = tanNx = 0.2623, A = 14.6976 deg

2A = 29.3952 deg

A + B = 2A = 29.3952 deg, tan 2A = **0.56336**

LHS of Problem 3a: $\tan N_{3e}(x)$ = **0.56325**

RHS of Problem 3a: [2 × 0.2623]/[1 – (0.2623 × 0.2623)] = **0.56340**

Hence **LHS = RHS.**

for Problem 3b:

$\tan N_{3e}(x)$ = [tan $2x$ + $sh\ 2x$]/[1 – tan $2x.sh\ 2x$]

tan A = tan $2x$ = – 1.732, A = – 60 deg

tan B = $sh\ 2x$ = 93.9558, B = 89.39 deg

A = – 60 deg, B = 89.3902 deg

A + B = 29.39 deg, tan(A + B) = **0.56324**

LHS of Problem 3b: $\tan N_{3e}(x)$ = **0.56325**

RHS of Problem 3b: = [–1.732 + 93.9547]

/ [1 – (– 1.732) (93.9547)] = **0.56326**

Hence **LHS = RHS.**

for Problem 4:

tan$x$ = [$\tan N_{3e}(x)$ – $\tan N_{2f}(x)$]/[1 – $\tan N_{3e}(x)$ $\tan N_{2f}(x)$]

tan A = $\tan N_{3e}(x)$ = 0.56325, A = 29.39 deg

tan B = $\tan N_{2f}(x)$ = 1.69025, B = 59.39 deg

A = 29.39 deg, B = 59.39 deg

A – B = – 30 deg, tan (A – B) = **– 0.577**

LHS of Problem 4: tan$x$ = **– 0.577**

RHS of Problem 4: = [0.56325 – 1.69025]

/ [1 + (0.56325) (1.69025)] = **– 0.577**

Hence **LHS = RHS.**

4.1.7 Complex Triple Nbic Function Ratio, R_{ntn}

In (2.3d), (2.5j), (2.24h) and (3.31), the expressions of the Complex circular function ratio (R_{nci}), Complex hyperbolic function ratio (R_{nhy}), Complex single Nbic function ratio (R_{nsn}), and Complex double Nbic function ratio (R_{ndn}), are given.

Following similar reasoning and procedure, the Complex triple Nbic function ratio, R_{ntn}, is derived as follows.

$$R_{ntn} = [\cos N_{3e}(nx) + i \sin N_{3e}(nx)] / [\cos N_{3e}(x) + i \sin N_{3e}(x)]^n \quad (4.17)$$

To simplify the numerator and the denominator, use (4.16), that is

$$\begin{aligned} \cos N_{3e}(x) &= \cos Nx \cos Nx - \sin Nx \sin Nx \\ &= \cos N^2 x - \sin N^2 x \\ &= \cos 2x - \sin 2x \; sh\, 2x \\ \sin N_{3e}(x) &= \sin Nx \cos Nx + \cos Nx \sin Nx \\ &= 2 \sin Nx \cos Nx \\ &= \sin 2x + \cos 2x \; sh\, 2x \end{aligned}$$

Thus

$$\begin{aligned} \cos N_{3e}(nx) &= \cos 2nx - \sin 2nx \; sh\, 2nx \\ i \sin N_{3e}(nx) &= i \sin 2nx + i \cos 2nx \; sh\, 2nx \\ \cos N_{3e}(x) &= \cos 2x - \sin 2x \; sh\, 2x \\ i \sin N_{3e}(x) &= i \sin 2x + i \cos 2x \; sh\, 2x \end{aligned}$$

Adding, the numerator of (4.17)

$$\begin{aligned} \cos N_{3e}(nx) + i \sin N_{3e}(nx) &= [\cos 2nx + i \sin 2nx] \\ &\quad \times [1 + i\; sh\, 2nx] = e^{2inx} [f^{\,inx}]^2 \end{aligned}$$

Adding, the denominator of (4.17)

$$\begin{aligned} [\cos N_{3e}(x) + i \sin N_{3e}(x)]^n &= [\cos 2x + i \sin 2x]^n \\ &\quad \times [1 + i\; sh\, 2x]^n = (e^{2ix})^n [f^{\,2ix}]^n \end{aligned}$$

So

$$\begin{aligned} R_{ntn} &= [\cos N_{3e}(nx) + i \sin N_{3e}(nx)] / [\cos N_{3e}(x) + i \sin N_{3e}(x)]^n \\ &= (1 + i\; sh\, 2nx) / (1 + i\; sh\, 2x)^n \\ &= [f^{\,inx}]^2 / [f^{\,2ix}]^n \end{aligned} \quad (4.18)$$

It is clearly seen that, **the Complex Nbic function ratios, R_{ntn} (for triple Nbic function), R_{ndn} (for double Nbic function), and, R_{nsn} (for single Nbic function), are identical.**

That is

$$R_{ntn} = R_{ndn} = R_{nsn} = [f^{\,inx}]^2 / [f^{\,2ix}]^n \quad (4.19)$$

4.1.8 Triple Nbic Function, $N_{3f}(x, y)$:

$N_{3f}(x, y)$:

The following type (F_2 type) of triple Nbic function are generated from the product of, $N_{2f}(x, y)$ and, $f^{\pm iy}$. That is, from (1.28f):

$$\begin{aligned} \cos N_{3f}(x, y) &= \cos N_{2f}(x, x)\; chy - \sin N_{2f}(x, x)\; shy \\ \sin N_{3f}(x, y) &= \sin N_{2f}(x, x)\; chy + \cos N_{2f}(x, x)\; shy \end{aligned} \quad (4.20a)$$

To simplify further

$$\begin{aligned} \cos N_{3f}(x, y) &= \cos N_{2f}(x, x)\; chy - \sin N_{2f}(x, x)\; shy \\ &= [\cos x - \sin x \; sh\, 2x]\; chy - [\sin x + \cos x \; sh\, 2x]\; shy \\ &= [\cos x \; chy - \sin x \; shy] \\ &\quad - sh2x.[\sin x \; chy + \cos x \; shy] \\ &= \cos N(x, y) - \sin N(x, y)\; sh\, 2x \\ \sin N_{3f}(x, y) &= \sin N_{2f}(x, x)\; chy + \cos N_{2f}(x, x)\; shy \\ &= [\sin x + \cos x \; sh\, 2x]\; chy + [\cos x - \sin x \; sh\, 2x]\; shy \end{aligned}$$

$$= [\sin x \; ch y + \cos x \; sh y] + sh\, 2x.[\cos x \; ch y - \sin x \; sh y]$$

$$= \sin N(x, y) + \cos N(x, y) \; sh\, 2x$$

$$\tan N_{3f}(x, y) = \frac{\sin N_{3f}(x, y)}{\cos N_{3f}(x, y)}$$

$$= \frac{\sin N(x, y) + \cos N(x, y).sh\, 2x}{\cos N(x, y) - \sin N(x, y).sh\, 2x}$$

Dividing both numerator and denominator by, $\cos N(x, y)$, and noting that,

$$\tan N(x, y) = [\sin N(x, y)] / [\cos N(x, y)], \text{ we get}$$

$$\tan N_{3f}(x, y) = \frac{\tan N(x, y) + sh\, 2x}{1 - \tan N(x, y).sh\, 2x} \qquad (4.20b)$$

This expression has the same form of, tan (A + B) formula, where, tan A and tan B are taken as $\tan N(x, y)$, and $sh\, 2x$, respectively.

4.1.9 Particular Case

for $y = x$

$$\sin N_{3f}(x, x) = \sin Nx + sh\, 2x \cos Nx$$

$$\cos N_{3f}(x, x) = \cos Nx - sh\, 2x \sin Nx$$

$$\tan N_{3f}(x) = \frac{\tan N(x) + sh\, 2x}{1 - \tan N(x).sh\, 2x}$$

$$\text{cosec} N_{3f}(x) = \frac{\sec N(x)}{\tan N(x) + sh\, 2x}$$

$$\sec N_{3f}(x) = \frac{\sec N(x)}{1 - \tan N(x).sh\, 2x} \qquad (4.21a)$$

$$\sin N_{3f}(x).\cos N_{3f}(x)$$

$$= \sin Nx.\cos Nx + sh\, 2x \cos N^2 x - sh^2\, 2x \sin N^2 x - sh^2\, 2x.\sin Nx.\cos Nx$$

$$= (1 - sh^2\, 2x) \sin Nx.\cos Nx + sh\, 2x (\cos N^2 x - \sin N^2 x)$$

Recollect the formulas, namely

$$\sin N_{3e}(x) = 2 \sin Nx \cos Nx$$

$$\cos N_{3e}(x) = \cos N^2 x - \sin N^2 x$$

$$\cos N^2 x + \sin N^2 x = ch\, 2x$$

and applying them suitably, the rearrangement results

$$2 \sin N_{3f}(x).\cos N_{3f}(x) = (1 - sh^2\, 2x) \sin N_{3e}(x) + 2\, sh\, 2x.\cos N_{3e}(x)$$

$$\cos N_{3f}^2(x) + \sin N_{3f}^2(x) = (\cos N^2 x + \sin N^2 x)(1 + sh^2\, 2x)$$

$$= (\cos N^2 x + \sin N^2 x)\, ch^2\, 2x$$

$$= ch^3\, 2x \qquad (4.21b)$$

$$\cos N_{3f}^2(x) - \sin N_{3f}^2(x)$$

$$= (\cos N^2 x - \sin N^2 x)(1 - sh^2\, 2x) - 4 \sin Nx.\cos Nx.sh\, 2x$$

$$= (1 - sh^2\, 2x) \cos N_{3e}(x) - 2\, sh\, 2x.\sin N_{3e}(x) \qquad (4.21c)$$

Thus the ratio

$$\frac{\cos N_{3f}^2(x)+\sin N_{3f}^2(x)}{\cos N_{3e}^2(x)+\sin N_{3e}^2(x)} = ch\ 2x \tag{4.21d}$$

$$\frac{\cos N_{3f}^2(x)-\sin N_{3f}^2(x)}{\cos N_{3e}^2(x)-\sin N_{3e}^2(x)} = \frac{(1-sh^2 2x).\cos N_{3e}(x)-2sh\ 2x.\sin N_{3e}(x)}{(1-sh^2 2x).\cos 4x-2sh\ 2x.\sin 4x} \tag{4.21e}$$

A comparison of the numerator and denominator of the RHS expression of (4.21e) shows that, $\sin N_{3e}(x)$, and $\cos N_{3e}(x)$ of the numerator are simply replaced by, sin $4x$, and cos $4x$, respectively, in the denominator.

Also the following ratio can be written in, tan-form as

$$\frac{2\sin N_{3f}(x).\cos N_{3f}(x)}{\cos N_{3f}^2(x)-\sin N_{3f}^2(x)} = \frac{(1-sh^2 2x).\sin N_{3e}(x)+2sh\ 2x.\cos N_{3e}(x)}{(1-sh^2 2x).\cos N_{3e}(x)-2sh\ 2x.\sin N_{3e}(x)} \tag{4.22a}$$

Divide the numerator and denominator of the LHS by, $\cos N_{3f}^2(x)$, and the numerator and denominator of the RHS by, $(1 - sh^2\ 2x).\cos N_{3f}^2(x)$, to get

$$\frac{2\tan N_{3f}(x)}{1-\tan N_{3f}^2(x)} = \frac{\tan N_{3e}(x)+\dfrac{2sh\ 2x}{1-sh^2 2x}}{1-\tan N_{3e}(x).\dfrac{2sh\ 2x}{1-sh^2 2x}} \tag{4.22b}$$

Notice that, both the LHS and RHS of this expression (4.22b), are of the same form of the plane trigonometric tan-formula.

The former one tallies with the formula

$$\tan 2A = [2 \tan A] / [1 - \tan^2 A]$$

where, $\tan A \Rightarrow \tan N_{3f}(x)$,

where as, the later one tallies with the formula

$$\tan (C + D) = [\tan C + \tan D] / [1 - \tan C \tan D]$$

where, $\tan C \Rightarrow \tan N_{3e}(x)$, $\tan D \Rightarrow 2\ sh\ 2x/(1 - sh^2\ 2x)$

Numerical Examples

In the following examples, some of the derived mathematical expressions are verified with numerical data:

Example 4.6 (Equation 4.21b)

Verify that, $\cos N_{3f}^2(x) + \sin N_{3f}^2(x) = ch^3\ 2x$

with, $x = 3.665191$ rad = 210 deg

Solution

Given that, $x = 3.665191$ rad (= 210 deg)

We have from (4.21a)

$$\cos N_{3f}(x) = \cos Nx - \sin Nx\ sh\ 2x$$
$$\sin N_{3f}(x) = \sin Nx + \cos Nx\ sh\ 2x$$

Now, for $x = 3.665191$ rad (= 210 deg)

$sh\ 2x = 762.9826$, $ch\ 2x =$ **762.9833**

$\cos Nx = -7.166622$, $\sin Nx = -26.676259$

$$\cos N_{3f}(x) = -7.166622 - (762.9826)(-26.676259) = 20346.35483$$
$$\sin N_{3f}(x) = -26.676259 + (762.9826)(-7.166622) = -5494.684146$$

For the given expression

$$\text{LHS} = [20346.35483]^2 + [-5494.684146]^2 = 444165708.8$$
$$[\text{LHS}]^{1/3} = [444165708.8]^{1/3} = \mathbf{762.9832586} = ch\ 2x$$

Hence, **LHS** $= ch^3 2x =$ **RHS**

Example 4.7 (Equation 4.22b)

Verify numerically the expression (4.22b)

$$\frac{2\tan N_{3f}(x)}{1-\tan N_{3f}^2(x)} = \frac{\tan N_{3e}(x)+\dfrac{2sh\ 2x}{1-sh^2 2x}}{1-\tan N_{3e}(x).\dfrac{2sh\ 2x}{1-sh^2 2x}}$$

with, $x = 3.665191$ rad = 210 deg

Solution

Take relevant data from Example 4.7.

$$\tan N_{3f}(x) = [\sin N_{3f}(x)]/[\cos N_{3f}(x)]$$
$$= -5494.684146/20346.35483 = -0.2700574$$

Calculate $\tan N_{3f}(x) = -0.5791$, $sh\ 2x = 762.9826$

Now, $$\text{LHS of (4.22b)} = \frac{2\times(-0.2700574)}{1-(-0.2700574)^2} = \frac{-0.540115}{0.927069} = \mathbf{-0.582605}$$

For RHS of (4.22b)

$$\frac{2sh\ 2x}{1-sh^2 2x} = \frac{2\times 762.9826}{1-(762.9826)^2} = -0.0026213$$

$$\text{LHS of (4.22b)} = \frac{-0.5791+(-0.0026213)}{1-(-0.5791)(-0.0026213)} = \mathbf{-0.5826053}$$

Hence, **LHS = RHS**

Example 4.8

Verify numerically the expression

$$\frac{\sin N_{3e}(x)}{\sin N_{3f}(x)} = \frac{\dfrac{\sin\ 2x}{sh\ 2x}+\cos\ 2x}{\dfrac{\sin N(x)}{sh\ 2x}+\cos N(x)}$$

CHAPTER 4

with, $x = 2.618$ rad $= 150$ deg

Solution

We have for, $x = 150$ deg $(= 2.618$ rad$)$

$\sin 2x = -0.866$, $\cos 2x = 0.5$, $sh\ 2x = 93.95465$

$\sin\text{N}x = -2.45895$, $\cos\text{N}x = -9.37622$

$\sin\text{N}_{3e}(x) = 46.1113$

Calculate

$$\sin\text{N}_{3f}(x) = \sin\text{N}x + sh\ 2x \cos\text{N}x$$
$$= -2.45895 + 93.95465 \times (-9.37622) = -883.398418$$

Then for the given expression

$$\text{LHS} = \frac{46.1113}{-883.398418} = \mathbf{-0.052198}$$

$$\text{RHS} = \frac{\dfrac{-0.866}{93.95465} + 0.5}{\dfrac{-2.45895}{93.95465} - 9.37622} = \mathbf{-0.052198}$$

Hence, **LHS = RHS**

Example 4.9 (Equation 4.21b)

Verify numerically the expression

$$2 \sin\text{N}_{3f}(x) . \cos\text{N}_{3f}(x) = (1 - sh^2\ 2x) \sin\text{N}_{3e}(x) + 2\ sh\ 2x . \cos\text{N}_{3e}(x)$$

with, $x = 3.665191$ rad $(= 210$ deg$)$

Solution

Given that, $x = 3.665191$ rad $(= 210$ deg$)$

Take some data from Example 4.6

$sh\ 2x = 762.9826$, $1 - sh^2\ 2x = -582141.4479$

$\sin\text{N}_{3f}(x) = -5494.684146$

$\cos\text{N}_{3f}(x) = 20346.35483$

Calculate from formula (4.17)

$\sin\text{N}_{3e}(x) = 382.3573$

$\cos\text{N}_{3e}(x) = -660.262$

Then

$$\textbf{LHS} = 2 \times (-5494.6841460)(20346.35483) = -\mathbf{223593586.6}$$

$$\textbf{RHS} = (-582141.4479)(382.3573) + 2(762.9826)(-660.262)$$
$$= -222586032.2 - 1007536.835 = -\mathbf{222593569.0}$$

Hence, **LHS = RHS**

❑❑❑

5

RECIPROCAL OF COMPLEX Nbic FUNCTIONS

5.1 VARIOUS TYPES OF RECIPROCAL OF COMPLEX Nbic FUNCTIONS

Recollect the various expressions of Complex Nbic functions and of Coci and Cohy-functions as discussed in Chapter 2.

Here, the reciprocal of a given complex function:

$$f = f(x) = f_1(x) + i\, f_2(x) = f_1 + if_2$$

is saught in the form of another complex function:

$$g = g(y) = g_1(y) - ig_2(y) = g_1 - ig_2$$

where

(f_1, g_1) and (f_2, g_2), are the **real** and **imaginary** parts, respectively, of the respective functions, with the intension to express y in terms of x, so that, $g = g(y)$, serves the solution for the reciprocal of $f = f(x)$.

That the function g(y), to be the reciprocal of the function $f(x)$, is ensured by the relation, $f(x).g(y) = 1$.

Reciprocal of Coci-function, e^{-ix} = (cos x + i sin x)

Let $\quad e^{-iy} = \cos y - i \sin y$, is the reciprocal of $e^{ix} = (\cos x + i \sin x)$

Then

$$\begin{aligned} \cos y - i \sin y &= 1/(\cos x + i \sin x) \\ &= (\cos x - i \sin x)/(\cos^2 x + \sin^2 x) \\ &= \cos x - i \sin x \end{aligned}$$

So, $\quad \cos y = \cos x$, and, $\sin y = \sin x \Rightarrow y = x$

Thus the reciprocal of $(\cos x + i \sin x)$ is $(\cos x - i \sin x)$ (5.1a)

That is, $\quad 1/e^{ix} = e^{-ix} \Rightarrow e^{ix}\, e^{-ix} = 1$ (5.1b)

Also the product of Coci-function and its reciprocal yields

$$(\cos y - i \sin y)\,(\cos x + i \sin x) = 1$$

or, $\quad (\cos y \cos x + \sin y \sin x) - i\,(\sin y \cos x - \cos y \sin x) = 1$

or, $\cos(y - x) - i \sin(y - x) = 1$

or, $e^{-i(y-x)} = e^{i(x-y)} = 1$ (5.1c)

Since here, $y = x$, $e^{i(x-y)} = e^0 = 1$ **(Checked)** (5.1d)

Reciprocal of Cohy-function, $f^{ix} = (ch\ x + i\ sh\ x)$

Since, $f^{-ix} = (chx - i\ shx)$, is not the reciprocal of $f^{ix} = (chx + i\ shx)$

Assume that

$$f^{-iy} = (chy - i\ shy), \text{ is the reciprocal of } f^{ix} = (chx + i\ shx)$$

so that, $f^{ix} f^{-iy} = 1$.

Then let

$$(chy - i\ shy) = 1/(chx + i\ shx) \quad (5.2a)$$

Multiply the numerator and denominator of RHS by $(chx + i\ shx)$, and simplify to get

$$(chy - i\ shy) = 1/(chx + i\ shx)$$
$$= (chx + i\ shx)/(1 + i\ sh\ 2x)$$

Multiply the numerator and denominator of RHS by $(1 - i\ sh\ 2x)$, and simplify to get

$$(chy - i\ shy) = [(chx + shx\ sh\ 2x) + i\ (shx - chx\ sh\ 2x)]\ /\ ch^2\ 2x$$
$$= [chx\ (1 + 2\ sh^2\ x) + i\ shx\ (1 - 2\ ch^2\ x)]\ /\ ch^2\ 2x$$
$$= (chx - i\ shx)\ /\ ch\ 2x$$

Separating **real** and **imaginary** parts

$$chy = chx\ /\ ch\ 2x$$
$$shy = shx\ /\ ch\ 2x \quad (5.2b)$$

Multiplying, shy and chy, and simplifying

$$shy\ chy = shx\ chx\ /\ ch^2\ 2x$$

from which

$$sh\ 2y = sh\ 2x\ /\ ch^2\ 2x = th\ 2x\ sech\ 2x \quad (5.2c)$$

thus

$$y = \frac{1}{2} sh^{-1}\ [sh\ 2x\ /\ ch^2\ 2x] = \frac{1}{2} sh^{-1}\ [th\ 2x\ sech\ 2x] \quad (5.2d)$$

This expression suggests the value of y in terms of the given x, so that the reciprocal of $(ch\ x + i\ shx)$, can be written down in terms of y.

5.1.1 Particular Case

Note that, in the case of complex circular function (Coci-function), the same procedure will lead to

$$y = x.$$

Also alternatively, directly this can be achieved by replacing hyperbolic terms, $(sh\ 2x)$ and $(ch\ 2x)$ simply by trigonometric terms, *viz.*, $(\sin 2x)$ and $(\cos 2x)$, respectively, in (5.2c). That is,

$$sh\ 2y = sh\ 2x\ /\ ch^2\ 2x = sh\ 2x\ /\ (ch^2\ x + sh^2\ x)^2$$

is rewritten with the suggested substitution, as

$$\sin 2y = \sin 2x/(\cos^2 x + \sin^2 x)^2 = \sin 2x$$

so that, $y = x$.

Here the function is:

$$e^{ix} = (\cos x + i \sin x)$$

and so, the required reciprocal is:

$$e^{-iy} = \cos y - i \sin y = \cos x - i \sin x = e^{-ix} \tag{5.3a}$$

Thus, for Coci function, as a particular case, we get

$$e^{ix} e^{-iy} \Rightarrow e^{ix} e^{-ix} = 1 \tag{5.3b}$$

Again, the product of two Cohy 1–forms, one is of x, and the other one of y, when expanded and rearranged, give the following forms of Cohy 2-functions:

In terms of Coci-function, $e^{-iy} = (\cos y - i \sin y)$

Let

$e^{-iy} = (\cos y - i \sin y)$, is the reciprocal of $f^{ix} = (chx + i\, shx)$

so that, $f^{ix} e^{-iy} = 1$.

Now

$$\cos y - i \sin y = 1/(chx + i\, shx) \tag{5.4a}$$

Then, $\cos y = chx\, /ch\, 2x$, and, $\sin y = shx\, /ch\, 2x$

Squaring and subtracting

$$\cos 2y = 1/ch^2\, 2x \Rightarrow y = \frac{1}{2} \cos^{-1} (1/ch^2\, 2x) \tag{5.4b}$$

Thus, $\cos y = chx/ch\, 2x$, $\sin y = shx/ch\, 2x$

The required reciprocal in terms of e^{-iy} is:

$$(chx - i\, shx)/ch\, 2x = f^{-iy}/ch\, 2x$$

Thus, $1/f^{ix} = e^{-iy} \Rightarrow f^{ix} f^{-iy} = ch\, 2x$,

where
$$y = \frac{1}{2} \cos^{-1} (1/ch^2\, 2x) \tag{5.4c}$$

Again

$$\cos y - i \sin y = 1\, /(chx + i\, shx)$$
$$\Rightarrow \cos y\, chx + \sin y\, shx - i(\sin y\, chx - \cos y\, shx) = 1$$
$$\Rightarrow \cos N(y, -x) - i \sin N(y, -x) = 1$$

This statement leads to the criterion:

$$e^{-iN/(y, -x)} = 1 \tag{5.4d}$$

In terms of Complex Single Nbic Functions (cosN Function), $e^{-iN/(y, y)} = \cos N(y, y) - i \sin N(y, y)$

$$1/f^{ix} = 1/(chx + i\, shx)$$
$$= \cos Ny - i \sin Ny = e^{-iN/\, y} \tag{5.5}$$
$$\Rightarrow (\cos Ny\, chx + \sin Ny\, shx) - i\, (\sin Ny\, chx - \cos Ny\, shx) = 1$$

By definition of double Nbic function, this statement leads to

$$\Rightarrow \cos N_{2f}\, (y, -x) - i \sin N_{2f}\, (y, -x) = 1$$

$$\Rightarrow e^{-iN_{2f}/(y, -x)} = 1 \tag{5.6}$$

Remembering that, $\cos N^2 y + \sin N^2 y = ch\ 2y$, we get

$$\cos Ny - i \sin Ny = 1/(chx + i\ shx) = (chx - i\ shx)/ch\ 2x$$

$$\Rightarrow \cos Ny = chx/ch\ 2x, \text{ and, } \sin Ny = shx/ch\ 2x$$

Squaring and adding

$$\cos N^2 y + \sin N^2 y = ch\ 2y = 1/ch\ 2x = \text{sech}\ 2x \tag{5.7a}$$

from which, $$y = \frac{1}{2}\ ch^{-1}\ (1/ch\ 2x) \tag{5.7b}$$

5.1.2 Reciprocal of Complex Single Nbic Functions (Cosn function), $e^{iN/x} = \cos Nx + i \sin Nx$

Then let $$(\cos Ny - i \sin Ny) = 1/(\cos Nx + i \sin Nx) \tag{5.8a}$$

Multiply the numerator and denominator of RHS by $(\cos Nx - i \sin Nx)$, and remembering that

$$\cos N^2 x + \sin N^2 x = ch\ 2x, \text{ we get on simplification}$$

$$(\cos Ny - i \sin Ny) = 1/(\cos Nx + i \sin Nx)$$

$$= (\cos Nx - i \sin Nx)/ch\ 2x$$

Separating **real** and **imaginary** parts

$$\cos Ny = \cos Nx/ch\ 2x$$

$$\sin Ny = \sin Nx/ch\ 2x \tag{5.8b}$$

So, $$\tan Ny = \tan Nx \tag{5.8c}$$

Squaring and adding

$$\cos N^2 y + \sin N^2 y = ch\ 2y$$

$$= (\cos N^2 x + \sin N^2 x)/ch^2\ 2x$$

$$= \text{sech}\ 2x$$

This suggests a criterion

$$ch\ 2y.ch\ 2x = 1$$

[The admissible range of values of $ch\ 2y$ and $ch\ 2x$ need to be remembered while applying this criterion]

Thus the reciprocal of $(\cos Nx + i \sin Nx)$ is:

$$(\cos Ny - i \sin Ny) = (\cos Nx - i \sin Nx)/ch\ 2x$$

$$\Rightarrow 1/e^{iN/x} = e^{-iN/x}/ch2x \Rightarrow e^{iN/x}\ e^{-i/N\ x} = ch\ 2x \tag{5.9a}$$

and, $$y = \frac{1}{2} ch^{-1}\ (1/ch\ 2x) \tag{5.9b}$$

This expression suggests the value of y in terms of given x, so that the reciprocal of $(\cos Nx + i \sin Nx)$ can be written down in terms of y.

Numerical Examples

The procedure is demonstrated with actual numerical calculation in the following example.

Example 5.1 (Equation 5.8a)

Take data, y = 0.31118 rad = 17.8293 deg

$\cos y$ = 0.095197, $\sin y$ = 0.30618, chy = 1.048808, shy = 0.316226

$2y$ = 0.62236, $ch\ 2y$ = 1.2

Using the definition (1.35a), get

$$\cos Ny = 0.95197 \times 1.048808 - 0.30618 \times 0.316226 = 0.901611675$$
$$\sin Ny = 0.30618 \times 1.048808 + 0.95197 \times 0.316226 = 0.622161698$$
$$\tan Ny = (\sin Ny)/(\cos Ny) = 0.69005506 = \tan Nx$$

We have, from reciprocal relation (5.8b)

$$\cos Ny = \frac{\cos N(x)}{ch\ 2x} = 0.901611675$$

$$\sin Ny = \frac{\sin N(x)}{ch\ 2x} = 0.622161698$$

From (5.8c)

$$\tan Nx = \tan Ny \quad \text{gives}$$

$$a = \frac{\sin N(x)}{\cos N(x)} = \frac{\sin N(y)}{\cos N(y)} = 0.69005506$$

$$a^2 = \frac{\sin N^2(x)}{\cos N^2(x)} = \frac{\sin N^2(y)}{\cos N^2(y)} = 0.476176$$

$$1 + a^2 = \frac{ch\ 2x}{\cos N^2(x)} = \frac{ch\ 2y}{\cos N^2(y)} = 1.476176$$

$$1/(1 + a^2) = \frac{\cos N^2(x)}{ch\ 2x} = 0.677426$$

But, $\dfrac{\cos N(x)}{ch\ 2x} = \cos Ny = 0.901611675$

From which on dividing,

$$\cos Nx = 0.677426/0.901611675 = 0.75135$$

Again

$$1/a^2 = \frac{\cos N^2(x)}{\sin N^2(x)} = 1/0.476176 = 2.1000638$$

$$1 + 1/a^2 = (1 + a^2)/a^2 = \frac{ch\ 2x}{\sin N^2(x)} = 3.1000638$$

$$a^2/(1 + a^2) = \frac{\sin N^2(x)}{ch\ 2x} = 0.322514$$

but, $\dfrac{\sin N(x)}{ch\ 2x} = \sin Ny = 0.622161898$

From which on dividing

$$\sin Nx = 0.322514/0.622161898 = 0.518376$$

Thus, the reciprocal of, (cosNx + i sinNx), *i.e.*, of, [0.75135 + 0.518376 i]
is given as: (cosNy – i sinNy) = [0.901611675 – 0.622161898 i]

Check of the numerical calculation:

On actual multiplication of the number and its reciprocal is

$$(0.75135 + 0.518376\, i)\,(0.901611675 - 0.622161898\, i)$$
$$= 0.999939624 + i\,(0.467461191 - 0.46737385) = 1 + i \times 0 = 1 \quad \textbf{(Verified)}$$

In terms of Coci function, e^{-iy} = (cos y – i sin y)

Let

$$(\cos y - i \sin y) = e^{-iy} \text{ be the reciprocal of, } (\text{cosN}x + i\, \text{sinN}x) \qquad (5.9c)$$

Thus with the help of, $(\text{cosN}^2 x + \text{sinN}^2 x) = ch\ 2x$, we get

$$(\cos y - i \sin y) = 1/(\cos y + i \sin y) = (\text{cosN}x - i\, \text{sinN}x)/ch\ 2x \qquad (5.9d)$$

Separating real and imaginary parts

$$\cos y = \frac{\text{cosN}(x)}{ch\ 2x}, \quad \sin y = \frac{\text{sinN}(x)}{ch\ 2x} \qquad (5.9e)$$

Now, y can be found out from

$$\tan y = (\sin y)/(\cos y) = (\text{sinN}x)/(\text{cosN}x) = \text{tanN}x$$

or, $$y = \tan^{-1}[\text{tanN}x] \qquad (5.9f)$$

The reciprocal of (cosNx + i sinNx) is given by (5.9d), that is

$$[\text{cosN}x - i\, \text{sinN}x]\,/\,ch\ 2x.$$

The relation (5.9f) can be alternatively arrived at by squaring cosNx and sinNx, and adding, and also subtracting them in view of (5.9e), and view it in the light of (1.48) and (1.45), that is

$$\text{cosN}^2 x - \text{sinN}^2 x = \cos 2x - \sin 2x\ sh\ 2x$$
$$2\, \text{cosN}x\, \text{sinN}x = \sin 2x + \cos 2x\ sh\ 2x \qquad (5.9g)$$

we have

$$\cos 2y = \cos^2 y - \sin^2 y = [\text{cosN}^2 x - \text{sinN}^2 x]\,/\,ch^2\, 2x$$
$$\sin 2y = 2 \sin y \cos y = 2[\text{cosN}x\, \text{sinN}x]\,/\,ch^2\, 2x \qquad (5.9h)$$

Substituting (5.9g) in (5.9h), we get

$$\cos 2y = [\cos 2x - \sin 2x\ sh\ 2x]\,/\,ch^2\, 2x$$
$$\sin 2y = [\sin 2x + \cos 2x\ sh\ 2x]\,/\,ch^2\, 2x$$

from which

$$\tan 2y = [\sin 2x + \cos 2x\ sh\ 2x]/[\cos 2x - \sin 2x\ sh\ 2x]$$
$$= [\tan 2x + sh\ 2x]/[1 - \tan 2x\ sh\ 2x]$$

The RHS of this expression according to (4.2m) can be well identified as = $\text{tanN}_{3e}(x)$.

So

$$\tan 2y = \text{tanN}_{3e}(x)$$
$$= \text{tanN}x/[1 - \text{tanN}^2 x] \qquad (5.9i)$$

If, tan y is given by (5.9h), (5.9i) confirms the relation for tan 2y.

5.1.3 Reciprocal of Complex Double Nbic Functions (Codn function), $e^{iN2/(x)} = \cos N_2(x) + i \sin N_2(x)$

The reciprocal of two types of Complex double Nbic functions, namely:

$\cos N_{2e}(x) + i \sin N_{2e}(x)$ and, $\cos N_{2f}(x) + i \sin N_{2f}(x)$, are described here.

For, **Complex double Nbic function, $e^{iN_{2e}/(x)} = \cos N_{2e}(x) + i \sin N_{2e}(x)$:**
Let

$$[\cos N_{2e}(y) - i \sin N_{2e}(y)] = 1 / [\cos N_{2e}(x) + i \sin N_{2e}(x)] \tag{5.10a}$$

Multiply the numerator and denominator of RHS by $[\cos N_{2e}(x) - i \sin N_{2e}(x)]$, and remembering (3.10a), *i.e*,

$$\cos N_{2e}^2(x) + \sin N_{2e}^2(x) = ch\ 2x$$

we get on simplification

$$[\cos N_{2e}(y) - i \sin N_{2e}(y)] = [\cos N_{2e}(x) - i \sin N_{2e}(x)] / ch\ 2x \tag{5.10b}$$

Separating **real** and **imaginary** parts

$$\cos N_{2e}(y) = [\cos N_{2e}(x)] / ch\ 2x$$
$$\sin N_{2e}(y) = [\sin N_{2e}(x)] / ch\ 2x \tag{5.10c}$$

So, $$\tan N_{2e}(y) = \tan N_{2e}(x) \tag{5.10d}$$

Squaring and adding

$$\begin{aligned}\cos N_{2e}^2(y) + \sin N_{2e}^2(y) &= ch\ 2y\\ &= [\cos N_{2e}^2(x) + \sin N_{2e}^2(x)]/ch^2\ 2x\\ &= \mathrm{sech}\ 2x\end{aligned}$$

Thus the reciprocal of $[\cos N_{2e}(x) + i \sin N_{2e}(x)]$ is:

$$\begin{aligned}1 / [\cos N_{2e}(x) + i \sin N_{2e}(x)] &= [\cos N_{2e}(y) - i \sin N_{2e}(y)]\\ &= [\cos N_{2e}(x) - i \sin N_{2e}(x)] / ch\ 2x\end{aligned} \tag{5.11}$$

That is $$1/e^{iN_{2e}/(x)} = e^{-iN_{2e}/(x)}/ch\ 2x$$

or, $$e^{iN_{2e}/(x)} . e^{-iN_{2e}/(x)} = ch\ 2x \tag{5.12a}$$

and, $$y = \frac{1}{2} ch^{-1} (1/ch\ 2x) \tag{5.12b}$$

This expression suggests the value of y in terms of given x, so that the reciprocal of $[\cos N_{2e}(x) + i \sin N_{2e}(x)]$ can be written down in terms of y.

For **Complex double Nbic function, $e^{iN_{2f}/(x)} = \cos N_{2f}(x) + i \sin N_{2f}(x)$:**

The procedure is exactly same as for the previous case, except that, the criterion governing the operation now is:

$$\cos N_{2f}^2(x) + \sin N_{2f}^2(x) = ch^2\ 2x \tag{5.13}$$

[which for the previous case was : $\cos N_{2e}^2(x) + \sin N_{2e}^2(x) = ch\ 2x$]

Now let

$$[\cos N_{2f}(y) - i \sin N_{2f}(y)] = 1/[\cos N_{2f}(x) + i \sin N_{2f}(x)] \tag{5.14a}$$

Multiply the numerator and denominator of RHS by, $[\cos N_{2f}(x) - i \sin N_{2f}(x)]$, and remembering (3.19a), that

$$\cos N_{2f}^2(y) + \sin N_{2f}^2(y) = ch^2\, 2y$$

we get on simplification

$$[\cos N_{2f}(y) - i \sin N_{2f}(y)] = 1/[\cos N_{2f}(x) + i \sin N_{2f}(x)]$$
$$= [\cos N_{2f}(x) - i \sin N_{2f}(x)] / ch^2\, 2x \quad (5.14b)$$

Separating **real** and **imaginary** parts

$$\cos N_{2f}(y) = [\cos N_{2f}(x)] / ch^2\, 2x$$
$$\sin N_{2f}(y) = [\sin N_{2f}(x)] / ch^2\, 2x \quad (5.14c)$$

So, $\tan N_{2f}(y) = \tan N_{2f}(x)$ (5.14d)

Squaring and adding

$$\cos N_{2f}^2(y) + \sin N_{2f}^2(y) = ch^2\, 2y$$
$$= [\cos N_{2f}^2(x) + \sin N_{2f}^2(x)] / ch^4\, 2x$$
$$= \operatorname{sech}^2 2x$$

Thus the reciprocal of, $[\cos N_{2f}(x) + i \sin N_{2f}(x)]$ is:

$$1 / [\cos N_{2f}(x) + i \sin N_{2f}(x)] = [\cos N_{2f}(y) - i \sin N_{2f}(y)]$$
$$= [\cos N_{2f}(x) - i \sin N_{2f}(x)] / ch^2\, 2x \quad (5.15)$$

That is $1 / e^{iN_{2f}(x)} = e^{-iN_{2f}(x)} / ch^2\, 2x$

or, $e^{iN_{2f}(x)} .\, e^{-iN_{2f}(x)} = ch^2\, 2x$ (5.16a)

and, $y = \dfrac{1}{2}\, ch^{-1}(1 / ch\, 2x)$ (5.16b)

This expression suggests the value of y in terms of given x, so that the reciprocal of $[\cos N_{2f}(x) + i \sin N_{2f}(x)]$, can be written down in terms of y.

Numerical Example

The procedure is demonstrated with actual numerical calculation in the following example.

Example 5.2 (Equation 5.14a)

Take $x = 0.31118 = 17.8293$ deg,

$\cos x = 0.95197$, $\sin x = 0.30618$, $chx = 1.048808$, $shx = 0.316226$

$2x = 0.663322$, $c = ch\, 2x = 1.2$

$\cos N_{2f}(x) = \cos x - \sin x\; sh\, 2x = 0.95197 - 0.30618 \times 0.663322$
$= 0.748874$

$\sin N_{2f}(x) = \sin x + \cos x\; sh\, 2x = 0.30618 + 0.95197 \times 0.663322$
$= 0.937643$

$\tan N_{2f}(x) = [\sin N_{2f}(x)] / [\cos N_{2f}(x)] = 1.25207 = \tan N_{2f}(y)$

So that the number, $[\cos N_{2f}(x) + i \sin N_{2f}(x)]$ is : $(0.748874 + 0.937643\, i)$

We have from reciprocal relation (5.14c) and (5.14d)

$$\cos N_{2f}(y) = \frac{\cos N_{2f}(x)}{ch^2 2x} = \frac{\cos N_{2f}(x)}{c^2}$$

$$\sin N_{2f}(y) = \frac{\sin N_{2f}(x)}{ch^2 2x} = \frac{\sin N_{2f}(x)}{c^2}$$

$$\tan N_{2f}(y) = \tan N_{2f}(x) \text{ gives}$$

$$a = \frac{\sin N_{2f}(x)}{\cos N_{2f}(x)} = \frac{\sin N_{2f}(y)}{\cos N_{2f}(y)} = 1.25207$$

$$a^2 = \frac{\sin N_{2f}^2(x)}{\cos N_{2f}^2(x)} = \frac{\sin N_{2f}^2(y)}{\cos N_{2f}^2(y)} = 1.567680$$

$$\sqrt{1+a^2} = \frac{ch\ 2x}{\cos N_{2f}(x)} = \frac{ch\ 2x}{\cos N_{2f}(y)} = \sqrt{2.567680} = 1.602398$$

$$\cos N_{2f}(y) = 1/[c.\sqrt{1+a^2}] = 1/(1.2 \times 1.602398) = 0.52006$$

$$\sin N_{2f}(y) = a/[c.\sqrt{1+a^2}] = 1.25207 \times 0.52006 = 0.651145$$

Thus the complex function is

$$[\cos N_{2f}(x) + i \sin N_{2f}(x)] = (0.748874 + 0.937643i)$$

Its reciprocal is given as :

$$[\cos N_{2f}(y) - i \sin N_{2f}(y)] = (0.52006 + 0.651145i)$$

The numerical check of the arithmetics is done through actual multiplication of the number and its reciprocal as:

$$(0.748874 + 0.937643\ i)\ (0.52006 + 0.651145\ i)$$
$$= 1.0000 + 5.05785 \times 10^{-6}\ i$$
$$= 1 + i \times 0 = 1 \qquad \textbf{(Verified)}$$

5.1.4 Reciprocal of Complex n-Function (Con-function), $n^{ix} = \text{sech}x + i\ \text{th}x$

Similar to the Coci–functions, statement (2.36d) shows that, the reciprocal of

$$n^{ix} = (\text{sech}\ x + i\ \text{th}x),\ \text{is}\ 1/n^{ix} = n^{-ix}\ [= (\text{sech}\ \ x - i\ \text{th}x)].$$

Also
$$n^{ix}\ n^{ix} = (cfx + i\ sfx)\ (cfx + i\ sfx) = cf^2x - sf^2x + 2i\ cfx\ sfx$$
$$= ch\ 2x - sh\ 2x = (chx - shx)^2$$
$$n^{ix}\ n^{-ix} = (cfx + i\ sfx)\ (cfx - i\ sfx) = cf^2\ x + sf^2\ x = 1 = n^0$$

This shows that the **reciprocal of n^{ix}** is n^{-ix}. That is, $n^{-ix} = 1/n^{ix}$ (5.17)

where, $cfx = chx$, and, $sfx = i\ shx$.

This is interesting to observe that, **powers do add up in Con-functions**, similar to the case of Coci-functions. But this is not the case with Cohy-functions.

5.1.5 Reciprocal of Complex Nbic Functions (General Procedure)

To obtain the reciprocal of the complex Nbic functions, a very simple procedure is shown, which can as well be considered as a general technique. The method is described as follows:

Let, $f = f(x)$ = given complex Nbic function

and, $g = g(x)$ = required reciprocal complex Nbic function, corresponding to $f(x)$

CHAPTER 5

Let

$$f = f_1(x) + i\, f_2(x) = f_1 + i\, f_2 \qquad (5.18a)$$

$$g = g_1(x) + i\, g_2(x) = g_1 + i\, g_2 \qquad (5.18b)$$

Since, f is reciprocal of g, and *vice versa*

$$f \cdot g = (f_1 + i\, f_2)\,(g_1 + i\, g_2) = 1$$

or,
$$(f_1 g_1 - f_2 g_2) + i\,(g_2 f_1 + f_2 g_1) = 1 + i\,(0) \qquad (5.19)$$

Separating the **real** and **imaginary** parts, we get two equations

$$f_1 g_1 - f_2 g_2 = 1$$
$$g_1 f_2 + f_1 g_2 = 0 \qquad (5.20)$$

The second one of (5.20) gives

$$\frac{g_1}{g_2} = -\frac{f_1}{f_2}$$

$$g_1 = -\frac{f_1}{f_2}\, g_2 \qquad (5.21a)$$

Substitute (5.21a) in the first one of (5.20) to yield

$$g_2 = -\frac{f_2}{f_1^2 + f_2^2} = -\frac{f_2}{f_3} \qquad (5.21b)$$

where,
$$f_3 = f_3(x) = f_1^2 + f_2^2 \qquad (5.21c)$$

The pair of equations (5.21a) and (5.21b), are written only in terms of the component parts of the given function $f(x)$. Hence, $g_2 = g_2(x)$, and $g_1 = g_1(x)$, from (5.21b) and (5.21a) respectively, are sufficient to write the reciprocal function, $g = g\,(x)$ completely, using (5.18b).

The applicability of this generalized method is demonstrated for the complex Nbic functions, including the cases of Coci and Cohy functions, as follows.

Application of the general procedure (5.21a–b)

For Reciprocal of Complex Circular function

The given function is

$$f = f_1 + i f_2 = \cos x + i \sin x$$

so that,
$$f_1 = \cos x,\; f_2 = \sin x$$
$$f_3 = f_1^2 + f_2^2 = \cos^2 x + \sin^2 x = 1 \qquad (5.22a)$$

$$g_2 = -\frac{f_2}{f_3} = -\sin x$$

$$g_1 = -\frac{f_1}{f_2}\, g_2 = -\frac{\cos x}{\sin x}\,(-\sin x) = \cos x$$

Thus, the required reciprocal function is

$$g = g_1 + i\, g_2 = \cos x - i \sin x \qquad (5.22b)$$

For Reciprocal of Complex Hyperbolic function

The given function is

$$f = f_1 + i\, f_2 = chx + i\ shx$$

so that,

$$f_1 = chx,\ f_2 = shx$$

$$f_3 = f_1^2 + f_2^2 = ch^2x + sh^2\, x = ch\ 2x \tag{5.22c}$$

$$g_2 = -\frac{f_2}{f_3} = -\frac{shx}{ch\ 2x}$$

$$g_1 = -\frac{f_1}{f_2} g_2 = -\frac{chx}{shx}\left(-\frac{shx}{ch\ 2x}\right) = \frac{chx}{ch\ 2x}$$

Thus, the required reciprocal function is

$$g = g_1 + i\ g_2 = [ch\ x - i\ shx] / ch\ 2x \tag{5.22d}$$

For Reciprocal of Complex Single Nbic function

The given function is

$$f = f_1 + i\, f_2 = \cos\text{N}x + i\ \sin\text{N}x$$

so that,

$$f_1 = \cos\text{N}x,\ f_2 = \sin\text{N}x$$

$$f_3 = f_1^2 + f_2^2 = \cos\text{N}^2x + \sin\text{N}^2x = ch\ 2x \tag{5.22e}$$

$$g_2 = -\frac{f_2}{f_3} = -\frac{\sin\text{N}(x)}{ch\ 2x}$$

$$g_1 = -\frac{f_1}{f_2} g_2 = \left(-\frac{\cos\text{N}(x)}{\sin\text{N}(x)}\right).\left(\frac{\sin\text{N}(x)}{ch\ 2x}\right) = \frac{\cos\text{N}(x)}{ch\ 2x}$$

Thus, the required reciprocal function is

$$g = g_1 + ig_2 = [\cos\text{N}x - i\ \sin\text{N}x] / ch\ 2x \tag{5.22f}$$

For Reciprocal of Complex Double Nbic function (E type)

The given function is

$$f = f_1 + i\, f_2 = \cos\text{N}_{2e}(x) + i\ \sin\text{N}_{2e}(x)$$

so that,

$$f_1 = \cos\text{N}_{2e}(x),\ f_2 = \sin\text{N}_{2e}(x)$$

$$f_3 = f_1^2 + f_2^2 = \cos\text{N}_{2e}^2(x) + \sin\text{N}_{2e}{}^2(x) = ch\ 2x \tag{5.22g}$$

$$g_2 = -\frac{f_2}{f_3} = -\frac{\sin\text{N}_{2e}(x)}{ch\ 2x}$$

$$g_1 = -\frac{f_1}{f_2} g_2 = \left(-\frac{\cos\text{N}_{2e}(x)}{\sin\text{N}_{2e}(x)}\right).\left(-\frac{\sin\text{N}_{2e}(x)}{ch\ 2x}\right) = \frac{\cos\text{N}_{2e}(x)}{ch\ 2x}$$

Thus, the required reciprocal function is

$$g = g_1 + i\ g_2 = [\cos\text{N}_{2e}(x) - i\ \sin\text{N}_{2e}(x)] / ch\ 2x \tag{5.22h}$$

For Reciprocal of Complex Double Nbic function (F type)

The given function is

$$f = f_1 + i\, f_2 = \cos\text{N}_{2f}(x) + i\ \sin\text{N}_{2f}(x)$$

so that,

$$f_1 = \cos\text{N}_{2f}(x),\ f_2 = \sin\text{N}_{2f}(x)$$

$$f_3 = f_1^2 + f_2^2 = \cos\text{N}^2{}_{2f}(x) + \sin\text{N}^2{}_{2f}(x) = ch^2\ 2x \tag{5.22i}$$

$$g_2 = -\frac{f_2}{f_3} = -\frac{\sin N_{2f}(x)}{ch^2 2x}$$

$$g_1 = -\frac{f_1}{f_2} g_2 = -\frac{\cos N_{2f}(x)}{\sin N_{2f}(x)}\left(-\frac{\sin N_{2f}(x)}{ch^2 2x}\right) = \frac{\cos N_{2f}(x)}{ch^2\ 2x}$$

Thus, the required reciprocal function is

$$g = g_1 + i\ g_2 = [\cos N_{2f}(x) - i\ \sin N_{2f}(x)]\ /\ ch^2\ 2x \qquad (5.22j)$$

For Reciprocal of Complex Triple Nbic function (E type)

The given function is

$$f = f_1 + i\ f_2 = \cos N_{3e}(x) + i\ \sin N_{3e}(x)$$

so that, $f_1 = \cos N_{3e}(x),\ f_2 = \sin N_{3e}(x)$

$$f_3 = f_1^2 + f_2^2 = \cos N_{3e}^2(x) + i\ \sin N_{3e}^2(x) = ch^3\ 2x \qquad (5.22k)$$

$$g_2 = -\frac{f_2}{f_3} = -\frac{\sin N_{3e}(x)}{ch^3 2x}$$

$$g_1 = -\frac{f_1}{f_2} g_2 = -\frac{\cos N_{3e}(x)}{\sin N_{3e}(x)}\left(-\frac{\sin N_{3e}(x)}{ch^3 2x}\right) = \frac{\cos N_{3e}(x)}{ch^3 2x}$$

Thus, the required reciprocal function is

$$g = g_1 + i\ g_2 = [\cos N_{3e}(x) - i\ \sin N_{3e}(x)]\ /\ ch^3\ 2x \qquad (5.22l)$$

For Reciprocal of Complex Triple Nbic function (F type)

The given function is

$$f = f_1 + i\ f_2 = \cos N_{3f}(x) + i\ \sin N_{3f}(x)$$

so that, $f_1 = \cos N_{3f}(x),\ f_2 = \sin N_{3f}(x)$

$$f_3 = f_1^2 + f_2^2 = \cos N_{3f}^2(x) + i\ \sin N_{3f}^2(x) = ch^2\ 2x \qquad (5.22m)$$

$$g_2 = -\frac{f_2}{f_3} = -\frac{\sin N_{3f}(x)}{ch^2 2x}$$

$$g_1 = -\frac{f_1}{f_2} g_2 = -\frac{\cos N_{3f}(x)}{\sin N_{3f}(x)}\left(-\frac{\sin N_{3f}(x)}{ch^2 2x}\right) = \frac{\cos N_{3f}(x)}{ch^2 2x}$$

Thus, the required reciprocal function is

$$g = g_1 + i g_2 = [\cos N_{3f}(x) - i\ \sin N_{3f}(x)]\ /\ ch^2\ 2x \qquad (5.22n)$$

In fact, it is clear that, the function, $f_3 = f_1^2 + f_2^2$, is the main key to the solution for the reciprocal functions. We see that, for reciprocal functions corresponding to the given complex functions:

f_3 = 1 = constant, for Complex circular function
f_3 = $ch\ 2x$, for Complex hyperbolic, and Complex double Nbic (E type) functions.
f_3 = $ch^2\ 2x$, for Complex triple Nbic (F type), and complex triple Nbic (F type) functions.
f_3 = $ch^3\ 2x$, for Complex triple Nbic (E type) function.

❑❑❑

6

CIRCULAR REPRESENTATION OF Nbic FUNCTIONS (Nbic CIRCLES)

6.1 VARIOUS TYPES OF Nbic CIRCLES

6.1.1 Circular Representation of Hyperbolic Functions

Hyperbolic circle, C_H

In a rectangular coordinate system (X,Y), if X = cos *c*, Y= sin *c*, then, $X^2 + Y^2 = 1$, represents circle of radius =1, having origin O at (0, 0).

Here *c* = the angle, (call it a circular angle), subtended by the radius with the X-axis, which is given by:

$$\tan c = \sin c/\cos c \tag{6.1}$$

But since the right side of the identity of hyperbolic functions, *viz.*,

$$ch^2\, x + sh^2\, x = ch\, 2x \tag{6.2}$$

is not equal to 1, and as such, it fails to represent the equation of a circle. But on substituting

$$X = \frac{chx}{\sqrt{ch\ 2x}},\ \text{and},\ Y = \frac{shx}{\sqrt{ch\ 2x}} \tag{6.3a}$$

or,

$$\left(\frac{chx}{\sqrt{ch\ 2x}}\right)^2 + \left(\frac{shx}{\sqrt{ch\ 2x}}\right)^2 = 1 \tag{6.3b}$$

for any point P in it, we arrive at the equation as: $X^2 + Y^2 = 1$.

The **geometrical angle** (gA) in terms of **tehy angle** H, is then defined as:

$$gA = H = \tan^{-1}(Y/X) \tag{6.4}$$

This equation, $X^2 + Y^2 = 1$, can now be identified as that of a circle, and will be termed as **Hyperbolic circle (C_H).**

The parameter *x* in (*shx*) or, in (*chx*) is a number, and can be converted to degree by multiplying it by (180/π), whenever required.

The Hyperbolic Circle with center at O (0,0) is not a full or closed circle, but can be termed as **open circle**, and is only a circular arc AB (or AC), of radius OP = 1, where P is at any location (X,Y) on the circumference of this circle.

The geometrical angle, gA (= teN angle) = H, on this circle made by the radius corresponding to a point P, namely:

$[X = \frac{chx}{\sqrt{ch\,2x}},\ Y = \frac{shx}{\sqrt{ch\,2x}}]$, with the horizontal axis, $X = \frac{chx}{\sqrt{ch\,2x}}$, is given by (6.4),

that is: $\tan H - \tan gA = [\frac{shx}{\sqrt{ch\,2x}}]/[\frac{chx}{\sqrt{ch\,2x}}] = shx/chx = thx$ (6.5a)

or, alternately by (2.39d), that is

$$\cos 2H.ch\,2x = 1 \qquad (6.5b)$$

The **geometrical angle** (gA) = H, (as can be obtained on the plot of the circle Y vs X), differs from the actual value of the parameter x (which can be converted to angle from radian measurement, if the need arises for the sake of comparison).

The **sweep angle** (sA) is defined as the measure of the teN angle of a point on the Nbic circle in excess of the corresponding value of x. It is thus the difference (or deviation) of the angle gA from that corresponding to the basic variable x. That is

$$sA = gA - x\ (deg) = H - x\ (deg) \qquad (6.6)$$

where in hyperbolic circles, H is the **tehy angle,** (which is a particular form of the **teN angle**).

Table 6.1 describes coordinates of P(X, Y) for various values of the basic variable x along with the corresponding gA values. It is clearly seen in this case that, gA (= H) < x. That is, sA = (gA – x), is negative. In this sense, Hyperbolic Circle can be called **negative circle** also.

Table 6.1 Hyperbolic Circle

Sr. No.	x (rad)	x (deg)	X	Y	gA = H (deg)	sA= H–x (deg)	gap angle (deg)
1 (start)	0	0	1.00000	0.00000	0.000	0.000	-
2	π/6	30	0.90136	0.43308	25.663	– 4.337	25.6630
3	π/3	60	0.78823	0.61538	37.980	– 22.020	12.3170
4	π/2	90	0.73698	0.67592	42.526	– 47.474	4.5460
5	2π/3	120	0.71774	0.69630	44.131	– 75.869	1.6050
6	5π/6	150	0.71086	0.70333	44.695	– 105.305	0.5640
7	π	180	0.70843	0.70579	44.892	– 135.108	0.1970
8	7π/6	210	0.70757	0.70664	44.962	– 165.038	0.0700
9	4π/3	240	0.70727	0.70694	44.987	– 195.013	0.0250
10	3π/2	270	0.70716	0.70705	44.995	– 225.005	0.0080
11	5π/3	300	0.70713	0.70709	44.998	– 255.002	0.0030
12	11π/6	330	0.70711	0.70710	44.999	– 285.0006	0.0014
13 (end)	2π	360	0.70711	0.70710	45.000	– 315.000	0.0006

The sweep angle sA = –ve, $[\text{gap angle}]_m = [(H)_m - (H)_{m-1}]$

In Euclidean circle (*i.e.*, ordinary circle), gA = x, so that, sA = 0. Hence, Euclidean circle can also be termed as **zero circle**.

Refer to **Figure 1.3** of Chapter 1, for general description of Nbic circles, which is valid for hyperbolic circles too. In **Figure 6.1** the Hyperbolic Circle is shown separately and is described in reference to **Figure 1.3.**

In a full circle of radius unity, and having the centre at (0, 0), a starting point B on the circumference at (1, 0) exactly coincide on the same point B, after travelling an angular distance 2π (360 deg), along the circumference.

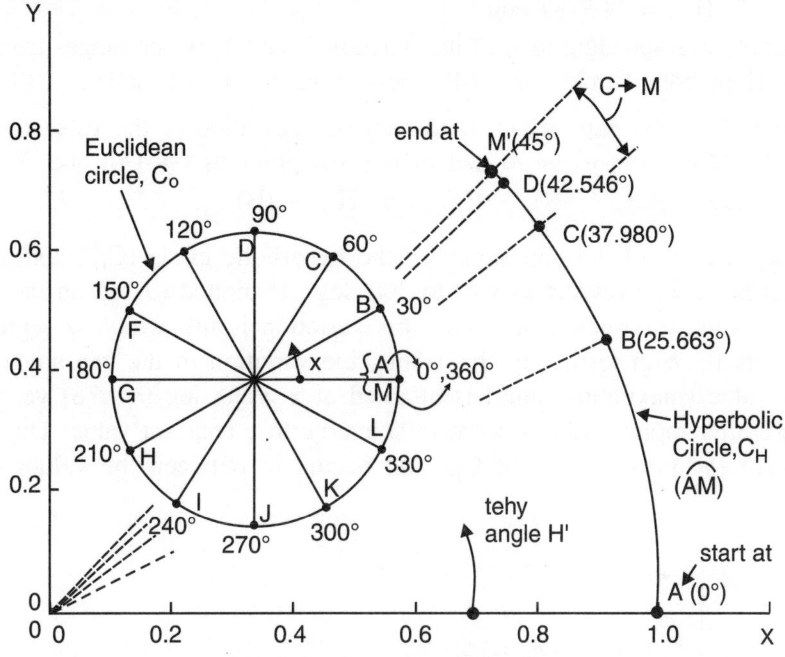

Fig. 6.1: Hyperbolic circle, C_H

Now here in Hyperbolic Circle, the starting point B (X = 1, Y = 0), [here it is point A] of the circle, on the symmetry line X axis, after travelling an angular distance corresponding to x = 2π (360 deg), along the circumference, does not end on the same point B(1, 0), but some where at the point C [here it is point M]. The geometrical angle H to locate this end point C, is obtained from the relation (6.5a), that is: tan H = t*h* x, with the value of the parameter, x = 2π = 6.2832 rad.

Thus, at the end point, tan H = t*h* (2π) = 1, so that, H = + 45 deg (+π/4).

Geometrically, this angle is in first quadrant, measured 45 deg anticlockwise from the horizontal axis X, which is represented by, $[(ch\ x)/(\sqrt{ch\ 2x})]$.

Similarly, considering a clockwise movement from 0, in the reverse direction, the point B ends at D, for which x = – 2π = – 6.2832 rad (= –360 deg), which yields H = – 45 deg (–π/4).

Thus the Hyperbolic Circle (C_H) is an open circular arc BC, [here it is AM] spreading from 0 deg to +45 deg, *i.e.*, 0 to π/4, (or, arc BD spreading from 0 deg to –45 deg, *i.e.*, 0 to –π/4).

The total arc length BC = π/4.

The coordinates of the end point M is (0.707, 0.707), in **Figure 6.1.**

At any intermediate point P on the circumference of this circle, say at x = 35 deg (corresponding to 0.610865 rad), the coordinates (X,Y) and angle H are:

$$X = [(ch\ x)/(\sqrt{ch\ 2x})] = 0.8782$$

$$Y = [(sh\ x)/(\sqrt{ch\ 2x})] = 0.4784 \quad (6.7a)$$

$$\tan H = th\ (Y/X)) = th\ (0.610865) = 0.54474, \text{ from which}$$

$$H = 28.5787 \text{ deg} \quad (6.7b)$$

Noteworthy is that, corresponding to x, in the function X and Y, which ranges from (0 to 2π), the angle H in the Hyperbolic Circle (arc AM), shows a range of (0 to π/4) *i.e.*, (0 to 45 deg).

The **gap angle** (or simply, **gap**) which is the angular deviation of the value of the m-th Nbic angle (teN angle) N in comparison to that of the just previous (m–1)th one. That is

$$[\text{gap angle}]_m = (N)_m - (N)_{m-1} = (H)_m - (H)_{m-1} \quad (6.7c)$$

In **Fig. 6.2** the variation of the gap angle for the hyperbolic circle (C_H), and all Nbic circles (C_N) considered here, with respect to x (0 to 360 deg), is plotted (based on the data of **Tables 6.1–6.6**). As obvious, for Euclidean circle, this variation is nil, which is represented by the 0-line (X axis) itself. With respect to this 0-line, the variation of the gap angle for C_H shows that, the peak value (maximum value) is attained at x = 30 deg (= π/6) very quickly from x = 0, and there after rapidly decays down to converge to a constant value. This value is equal to zero (= 0) for the present case of C_H, and occurs in between the values of x = 180 to 210 deg.

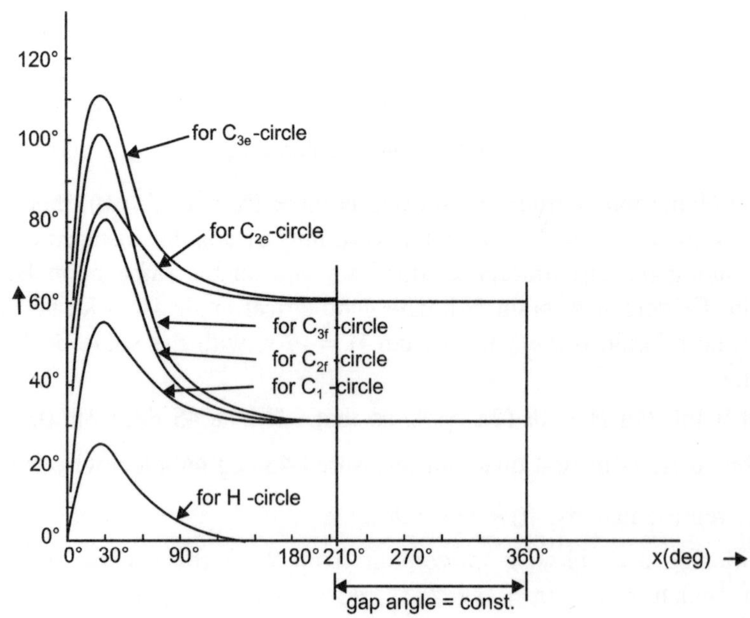

Fig. 6.2: Gap angle for various Nbic circles

Whether it is hyperbolic circle or Nbic circle, all the functions of the gap angle with respect to x, follow the same trend, as is clearly noticed in **Fig. 6.2**. Particularly to mention that in C_H, the value of x commencing from which the constant value of the gap angle reaches a constant value, is in between x = 180 to 210 deg, (say at 210 deg).

In the following data, the parameter x (in rad, and deg), and the corresponding angle H (in deg) in the Hyperbolic Circle (C_H) are given in the following order,

[x rad, x deg $\Rightarrow$ H deg] as:

[0.0, 0.0 $\Rightarrow$ 0.0], [0.2, 11.459 $\Rightarrow$ 11.165], [0.4, 22.918 $\Rightarrow$ 20.804],

[0.6, 34.378 $\Rightarrow$ 28.238], [0.8, 45.837 $\Rightarrow$ 33.586], [1.0, 57.296 $\Rightarrow$ 37.293],

[1.2, 68.755 $\Rightarrow$ 39.816], [1.4, 80.214 $\Rightarrow$ 41.520], [1.6, 91.674 $\Rightarrow$ 42.666],

[1.8, 103.133 $\Rightarrow$ 43.435], [2.0, 114.592 $\Rightarrow$ 43.951], [2π, 360 $\Rightarrow$ 45.0]

This can be demonstrated with an example as follows.

Numerical Example

Example 6.1 (Equation 6.5b)

Calculate the geometrical angle H, if x is given, or vice versa, using relation (6.5b), *i.e.*, cos 2H. ch $2x$ = 1.

Solution

(a) When x = 0.9 (51.5664 deg),

chx = 3.10747, cos 2H = 0.3218, H = 35.6139 deg.

(b) When H = 28.93 deg,

Use (6.5a), *i.e.*, thx = tan H = 0.55275 (from actual values of shx and chx corresponding to the point P),

get x = 0.62236 radian, which in terms of angle = 35.6586 deg.

(c) When x = – 2.0944 rad (–120 deg), ch $2x$ = 32.9791,

cos 2H = 0.030322, H = 44.13 deg.

(d) When x = 360 deg = 4π, H will show an angle given by

cos 2H.ch $2x$ = 1, tanH = thx = 0.0008, H = 1.1 rad = 63.0256 deg,

to give H = 44.9998 deg (= 45 deg).

6.1.2 Circular Representation of Single Nbic Functions

Single Nbic circle, C_1

Similar to the expression involving hyperbolic functions, *viz.*, $ch^2\,x + sh^2\,x = ch\,2x$, the Nbic functions also follow the same form, that is

$$\cos N^2 x + \sin N^2 x = ch\ 2x \qquad (6.8a)$$

and as such, it cannot represent the equation of a circle. But on substituting

$$X = [\cos Nx]/(\sqrt{ch\ 2x})$$

$$Y = [\sin Nx]/(\sqrt{ch\ 2x}) \qquad (6.8b)$$

for any point P in it, we arrive at the equation: $X^2 + Y^2 = 1$.

In view of the substitution (6.8b), (6.8a) takes the form

$$[\cos Nx / (\sqrt{ch\ 2x})]^2 + [\sin Nx / (\sqrt{ch\ 2x})]^2 = 1 \qquad (6.8c)$$

This can be identified as that of a circle of radius unity, and will be termed as **Single Nbic Circle** (or in short, **C_1 circle**).

The parameter x in (shx) or (chx) is a number, and can be converted to degree by multiplying it by (180/π), whenever required.

In **Table 6.2,** the coordinates of P(X, Y) for various values of the basic variable x along with the corresponding gA values as N_1, are given.

Table 6.2 Single Nbic Circle, C_1

Sr. No.	x (rad)	x (deg)	X	Y	gA = N_1 (deg)	sA= $N_1 - x$ (deg)	gap angle (deg)
1 (start)	0	0	1.00000	0.00000	0	0.000	–
2	π/6	30	0.56406	0.82573	55.65997	25.65997	55.65997
3	π/3	60	–0.13882	0.99032	97.97858	37.97858	42.31861
4	π/2	90	–0.67592	0.73698	132.52382	42.52382	34.54524
5	2π/3	120	–0.96189	0.27344	164.13331	44.13331	31.60949
6	5π/6	150	–0.96729	–0.25368	194.70740	44.70740	30.57409
7	π	180	–0.70843	–0.70579	224.89466	44.89466	30.18726
8	7π/6	210	–0.25945	–0.96576	254.96045	44.96045	30.06579
9	4π/3	240	0.25860	–0.96599	284.98678	44.98678	30.02633
10	3π/2	270	0.70705	–0.70716	314.99595	44.99595	30.00917
11	5π/3	300	0.96592	–0.25885	344.99514	44.99514	29.99919
12	11π/6	330	0.96593	0.25881	374.99933	44.99933	30.00419
13(end)	2π	360	0.70711	0.70710	405.000	45.00000	30.00067

The sweep angle sA = + ve, $[\text{gap angle}]_m = [(N_1)_m - (N_1)_{m-1}]$

>>

In **Figure 6.3,** the plot of Y versus X according to (6.8b) for Single Nbic Circle, C_1 is given.

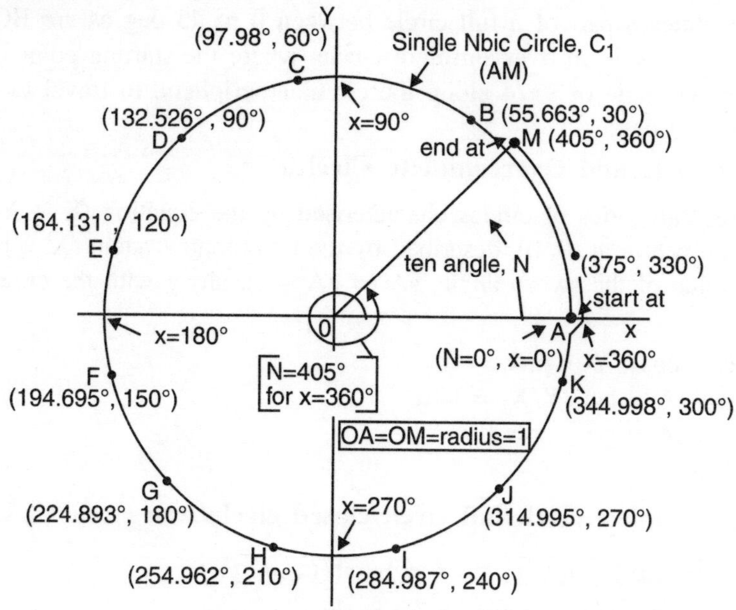

Fig. 6.3: Single Nbic circle, C_1

The angle, subtended by the radius corresponding to a point P at X = $[\cos Nx]/(\sqrt{ch\ 2x})]$, Y = $[\sin Nx]/(\sqrt{ch\ 2x})]$, on this circle with the horizontal axis, X = $[\cos Nx]/(\sqrt{ch\ 2x})]$, is the geometrical angle, gA = (teN angle) N_1.

Contrary to the Hyperbolic circle, it can be noticed that, the starting point A(1, 0) in single Nbic circle on the X axis, after travelling an angular distance corresponding to $x = 2\pi$ (360 deg), along the circumference, does not end on the same point A(1, 0), but continues to travel further in excess of 2π, along same periphery to an additional sweep angle, and ends some where at the point M (refer to **Figure 6.3**).

The single Nbic circle is characterized by circular helix form having circumferential length of $\frac{9}{4}\pi$, which is $\frac{1}{4}\pi$ in excess of 2π.

The sweep angle (sA) appears to be stable at 45 deg right from x = 210 deg onwards till x = 360 deg.

It is clearly seen in this case that, $N_1 > x$. That is, sA = $(N_1 - x)$, is (+ve). In this sense, single Nbic circle can be called, **positive circle** also.

Thus, the single Nbic circle (when viewed in reference to **Figure 1.3**) **neither** is a complete circle, where the starting point B and the end point C, corresponding to 0 deg and 360 deg, respectively, coincide on the same point, as in the case of **Euclidean circle** (to be denoted by $\mathbf{C_0}$), **nor** as an open or incomplete circle, where the starting point B (corresponding to 0 deg), sweeps an angle of $\pi/4$ along the circular periphery, to travel to the end point C (corresponding to 360 deg), constitutes a part of a full circle between 0 to 45 deg as arc BC, as in the case of Hyperbolic Circle; but is an **overcomplete circle**, where the starting point B (corresponding to 0 deg), sweeps an angle of $9\pi/4$ along the circular periphery, to travel to the end point C (corresponding to 360 deg).

Complete, Incomplete and Overcomplete Circles

In this sense, three categories of circles characterised by the equation, $X^2 + Y^2 = 1$, possessing radius of unity, and centre at (0, 0), described by the coordinate system (X, Y), can be done on the basis of the value of the sweep angle, sA (= gA – x), along with the criterion, gA = $\tan^{-1}$ (Y/X), as follows:

For $X = \cos x$, $Y = \sin x$,

$$\tan gA = \tan (Y/X) = \tan x$$

$$sA = gA - x = 0 \tag{6.9a}$$

They are termed as

Euclidean circle/complete circle/full circle/closed circle/zero circle/C_0 circle

For $X = [chx]/(\sqrt{ch\ 2x})$, $Y = [shx]/(\sqrt{ch\ 2x})$

$$\tan gA = \tan H = \tan (Y/X) = thx$$

$$sA = gA - x = (-\text{ ve}) \tag{6.9b}$$

They are termed as

Incomplete circle/open circle/inferior circle/negative circle/C_H circle

For $X = [\cos Nx]/(\sqrt{ch\ 2x})$, $Y = [\sin Nx]/(\sqrt{ch\ 2x})$

$$\tan gA = \tan S_1 = \tan (Y/X) = \tan Nx = \frac{\tan x + thx}{1 - \tan x.thx}$$

$$sA = gA - x = (+\text{ve}) \tag{6.9c}$$

They are termed as

Overcomplete circle/superior circle/positive circle/C_N circle

Noteworthy is that, in single Nbic circles, corresponding to x in the function X and Y, which ranges from (0 to 2π), the angle, N_1 in the Circle (arc A to M), shows a range of (0 to $9\pi/4$) *i.e.*, (0 to 405 deg).

This geometrical angle (teN angle) differs from the actual value of the parameter x (which can be converted to angle in degree measure), and can be calculated using (1.34), that is

$$\tan N_1 = \tan Nx = \frac{\tan x + thx}{1 - \tan x.thx} \tag{6.10}$$

This can be demonstrated with an example as follows.

Numerical Example

Example 6.2

Calculate the geometric angle N_1 for

(a) x = + 120 deg = 2.0944 rad, and
(b) x = + 210 deg = 3.665 rad

Solution

(a)

$$x = +120 \text{ deg} = 2.0944 \text{ rad}$$

$$\sin x = 0.866, \cos x = -0.5,$$

$$shx = 3.99869,\ chx = 4.121836,\ thx = 0.970124$$

$$\sin N(x, x) = \sin x\ chx + \cos x\ shx$$

$$= (0.866)(4.121836) + (-0.5)(3.99869) = 1.5703$$

$$\cos N(x, x) = \cos x\ chx - \sin x\ shx$$

$$= (-0.5)(4.121836) - (0.866)(3.99869) = -5.5239$$

$$\tan N(x, x) = [\sin N(x, x)] / [\cos N(x, x)]$$

$$= 1.5703/-5.5239 = \mathbf{-0.28427}$$

alternatively, using (1.34)

$$\tan N(x, x) = \frac{\tan x + thx}{1 - \tan x.thx} = \frac{-1.732 + 0.970}{1 + 1.732 \times 0.970} = \mathbf{-0.284324}$$

$$\tan N_1 = 0.284324 \Rightarrow N_1 = -15.87 \text{ deg } (-0.277 \text{ rad})$$

Geometrically, this angle is in second quadrant, measured 15.86669 deg clockwise from the horizontal axis-X, represented by, $[\cos N(x, x) / (\sqrt{ch\ 2x})]$. In reference to the 0 – line, *i.e.*, OX line, this angle shows, 180 – 15.86669 = 164.13331 deg (**Table 6.2**, item 5).

(b)

$$x = +210 \text{ deg} = 3.665 \text{ rad}$$

$$\sin x = -0.5,\ \cos x = -0.866,\ \tan x = 0.57735$$

$$shx = 19.51901,\ chx = 19.54461,\ thx = 0.99869$$

$$\sin N(x, x) = \sin x\ chx + \cos x\ shx$$

$$= (-0.5)(19.54461) + (-0.866)(19.51901) = -26.676259$$

$$\cos N(x, x) = \cos x\ chx - \sin x\ shx$$

$$= (-0.866)(19.54461) - (-0.5)(19.51901) = -7.166622$$

$$\tan N(x, x) = [\sin N(x, x)] / [\cos N(x, x)]$$

$$= -26.676259/-7.166622 = \mathbf{3.722292}$$

Alternatively, using formula (1.34)

$$\tan N(x, x) = \frac{\tan x + thx}{1 - \tan x.thx} = \frac{0.57735 + 0.99869}{1 - 0.57735 \times 0.99869} = \mathbf{3.7222873}$$

$$\tan N_1 = 3.7222873 \Rightarrow N_1 = 74.9624 \text{ deg } (-0.277 \text{ rad})$$

Thus take

$$\tan N_1 = \tan N(x, x) \Rightarrow N_1 = +74.9624 \text{ deg.}$$

Geometrically, this angle is in first quadrant, measured 74.9624 deg anticlockwise from the horizontal axis-X, represented by, $[\cos N(x, x) / (\sqrt{ch\ 2x})]$.

6.1.3 Circular Representation of Double Nbic Functions (E type, F type)

Double Nbic circle (E type), C_{2e}, and Double Nbic circle (F type), C_{2f}

Similar to the expression involving hyperbolic functions (6.2) *viz.*, $ch^2x + sh^2x = ch\ 2x$, and single Nbic functions (6.8a) *viz.*, $\cos N^2\ x + \sin N^2\ x = ch\ 2x$, it can be noted that both the following expressions corresponding to the expressions (3.10a), and (3.19a) *viz*:

double Nbic functions (E type): $\cos N^2_{2e}(x) + \sin N^2_{2e}(x) = ch\ 2x$ (6.11)

double Nbic functions (F type): $\cos N^2_{2f}(x) + \sin N^2_{2f}(x) = ch^2\ 2x$ (6.12)

cannot represent the equation of a circle. But, if suitably rearranged, each with proper set of coordinate system (X,Y), it can be seen that both will follow the circular form of equation, that is, $X^2 + Y^2 = 1$. Substituting then:

for double Nbic functions (E type):

$$X = [\cos N_{2e}(x)]/(\sqrt{ch\ 2x})$$

$$Y = [\sin N_{2e}(x)]/(\sqrt{ch\ 2x}) \qquad (6.13a)$$

so that the teN angle N_{2e} is given by using (3.9b), as

$$\begin{aligned}\tan N_{2e}(x) &= \tan\ (Y/X)\\ &= [\sin N_{2e}(x)\ /(\sqrt{ch\ 2x})]/[\cos N_{2e}(x)/(\sqrt{ch\ 2x})]\\ &= [\sin N_{2e}(x)]/[\cos N_{2e}(x)]\\ &= \tan N_{2e}(x)\\ &= [\tan\ 2x + thx]/[1 - \tan\ 2x\ thx] \qquad (6.13b)\end{aligned}$$

and

for double Nbic functions (F type):

$$X = [\cos N_{2f}(x)]/(ch\ 2x)$$

$$Y = [\sin N_{2f}(x)]/(ch\ 2x) \qquad (6.14a)$$

so that the teN angle N_{2f} is given by using (3.21b), as

$$\begin{aligned}\tan N_{2f}(x) &= \tan\ (Y/X)\\ &= [\sin N_{2f}(x)/ch\ 2x]\ /\ [\cos N_{2f}(x)\ /\ ch\ 2x]\\ &= [\sin N_{2f}(x)]\ /\ [\cos N_{2f}(x)]\\ &= \tan N_{2f}(x)\\ &= [\tan\ 2x + thx]\ /\ [1 - \tan\ 2x\ thx]\\ &= [\tan x + sh\ 2x]\ /\ [1 - \tan x\ sh\ 2x] \qquad (6.14b)\end{aligned}$$

both the expressions (6.11and 6.12) satisfy the circular form of equation, $X^2 + Y^2 = 1$.

Thus in view of the substitution (6.13a), (6.11) takes the form

$$[\cos N_{2e}(x)\ /\ (\sqrt{ch\ 2x})]^2 + [\sin N_{2e}(x)\ /\ (\sqrt{ch\ 2x})]^2 = 1 \qquad (6.15)$$

This equation is that of **double Nbic circle (E type),** or in short, **C_{2e} circle** of radius unity **(Figure 6.4**).

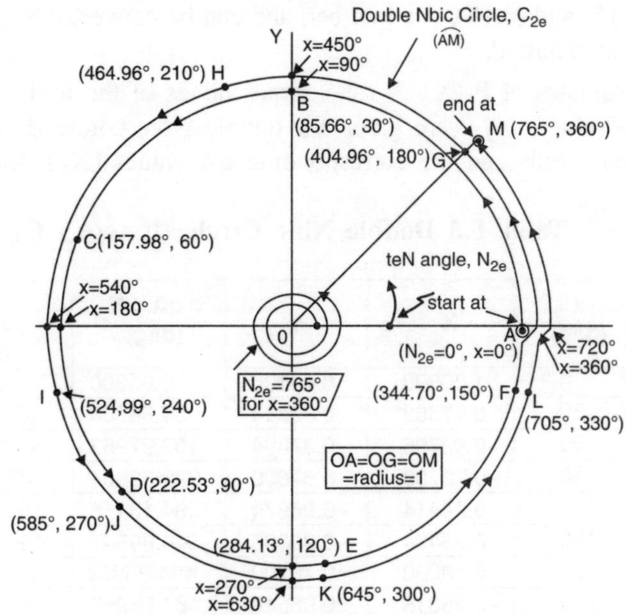

Fig. 6.4 : Double Nbic circle, C_{2e}

Expression (6.12) on the strength of (6.14a), reappears in the form

$$[\cos N_{2f}(x) / (ch\ 2x)]^2 + [\sin N_{2f}(x) / (ch\ 2x)]^2 = 1 \tag{6.16}$$

This equation is that of **double Nbic Circle (F type),** or in short, $\mathbf{C_{2f}}$ **circle** of radius unity (**Figure 6.5**).

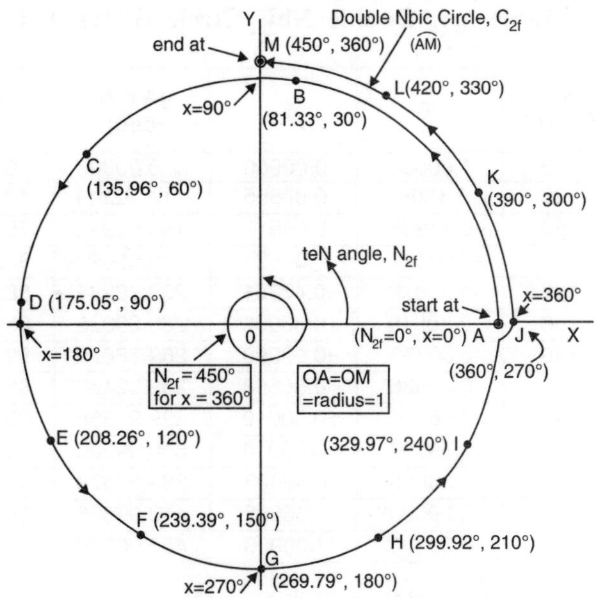

Fig. 6.5 : Double Nbic circle, C_{2f}

The parameter x in (6.15) and (6.16), is a number, and can be converted to degree by multiplying it by (180/π), whenever required.

In **Table 6.3**, the coordinates of P (X, Y) for various values of the basic variable x along with the corresponding gA values (N_{2e}), are given for double Nbic Circle (E type), whereas **Table 6.4**, presents the similar items and the corresponding gA values (N_{2f}), for double Nbic Circle (F type).

Table 6.3 Double Nbic Circle (E type), C_{2e}

Sr. No.	x (rad)	x (deg)	X	Y	gA = N_{2e} (deg)	sA= N_{2e} − x (deg)	Gap angle (deg)
1 (start)	0	0	1.00000	0.00000	0.00000	0.00000	–
2	π/6	30	0.07562	0.99714	85.66304	55.66304	85.66304
3	π/3	60	–0.92705	0.37494	157.97963	97.97963	72.31659
4	π/2	90	–0.73700	–0.67600	222.52804	132.52804	64.54841
5	2π/3	120	0.24414	–0.96974	284.13118	164.13118	61.60314
6	5π/6	150	0.96454	–0.26395	344.69547	194.69547	60.56429
7	π	180	0.70800	0.70700	404.95951	224.95951	60.26404
8	7π/6	210	–0.25818	0.96610	464.96205	254.96205	60.00254
9	4π/3	240	–0.96587	0.25904	524.98689	284.98689	60.02484
10	3π/2	270	–0.70700	–0.70700	585.00000	315.00000	60.01311
11	5π/3	300	0.25879	–0.96593	644.99844	344.99844	59.99844
12	11π/6	330	0.96592	–0.25883	704.99935	374.99935	60.00091
13(end)	2π	360	0.707000	0.70700	765.00000	405.00000	60.00065

The sweep angle sA = +ve, $[\text{gap angle}]_m = [(N_{2e})_m - (N_{2e})_{m-1}]$

Table 6.4 Double Nbic Circle (F type), C_{2f}

Sr. No.	x (rad)	x (deg)	X	Y	gA = N_{2f} (deg)	sA = N_{2f} −x (deg)	gap angle (deg)
1 (start)	0	0	1.00000	0.00000	0.00000	0.00000	–
2	π/6	30	0.15081	0.98856	81.32611	51.32611	81.32611
3	π/3	60	–0.71885	0.69517	135.95942	75.95942	54.63331
4	π/2	90	–0.99627	0.08627	175.05093	85.05093	39.09151
5	2π/3	120	–0.88079	–0.47351	208.26234	88.26234	33.20304
6	5π/6	150	–0.50919	–0.86066	239.39028	89.39028	31.12794
7	π	180	–0.00373	–0.99999	269.78629	89.78629	30.39601
8	7π/6	210	0.49886	–0.86668	299.92468	89.92468	30.13839
9	4π/3	240	0.86580	–0.50040	329.97369	89.97369	30.04901
10	3π/2	270	0.99999	–0.00016	360.00000	90.00000	30.02631
11	5π/3	300	0.86605	0.49995	389.99678	89.99678	29.99678
12	11π/6	330	0.50000	0.86602	419.99885	89.99885	30.00207
13 (end)	2π	360	0.00000	1.00000	450.00000	90.00000	30.00115

The sweep angle sA = +ve, $[\text{gap angle}]_m = [(N_{2f})_m - (N_{2f})_{m-1}]$

It can be noticed from **Table 6.3** and **Table 6.4** that, for both the Nbic circles, [double Nbic Circle (E type), as well as double Nbic Circle (F type)], the starting point A(1, 0) on Nbic Circles on the X-axis, after travelling an angular distance corresponding to $x = 2\pi$ (360 deg), along the circumference, does not end on the same point A(1, 0), but continues to travel further along same periphery, and ends some where at the point M.

For the former circle $\mathbf{C_{2e}}$, the end point M rests in the first quadrant, after completing two full turns covering an angular distance of $\pi/4$ in excess of 4π, and the corresponding gA (= N_{2e}) is at $17\pi/4$. Records (**Table 6.3**) show that, at this end point M, the sweep angle sA = $(N_{2e} - x)$ is $9\pi/4$.

The variation of the gap angle with respect to x, (refer to **Figure 6.2** in the light of **Table 6.3**) for $\mathbf{C_{2e}}$ circle reveals that, the gap angle approches to be stable at 60 deg, and attains stability right from $x = 210$ deg onwards till $x = 360$ deg.

Since, $N_{2e} > x$. that is, sA = $(N_{2e} - x)$ is positive, double Nbic circles (E type) are then **positive circles**.

For the later circle $\mathbf{C_{2f}}$, the end point M (corresponding to 2π) is located finally on the Y-axis itself in the first quadrant, after circling one full turn, which anglewise, is $\pi/2$ in excess of 2π.

In this case, the sweep angle sA, can be seen to converge towards the value of 90 deg right from $x = 210$ deg onwards till $x = 360$ deg. The data of **Table 6.4** confirms this observation. The variation of the gap angle with respect to x, (refer to **Figure 6.2** in the light of **Table 6.4**) for $\mathbf{C_{2f}}$ circle reveals that, the gap angle approches to be stable at 30 deg, and attains stability right from $x = 210$ deg onwards till $x = 360$ deg. The trend of the gap angle as the function of x, is identical with those of the other Nbic circles, the peak value is 81.33 deg, which occurs at $x = \pi/6$.

The Nbic angle N_{2f} for the end point M is $5\pi/2$ (450 deg), which is $\pi/2$ (90 deg) in excess of 2π. That is, sA = $(S_1 - x)$, is (+ve), **double Nbic circle (F type)** then complies with **positive circle**.

Typical records in **Table 6.3** and **Table 6.4** with the help of numerical calculation, are shown in the following examples.

Numerical Example:

Example 6.3

Obtain the values of gA, sA and gap angles, for:

$$x = 300 \text{ deg for } C_{2e} \text{ circle.}$$
$$x = 150 \text{ deg for } C_{2f} \text{ circle.}$$

Refer to the data from **Table 6.3.**

For $\mathbf{C_{2e}}$ circle:

$$x = 300 \text{ deg} = 5.235988 \text{ rad}$$
$$\sin x = -0.866025, \cos x = 0.5$$

$$shx = 93.95465,\ chx = 93.95998,\ \sqrt{ch\ 2x} = 132.875741$$

$$\begin{aligned}\text{sinN}x &= \sin x\ chx + \cos x\ shx\\ &= (-0.866025)\ (93.95998) + (0.5)\ (93.95465) = -\ 34.394366\\ \text{cosN}x &= \cos x\ chx - \sin x\ shx\\ &= (0.5)\ (93.95998) - (-\ 0.866025)\ (93.95465) = 128.347065\end{aligned}$$

Recall (3.9a)

$$\begin{aligned}\text{sinN}_{2e}(x, x) &= \text{sinN}x \cos x + \text{cosN}x \sin x\\ &= (-34.394366)\ (0.5) + (128.347065)\ (-\ 0.866025)\\ &= -\ 128.348950\\ \text{cosN}_{2e}(x, x) &= \text{cosN}x \cos x - \text{sinN}x \sin x\\ &= (128.347065)\ (0.5) - (-\ 34.394366)\ (-\ 0.866025) = 34.387152\end{aligned}$$

$$X = [\text{cosN}_{2e}(x, x)] / \sqrt{ch\ 2x} = 34.387152 / 132.875741 = 0.258792$$

$$Y = [\text{sinN}_{2e}(x, x)] / \sqrt{ch\ 2x} = -\ 128.348950 / 132.875741 = -\ 0.965932$$

$$X^2 + Y^2 = (0.258792)^2 + (-\ 0.965932)^2 = 1$$

$$\begin{aligned}gA = N_{2e} &= \tan^{-1}(Y/X)\\ &= \tan^{-1}\ [(-128.348950)/(34.387152)] = \tan^{-1}(-\ 3.732468)\\ &= -\ 75.0016 \text{ deg}\end{aligned}$$

Since, X is positive and Y is negative, angle gA (= N_{2e}) will lie in the 4-th quadrant. It is important to note that, just after x = 150 deg (0.964539 rad), the corresponding teN angle N_{2e} completes one complete turn of 360 deg, and crosses the zero line OX once. During the next turn in continuation, this value of –75.0016 deg (corresponding to x = 300 deg), is obtained in the 4-th quadrant.

So measuring from OX line (anticlockwise +ve),

Geometrical angle gA (= teN angle N_{2e}) = 360 + (360 – 75.0016) = 644.99844 = 645 deg.

Sweep angle sA = gA – x = 645 – 300 = 345 deg

Gap angle (at x = 300 deg)

= present value of gA (at x = 300 deg)
– previous value of gA (at x = 270 deg)
= 645 – 585 = 60 deg

These values are entered in the 11th row of **Table 6.3**.

Example 6.4

Obtain the values of gA, sA and gap angles, for x = 90 deg for C_{2f} circle.

Refer to the data from **Table 6.4**.

For C_{2f} circle:

$$\begin{aligned}x &= 90 \text{ deg} = 1.570796 \text{ rad}\\ \sin x &= 1,\ \cos x = 0\\ shx &= 2.301299,\ chx = 2.509178,\ sh\ 2x = 11.548738,\ ch\ 2x = 11.59195\end{aligned}$$

$$\begin{aligned} \sin N x &= \sin x \, ch x + \cos x \, sh x \\ &= (1)\,(2.509178) + (0)\,(2.301299) = 2.509178 \\ \cos N x &= \cos x \, ch x - \sin x \, sh x \\ &= (0)\,(2.509178) - (1)\,(2.301299) = -\,2.301299 \end{aligned}$$

Recall (3.22a)

$$\begin{aligned} \sin N_{2f}(x) &= \sin N x \, ch x + \cos N x \, sh x = \sin x + \cos x \, sh\, 2x \\ \cos N_{2f}(x) &= \cos N x \, ch x - \sin N x \, sh x = \cos x - \sin x \, sh\, 2x \end{aligned}$$

Use either

$$\begin{aligned} \sin N_{2f}(x) &= \sin N x \, ch x + \cos N x \, sh x \\ &= (2.509178)\,(2.509178) + (\,-2.301299)\,(2.301299) = 1 \\ \cos N_{2f}(x) &= \cos N x \, ch x - \sin N x \, sh x \\ &= (-\,2.301299)\,(2.509178) - (2.509178)\,(2.301299) \\ &= -\,11.548738 \end{aligned}$$

Or alternatively, use

$$\begin{aligned} \sin N_{2f}(x) &= \sin x + \cos x \, sh\, 2x \\ &= 1 + (0)\,(11.548738) = 1 \\ \cos N_{2f}(x) &= \cos x - \sin x \, sh\, 2x \\ &= 0 - (1)\,(11.548738) = -\,11.548738 \end{aligned}$$

$$X = [\cos N_{2f}(x)]/(ch\, 2x) = -\,11.548738/11.59195 = -\,0.996272$$

$$Y = [\sin N_{2f}(x)]/(ch\, 2x) = 1/11.59195 = 0.086267$$

$$X^2 + Y^2 = (-\,0.996272)^2 + (0.086267)^2 = 1$$

$$\begin{aligned} gA = N_{2f} &= \tan^{-1}(Y/X) \\ &= \tan^{-1}[(0.086267)/(-\,0.996272)] \\ &= \tan^{-1}(-0.08659) = -\,4.948887 \text{ deg} \end{aligned}$$

Since, X is negative and Y is positive, angle gA (= N_{2f}) will lie in the 2-nd quadrant. It is important to note that, the corresponding teN angle N_{2f} does not complete even a turn of 180 deg. This value of –4.948887 deg (corresponding to x = 90 deg) obtained in the 2-nd quadrant when measured from OX line (anticlockwise positive), reads

Geometrical angle gA (= teN angle N_{2e}) = 180 – 4.948887
= 175.051113 (=175.05093) deg.

Sweep angle sA = gA – x =175.05093 – 90 = 85.05093 deg

Gap angle (at x = 90 deg)
= present value of gA (at x = 90 deg)
– previous value of gA (at x = 60 deg)
= 175.05093 – 135.95942 = 39.09151 deg

These values are entered in the 4-th row of **Table 6.4**.

6.1.4 Circular Representation of Triple Nbic Functions (E type, F type)

Triple Nbic Functions (E type), C_{3e}, and Triple Nbic Functions (F type), C_{3f}

Refer to the expressions (4.8a) and (4.21b) related to:

Triple Nbic functions (E type) *viz.*, $\cos N^2_{3e}(x) + \sin N^2_{3e}(x) = ch^2\, 2x$ (6.17)

Triple Nbic functions (F type) *viz.*, $\cos N^2_{3f}(x) + \sin N^2_{3f}(x) = ch^3\, 2x$ (6.18)

Neither (6.17) nor (6.18) speaks for the equation of a circle. Both of these two equations, however, can be converted to that of a circle, of the type, $X^2 + Y^2 = 1$, with proper adjustment on expressing each of them under suitable coordinate system (X,Y). Thus on substituting:

for Triple Nbic functions (E type):

$$X = [\cos N_{3e}(x)]/(ch\ 2x)$$
$$Y = [\sin N_{3e}(x)]/(ch\ 2x) \qquad (6.19)$$

and

for Triple Nbic functions (F type):

$$X = [\cos N_{3f}(x)]/(\sqrt[3/2]{ch\ 2x})$$
$$Y = [\sin N_{3f}(x)]/(\sqrt[3/2]{ch\ 2x}) \qquad (6.20)$$

both the expressions (6.19, 6.20) conform to the circular form of equation, $X^2 + Y^2 = 1$.

Thus in view of the substitution (6.19), (6.17) takes the form

$$[\cos N_{3e}(x) / (ch\ 2x)]^2 + [\sin N_{3e}(x)/(ch\ 2x)]^2 = 1 \qquad (6.21)$$

This equation is that of **Triple Nbic Circle (E type)**, or in short, $\mathbf{C_{3e}}$ **circle,** of radius unity **(Figure 6.6).**

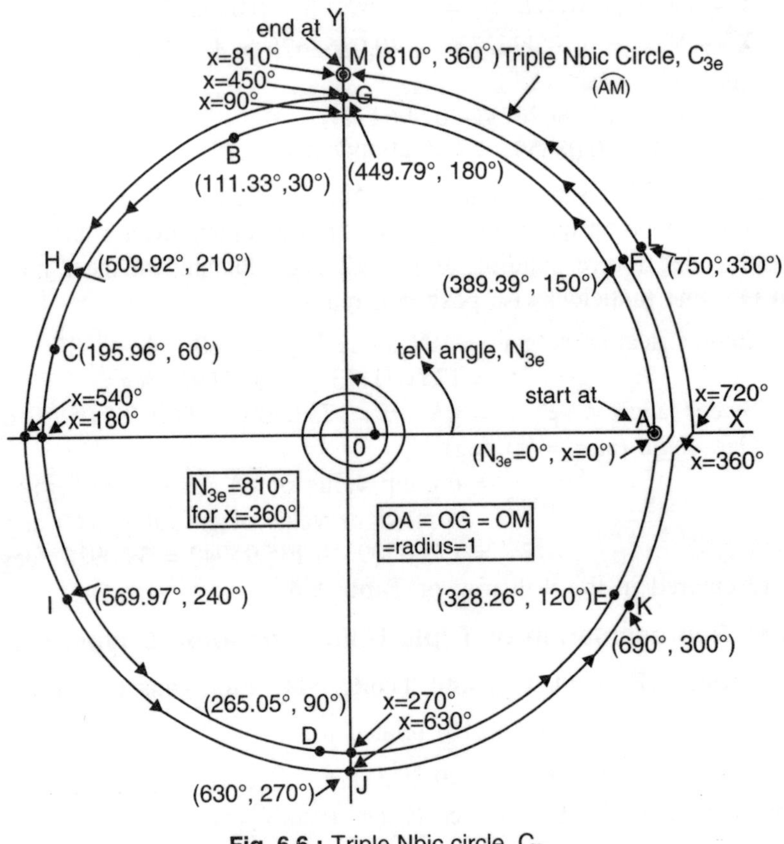

Fig. 6.6 : Triple Nbic circle, C_{3e}

Now, (6.18) on the strength of (6.20), reappears in the form

$$[\cos N_{3f}(x)\ /(\sqrt[3/2]{ch\ 2x}\ )]^2 + [\sin N_{3f}(x)\ /(\sqrt[3/2]{ch\ 2x}\ )]^2 = 1 \qquad (6.22)$$

This equation is that of **Triple Nbic Circle (F type)**, or in short, **C_{3f} circle,** of radius unity (**Figure 6.7**).

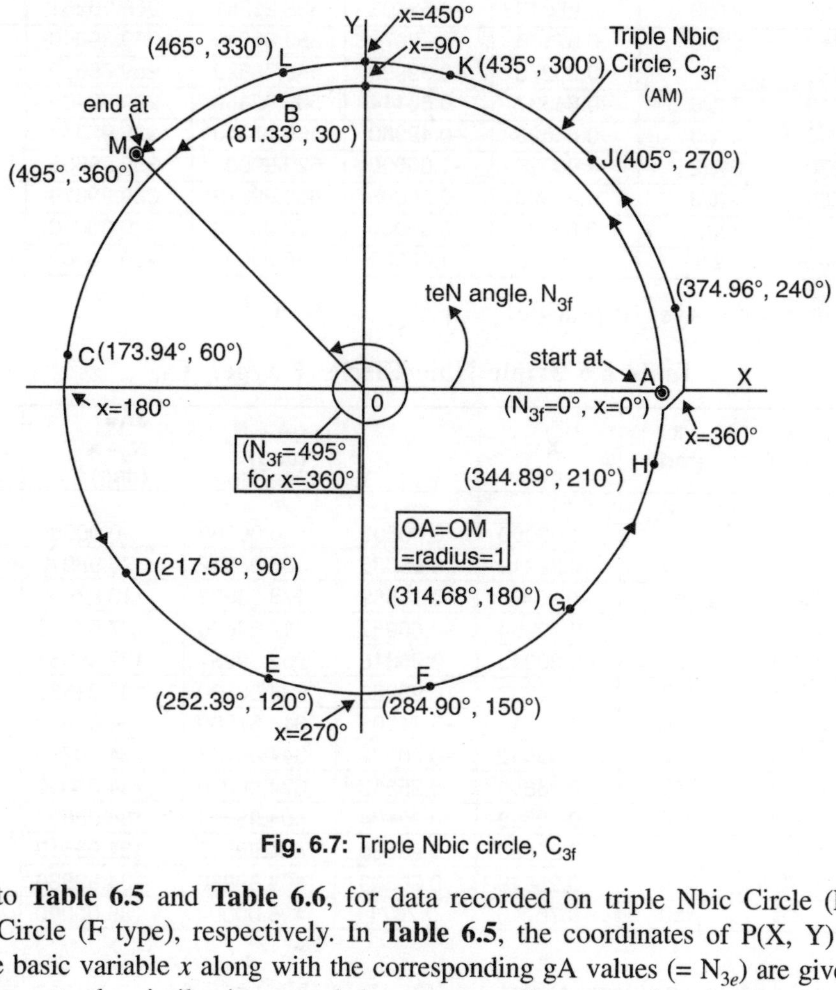

Fig. 6.7: Triple Nbic circle, C_{3f}

Refer to **Table 6.5** and **Table 6.6**, for data recorded on triple Nbic Circle (E type) and triple Nbic Circle (F type), respectively. In **Table 6.5**, the coordinates of P(X, Y) for various values of the basic variable x along with the corresponding gA values (= N_{3e}) are given, whereas, **Table 6.6** presents the similar items and the corresponding gA values (= N_{3f}).

Table 6.5 Triple Nbic Circle (E type), C_{3e}

Sr. No.	x (deg)	x (rad)	X	Y	gA = N_{3e} (deg)	sA= $N_{3e} - x$ (deg)	gap angle (deg)
1 (start)	0	0	1.00000	0.00000	0.00000	0.00000	–
2	30	π/6	–0.36367	0.93153	111.32571	81.32571	111.32571
3	60	π/3	–0.96146	–0.27496	195.95956	135.95956	84.63385
4	90	π/2	–0.08627	–0.99627	265.05110	175.05110	69.09154
5	120	2π/3	0.85047	–0.52603	328.26251	208.26251	63.21141
6	150	5π/6	0.87130	0.49076	389.39040	239.39040	61.12789
7	180	π	0.00373	0.99999	449.78629	269.78629	60.39589
8	210	7π/6	–0.86537	0.50114	509.92466	299.92466	60.13837
9	240	4π/3	–0.86626	–0.49960	569.97343	329.97343	60.04877
10	270	3π/2	–0.00016	–1.00000	629.99083	359.99083	60.01740
11	300	5π/3	0.86600	–0.50005	689.99679	389.99679	60.00590
12	330	11π/6	0.86603	0.50000	750.00000	420.00000	60.00321
13 (end)	360	2π	0.00000	1.00000	810.00000	450.00000	60.00000

The sweep angle sA = +ve, $[\text{gap angle}]_m = [(N_{3e})_m - (N_{3e})_{m-1}]$

Table 6.6 Triple Nbic Circle (F type), C_{3f}

Sr. No.	x (deg)	x (rad)	X	Y	gA = N_{3f} (deg)	sA= $N_{3f} - x$ (deg)	gap angle (deg)
1 (start)	0	0	1.00000	0.00000	0.00000	0.00000	–
2	π/6	30	–0.29219	0.95636	106.98912	76.98912	106.98912
3	π/3	60	–0.99441	0.10559	173.93903	113.93903	66.94991
4	π/2	90	–0.79254	–0.60982	217.57670	127.57670	43.63767
5	2π/3	120	–0.30248	–0.95316	252.39357	132.39357	34.81687
6	5π/6	150	0.24337	–0.96993	284.08530	134.08530	31.69173
7	π	180	0.70313	–0.71106	314.67900	134.67900	30.59370
8	7π/6	210	0.96542	–0.26072	344.88737	134.88737	30.20837
9	4π/3	240	0.96610	0.25815	374.96050	134.96050	30.07313
10	3π/2	270	0.70728	0.70694	404.98614	134.98614	30.02564
11	5π/3	300	0.25890	0.96590	434.99513	134.99513	30.00899
12	11π/6	330	–0.25879	0.96593	464.99830	134.99830	30.00317
13(end)	2π	360	–0.70710	0.70711	495.00000	135.00000	30.00170

The sweep angle sA = +ve, $[\text{gap angle}]_m = [(N_{3f})_m - (N_{3f})_{m-1}]$

For the triple Nbic Circle (E type), starting from A (x = 0 deg), the locus of the point P takes two complete turns and additional quarter turn, on the circular periphery of radius unity, to reach the end point M. That is to say that, at this end point M, gA (= N_{3e}) = $4\pi + \pi/2 = 9\pi/2$.

The sweep angle sA (which is the measure of the teN angle of a point on the Nbic circle in excess of the corresponding value of x) at M is, gA – x = $N_{3e} - x = 5\pi/2$.

Figure 6.2 reveals that the peak value of the gap angle is 111.32 deg which, like all other Nbic circles (including hyperbolic circles), occurs at x = 30 deg (π/6). Striking similarity exists between C_{3e} circles and C_{2e} circles in regard to the occurrence of the stable value of the gap angle. It is of same value 60 deg, and commences to converge at x = 210 deg and remains constant up to x = 360 deg.

The Nbic angle N_{3e} for the end point M is 9π/2 (810 deg), which is π/2 (90 deg) in excess of 4π. This ensures that, sA is positive. **Triple Nbic circle (E type)** then complies with **positive circle**.

For the triple Nbic Circle (F type), the end point M (corresponding to 2π) is reached finally in the third quadrant, which is π/4 in excess of 3π. This is measured by, gA = N_{3f} = 13π/4 (= 495 deg). In this case, the sweep angle sA at the end point M = 5π/4 (= 135 deg).

Figure 6.2 reveals that the peak value of the gap angle is 106.98 deg which, like all other Nbic circles (including hyperbolic circles), occurs at x = 30 deg (π/6). Striking similarity exists between C_{3f} circles and C_{2f} circles in regard to the occurrence of the stable value of the gap angle. It is of same value 30 deg, and commences to converge at x = 210 deg and remains constant till the end point M (at x = 360 deg) is reached. The data of **Table 6.6** confirms this observation.

Since the sweep angle, sA is positive, **triple Nbic circle (F type)**, or in short, C_{3f} circles, are **positive circle**.

Typical records in **Table 6.5** and **Table 6.6** with the help of numerical calculation, are shown in the following examples.

Numerical Example

Example 6.5

Obtain the values of gA, sA and gap angles, for x = 300 deg for C_{3e} circle. Refer to the data from **Table 6.5**.

For C_{3e} circle:

Recall data from Example 6.3

$$\begin{aligned}
x &= 300 \text{ deg} = 5.235988 \text{ rad} \\
\sin x &= -\ 0.866025,\ \cos x = 0.5, \\
\sin 2x &= -\ 0.866025,\ \cos 2x = -\ 0.5 \\
shx &= 93.95465,\ chx = 93.95998, \\
sh\ 2x &= 17655.95,\ ch\ 2x = 17655.95 \\
\sin Nx &= -\ 34.394366,\ \cos Nx = 128.347065
\end{aligned}$$

Recall (4.16) for, sets of $\sin N_{3e}(x)$ and $\cos N_{3e}(x)$ formulas, *viz*

$$\begin{aligned}
\sin N_{3e}(x) &= \sin 2x + \cos 2x\ sh\ 2x \\
&= 2 \sin Nx \cos Nx \\
&= \sin N_{2f}(x) \cos x + \cos N_{2f}(x) \sin x \\
\cos N_{3e}(x) &= \cos 2x - \sin 2x\ sh\ 2x \\
&= \cos N^2 x - \sin N^2 x \\
&= \cos N_{2f}(x) \cos x - \sin N_{2f}(x) \sin x
\end{aligned}$$

Use first set of formula

$$\sin N_{3e}(x) = \sin 2x + \cos 2x \; sh\ 2x$$
$$= -\ 0.866025 + (-\ 0.5\ )\ (17655.95) = -\ 8828.841025$$
$$\cos N_{3e}(x) = \cos 2x - \sin 2x \; sh\ 2x$$
$$= -\ 0.5 - (-\ 0.866025)\ (17655.95) = 15289.9941$$

Use second set of formula

$$\sin N_{3e}(x) = 2 \sin Nx \cos Nx$$
$$= 2\ (-\ 34.394366)\ (128.347065) = -\ 8828.6831857$$
$$\cos N_{3e}(x) = \cos N^2 x - \sin N^2 x$$
$$= (128.347065)^2 - (-\ 34.394366)^2 = 15289.99668$$

Use third set of formula

$$\sin N_{3e}(x) = \sin N_{2f}(x) \cos x + \cos N_{2f}(x) \sin x$$
$$= (8827.111)\ (0.5) + (15291.0)\ (-\ 0.866025) = -\ 8828.832775$$
$$\cos N_{3e}(x) = \cos N_{2f}(x) \cos x - \sin N_{2f}(x) \sin x$$
$$= (15291.0)\ (0.5) - (8827.111)\ (-\ 0.866025) = 15289.9988$$

Take first set values

$$\sin N_{3e}(x) = -\ 8828.841025, \quad \cos N_{3e}(x) = 15289.9941$$

for triple Nbic functions (E type):

Recall (6.19)

$$X = [\cos N_{3e}(x)]\ /(ch\ 2x) = 15289.9941/17655.95 = 0.865997$$
$$Y = [\sin N_{3e}(x)]\ /(ch\ 2x) = -\ 8828.841025/17655.95 = -0.500049$$
$$X^2 + Y^2 = (0.865997)^2 + (-\ 0.500049)^2 = 1$$
$$gA = N_{3e} = \tan^{-1}(Y/X)$$
$$= \tan^{-1}\ [(-\ 0.500049)/(0.865997)]$$
$$= \tan^{-1}\ (-0.577426) = -\ 30.003245 \text{ deg}$$

Since, X is positive and Y is negative, angle gA (= N_{3e}) will lie in the 4-th quadrant. It is important to note that, after x = 120 deg, the corresponding teN angle N_{3e} completes one complete turn of 360 deg, and crosses the zero line OX once. During the next turn in continuation, this value of – 30.003245 deg (corresponding to x = 300 deg), is obtained in the 4-th quadrant.

So measuring from OX line (anticlockwise +ve),

Geometrical angle gA (= teN angle N_{3e})
$$= 360 + (360 - 30.003245)$$
$$= 689.996755 = 690 \text{ deg.}$$

Sweep angle sA = gA – x = 689.996755 – 300
$$= 389.996755 = 390 \text{ deg}$$

Gap angle (at x = 90 deg)
= present value of gA (at x = 300 deg)
– previous value of gA (at x = 270 deg)
= 689.99676 – 629.99083 = 60.00593 = 60 deg

These values are entered in the 11-th row of **Table 6.5**.

Example 6.6

Obtain the values of gA, sA and gap angles, for x = 210 deg for C_{3f} circle. Refer to the data from Example 4.6

For C_{3f} circle:

$$x = 210 \text{ deg} = 3.665191 \text{ rad}$$

$$sh\ 2x = 762.981962,\ ch\ 2x = 762.982617, \sqrt{ch^3\ 2x} = 21075.21266$$

$$\text{sinN}x = -\ 26.676259,\ \text{cosN}x = -\ 7.166622$$

We have from (4.21a)

$$\text{cosN}_{3f}\ (x) = \text{cosN}x - \text{sinN}x\ sh\ 2x$$

$$\text{sinN}_{3f}\ (x) = \text{sinN}x + \text{cosN}x\ sh\ 2x$$

So,

$$\text{cosN}_{3f}\ (x) = -\ 7.166622 - (-\ 26.676259)\ (762.981962) = 20346.33781$$

$$\text{sinN}_{3f}\ (x) = -\ 26.676259 + (-\ 7.166622)\ (762.981962) = -\ 5494.679573$$

for triple Nbic functions (F type):

Recall (6.20)

$$X = [\text{cosN}_{3f}\ (x)]\ /\ (\sqrt{ch^3 2x}\,) = 20346.33781/21075.21266 = 0.965416$$

$$Y = [\text{sinN}_{3f}\ (x)]\ /\ (\sqrt{ch^3\ 2x}\,) = -\ 5494.679573/21075.21266 = -\ 0.260718$$

$$X^2 + Y^2 = (0.965416)^2 + (-\ 0.260718)^2 = 1$$

$$\text{gA} = \text{N}_{3f} = \tan^{-1}(Y/X) = \tan^{-1}\ [(-\ 0.260718)\ /\ (0.965416)] = \tan^{-1}\ (-\ 0.270058) = -\ 15.112655 \text{ deg}$$

Since, X is positive and Y is negative, angle gA (= N_{3f}) will lie in the 4-th quadrant. It is important to note that, the corresponding teN angle N_{3f} has not made complete one complete turn of 360 deg, and so does not cross the zero line OX even once. This value of –15.112655 deg (corresponding to x = 210 deg), is obtained in the 4-th quadrant.

So measuring from OX line (anticlockwise +ve),

Geometrical angle gA (= teN angle N_{3f})
= 360 – 15.112655
= 344.887345 deg

Sweep angle sA = gA – x = 344.887345 – 300 = 134.887345 deg

Gap angle (at x = 210 deg)
= present value of gA (at x = 210 deg)
– previous value of gA (at x = 180 deg)
= 344.887345 – 314.679003 = 30.208342 deg

These values are entered in the 8-th row of **Table 6.6.**

Collecting information together

In **Table 6.7** important information regarding various Nbic circles are condensed.

Table 6.7 Information for Different Nbic Circles

Description of Circle	Eqn. No.	Starting point A (X,Y)	End point M (X,Y)	gA (= N) at M = k .π/4	sA at M = m .π/4	Gap angle[b] Stable value = n .π/6	Remarks for circle type
				k	m	n	
C_H	(6.3b)	(1, 0)	(a, a)	1	–7	0	Inferior
C_0	—	(1, 0)	(1, 0)	8	0	0	complete
C_1	(6.8c)	(1, 0)	(a, a)	9	1	1	superior
C_{2f}	(6.16)	(1, 0)	(0, 1)	10	2	1	Do
C_{3f}	(6.22)	(1, 0)	(–a, a)	13	5	1	Do
C_{2e}	(6.15)	(1, 0)	(a, a)	17	9	2	Do
C_{3e}	(6.21)	(1, 0)	(0, 1)	18	10	2	Do

$a = 1/\sqrt{2} = 0.7071$

b = gap angle stable value starts at

$x = 7\pi/6$ onwards, and maintains till $x = 2\pi$

π (rad) = 180°

It is interesting to see the characteristic features of the suggested form of various Nbic circles, regarding the Nbic angle (teN angle, N) expressed in terms of the general angle (geometrical angle) gA, the sweep angle sA, and the gap angle (along with their variation with respect to the circular angle at the end point M), as provided in **Table 6.7**.

All the types of Nbic circles including the Euclidean circle and hyperbolic circle, is seen to follow a nice chain of ascending form of ladder, when arranged suitably. **Table 6.7** is slightly reshuffled during presentation to highlight this mode of arrangement for achieving the intended order. The mode of arrangement of the Nbic circle types presented, is as follows:

$$[C_H], [C_0], [C_1, (C_{2f}, C_{3f}), (C_{2e}, C_{3e})].$$

6.1.5 New Categorization for Circles

In consequence to this pattern of arrangement, all the **k-s, m-s** and **n-s** show values in increasing order, as the circle types are designed with the progressively higher order Nbic functions. It is thus possible to include the Euclidean circle C_0, and the hyperbolic circle C_H, under the same umbrella of Nbic circles C_N, namely [C_{N1} (or simply C_N), (C_{2f}, C_{3f}), (C_{2e}, C_{3e})]:

Thus according to this **new categorization for circles** we identify:

Hyperbolic Circle: C_H circle/negative circle/inferior circle/incomplete circle

Euclidean Circle: C_0 circle/zero circle/full circle/complete circle

Nbic Circles: C_N circle/positive circle/superior circle/overcomplete circle.

Note:

All the Nbic circles (**C_0, C_H** and **C_N** circles) are represented by the equation of the type, $X^2 + Y^2 = 1$, in the suitably designed coordinate system (X,Y), and satisfy the Pythagorean criterian, which is derived from the geometrical property of the right angled triangle OAP [where, OP = radius = 1, O is the centre at (0, 0), and P is on the circular periphery at (X, Y), A is the foot of the perpendicular on the X axis drawn from P].

Looking from a different angle, it can be noticed that, the Pythagorean criterian is derived from the geometrical figure of the triangle OAP, which lies inside a circle such that, out of three vertices, one vertex point (O) is at the centre, and one vertex point (P) is on the circumference, and the third vertex point (A), of course, is well within the circular domain.

Considering in an almost similar situation, the geometrical property of a particular type of geometric figure *viz., trapezium,* positioned inside a circle, three vertices of which are on the periphery, while the fourth vertex is positioned at the centre, gives rise to a simple Theorem, with the help of which, the year old unsolved problem of **trisecting an angle**, which remains unsolved since Archimedes till to day, using geometric means, has been attempted.

In **Appendix C**, this Theorem (**Bairagi's Theorem**) proposed by the author of this book, is presented, and discussed.

❑❑❑

7

APPLICATION OF SINGLE Nbic FUNCTIONS TO STRUCTURAL PROBLEMS

7.1 APPLICATION OF SINGLE Nbic FUNCTIONS TO STRUCTURAL PROBLEMS [BEAMS ON ELASTIC FOUNDATION]

7.1.1 Basic Functions

Let

$$\begin{aligned} A_{kx} &= \sin kx \; ch \; kx \\ B_{kx} &= \sin kx \; sh \; kx \\ C_{kx} &= \cos kx \; sh \; kx \\ D_{kx} &= \cos kx \; ch \; kx \end{aligned} \qquad (7.1)$$

Remember that:

Symmetrical function $f(x)$: $f(-x) = + f(x)$

Anti-symmetrical function $f(x)$: $f(-x) = - f(x)$

Then

$\mathbf{B}_{kx}$ $(= \sin kx \; sh \; kx) \Rightarrow$ Symmetrical functions

$\mathbf{D}_{kx}$ $(= \cos kx \; ch \; kx) \Rightarrow$ Symmetrical functions

$\mathbf{A}_{kx}$ $(= \sin kx \; ch \; kx) \Rightarrow$ Anti-symmetrical functions

$\mathbf{C}_{kx}$ $(= \cos kx \; sh \; kx) \Rightarrow$ Anti-symmetrical functions

Then writing in abbreviated form

$$\begin{aligned} &\text{sinN}(kx, kx) = \text{sinN } kx = A_{kx} + C_{kx} && \Rightarrow \text{anti-symmetrical function} \\ &\text{sinN}(kx, -kx) = A_{kx} - C_{kx} && \Rightarrow \text{anti-symmetrical function} \\ &\text{cosN}(kx, kx) = \text{cosN } kx = D_{kx} - B_{kx} && \Rightarrow \text{symmetrical function} \\ &\text{cosN}(kx, -kx) = D_{kx} + B_{kx} && \Rightarrow \text{symmetrical function} \end{aligned} \qquad (7.2)$$

7.1.2 Differentiation Rules

$$\begin{aligned} A'_{kx} &= (k)\,[D_{kx} + B_{kx}] = (k)\,\text{cosN}(kx, -kx) \\ B'_{kx} &= (k)\,[C_{kx} + A_{kx}] = (k)\,\text{sinN}(kx, kx) \\ C'_{kx} &= (k)\,[D_{kx} - B_{kx}] = (k)\,\text{cosN}(kx, kx) \end{aligned}$$

$$D'_{kx} = (k)\,[C_{kx} - A_{kx}] = (-k)\,\text{sinN}(kx, -kx) \qquad (7.3a)$$

where, symbol ( )$'$, and ( )$''$, are used to represent, $\frac{d}{dx}$ ( ), and $\frac{d^2}{dx^2}$ ( ), etc.

Thus for

$$\begin{aligned}
\text{sinN}(kx, kx) &= C_{kx} + A_{kx} \\
\text{sinN}'(kx, kx) &= C'_{kx} + A'_{kx} = (k)\,[\text{cosN}(kx, kx) + \text{cosN}(kx, -kx)] \\
&= (2k)\,D_{kx} \\
\text{sinN}''(kx, kx) &= -(2k^2)\,[A_{kx} - C_{kx}] = -(2k^2)\,\text{sinN}(kx, -kx) \\
\text{sinN}'''(kx, kx) &= -(2k^3)\,[\text{cosN}(kx, -kx) - \text{cosN}(kx, kx)] = (-4k^3)\,B_{kx} \\
\text{sinN}^{iv}(kx, kx) &= (-4k^3)\,\text{sinN}(kx, kx)
\end{aligned}$$

$$\mathbf{sinN^{iv}\,kx + 4k^4\,sinN\,kx = 0} \qquad (7.3b)$$

for

$$\begin{aligned}
\text{sinN}(kx, -kx) &= A_{kx} - C_{kx} \\
\text{sinN}'(kx, -kx) &= A'_{kx} - C'_{kx} = (2k)\,B_{kx} \\
\text{sinN}''(kx, -kx) &= (2k^2)\,[A_{kx} + C_{kx}] = (2k^2)\,\text{sinN}(kx, kx) \\
\text{sinN}'''(kx, -kx) &= (4k^3)\,D_{kx} \\
\text{sinN}^{iv}(kx, -kx) &= (-4k^4)\,\text{sinN}(kx, -kx)
\end{aligned}$$

$$\mathbf{sinN^{iv}(kx, -kx) + 4k^4\,sinN(kx, -kx) = 0} \qquad (7.3c)$$

for

$$\begin{aligned}
\text{cosN}(kx, kx) &= D_{kx} - B_{kx} \\
\text{cosN}'(kx, kx) &= (k)\,[(C_{kx} - A_{kx}) - (C_{kx} + A_{kx})] = (-2k)\,A_{kx} \\
\text{cosN}''(kx, kx) &= -(2k^2)\,[D_{kx} + B_{kx}] = -(2k^2)\,\text{cosN}(kx, -kx) \\
\text{cosN}'''(kx, kx) &= -(2k^3)\,[\text{sinN}(kx, kx) - \text{sinN}(kx, -kx)] = (-4k^3)\,C_{kx} \\
\text{cosN}^{iv}(kx, kx) &= (-4k^4)\,\text{cosN}(kx, kx)
\end{aligned}$$

$$\mathbf{cosN^{iv}(kx, kx) + 4k^4\,cosN(kx, kx) = 0} \qquad (7.3d)$$

for

$$\begin{aligned}
\text{cosN}(kx, -kx) &= D_{kx} + B_{kx} \\
\text{cosN}'(kx, -kx) &= (2k)\,C_{kx} \\
\text{cosN}''(kx, -kx) &= (2k^2)\,[D_{kx} - B_{kx}] = (2k^2)\,\text{cosN}(kx, kx) \\
\text{cosN}'''(kx, -kx) &= (-4k^3)\,A_{kx} \\
\text{cosN}^{iv}(kx, -kx) &= (-4k^4)\,\text{cosN}(kx, -kx)
\end{aligned}$$

$$\mathbf{cosN^{iv}(kx, -kx) + 4k^4\,cosN(kx, -kx) = 0} \qquad (7.3e)$$

Thus, omitting here the argument (kx), without loss of understanding, we write

$$\begin{aligned}
& A' = B + D,\ A'' = 2C,\ A''' = 2[D - B] = 2C',\ A^{iv} = -4A \\
\Rightarrow\ & A^{iv} + 4A = 0 \\
& B' = A + C,\ B'' = 2D,\ B''' = 2[A - C] = 2D',\ B^{iv} = -4B \\
\Rightarrow\ & B^{iv} + 4B = 0 \\
& C' = D - B,\ C'' = -2A,\ C''' = -2[D + B] = 2A',\ C^{iv} = -4C \\
\Rightarrow\ & C^{iv} + 4C = 0 \\
& D' = C - A,\ D'' = -2B,\ D''' = 2[A + C] = -2B',\ D^{iv} = -4D \\
\Rightarrow\ & D^{iv} + 4D = 0
\end{aligned} \qquad (7.4)$$

CHAPTER 7

7.1.3 Beams on Elastic Foundation Equation

A long beam of constant flectural rigidity (EI), extending from $-\infty$ to $+\infty$, rests on an elastic foundation (possessing modulus of subgrade reaction represented by the parameter k), is carrying a uniformly distributed loading of constant intensity (q) including its self weight. The foundation is assumed to behave according to Winkler's hypothesis.

Since the beam is infinitely long both sides, it is immaterial where exactly would be its mid point. The mid point O can be safely taken anywhere more or less at the middle zone of the beam. At this point O, fix the origin of the coordinates (x, y). Along the right side of the beam, x is +ve, whereas, y is +ve downwards.

The governing fourth order differential equation for deflection $y = y(x)$ of beams on elastic foundation is :

$$y^{iv} + 4k^4 y = q/(\mathrm{EI}) \tag{7.5}$$

7.1.4 Solution Using Single Nbic Function

Applying the proposed single Nbic functions, for the solution of this problem the **homogeneous solution** of the equation is sought as

$$\begin{aligned} y = {} & A_1 \operatorname{sinN}(kx, kx) + B_1 \operatorname{sinN}(kx, -kx) \\ & + C_1 \operatorname{cosN}(kx, kx) + D_1 \operatorname{cosN}(kx, -kx) \end{aligned} \tag{7.6a}$$

where, A_1, B_1, C_1, D_1 are four arbitrary constants.

Using differentiation rules (7.3a)

$$\begin{aligned} y' &= A_1 (2k) D_{kx} + B_1 (2k) B_{kx} + C_1 (-2k) A_{kx} + D_1 (2k) C_{kx} \\ &= (2k) [A_1 D_{kx} + B_1 B_{kx} - C_1 A_{kx} + D_1 C_{kx}] \\ y'' &= (2k^2) [- A_1 \operatorname{sinN}(kx, - kx) + B_1 \operatorname{sinN}(kx, kx) \\ & \quad - C_1 \operatorname{cosN}(kx, - kx) + D_1 \operatorname{cosN}(kx, kx)] \\ y''' &= (4k^3) [- A_1 B_{kx} + B_1 D_{kx} - C_1 C_{kx} - D_1 A_{kx}] \\ y^{iv} &= (-4k^4) [A_1 \operatorname{sinN}(kx, kx) + B_1 \operatorname{sinN}(kx, - kx) \\ & \quad + C_1 \operatorname{cosN}(kx, kx) + D_1 \operatorname{cosN}(kx, - kx)] \\ &= (4k^4)\, y \end{aligned} \tag{7.6b}$$

So

$$\mathbf{y^{iv} + 4k^4 y = 0} \tag{7.7}$$

For **Symmetric Problem**, (7.6) in the light of (7.2), reduces to

$$y = C_1 \operatorname{cosN}(kx, kx) + D_1 \operatorname{cosN}(kx, - kx) \tag{7.8}$$

Note that, the hyperbolic part [*viz.*, $ch(\mathrm{kx})$ and $sh(\mathrm{kx})$] in each of the Nbic functions consists of e^{kx} and e^{-kx}. At $x = +\infty$, the displacement, slope, moment, shear etc., all cannot be felt at all, and hence the terms containing e^{kx}, can be taken as zero, leaving only terms containing e^{-kx}. Then it will be sufficient to analyse the right hand side of the beam (*i.e.*, corresponding to, $x = 0$ to $+\infty$) only, using (7.8) involving two arbitrary constants, C_1 and D_1.

Thus for infinitely long beam, the equation for the deflection $y = y(x)$, becomes

$$\begin{aligned} y &= C_1 \operatorname{cosN}(kx, kx) + D_1 \operatorname{cosN}(kx, - kx) \\ &= (C_1 + D_1) \cos kx \,.ch\, kx + (-C_1 + D_1) \sin kx \,.sh\, kx \\ &= (D_1 + C_1) D_{kx} + (D_1 - C_1) B_{kx} \end{aligned} \tag{7.9}$$

Let

$$D^*_{kx} = e^{-kx}\cos kx$$
$$B^*_{kx} = e^{-kx}\sin kx$$
$$A^*_{kx} = D^*_{kx} + B^*_{kx} = e^{-kx}[\cos kx + \sin kx]$$
$$C^*_{kx} = D^*_{kx} - B^*_{kx} = e^{-kx}[\cos kx - \sin kx] \tag{7.10a}$$

so that, we have

$$(D^*_{kx})' = (e^{-kx}\cos kx)' = -k\,A^*_{kx}$$
$$(B^*_{kx})' = (e^{-kx}\cos kx)' = k\,C^*_{kx} \tag{7.10b}$$

With these, get for $x = +\infty$, with **condition that: $e^{kx} = 0$**

$$B_{kx} = \sin kx\,.sh\,kx \Rightarrow -\frac{1}{2}e^{-kx}\sin kx = -\frac{1}{2}B^*_{kx}$$

$$D_{kx} = \cos kx\,.ch\,kx \Rightarrow \frac{1}{2}e^{-kx}\cos kx = \frac{1}{2}D^*_{kx} \tag{7.10c}$$

Now recall (7.9) with (7.10c)

$$y = (D_1 + C_1)D_{kx} + (D_1 - C_1)B_{kx}$$
$$= (D_1 + C_1)\frac{1}{2}D^*_{kx} - (D_1 - C_1)\frac{1}{2}B^*_{kx}$$

$$y' = -\frac{1}{2}k(D_1 + C_1)A^*_{kx} - \frac{1}{2}k(D_1 - C_1)C^*_{kx} \tag{7.11}$$

To evaluate the two arbitrary constants C_1 and D_1, two boundary conditions, namely, the deflection and shear criteria, are applied as

Boundary Condition–1 (deflection criterion) that:

$$y' = 0 \text{ at } x = 0$$

With A^*_{kx} and C^*_{kx}, both are = 1, for $x = 0$, which gives, $D_1 = 0$ (7.12)

Hence

$$y = C_1\cos N(kx, kx) = C_1[D_{kx} - B_{kx}]$$
$$= \frac{1}{2}C_1[D^*_{kx} + B^*_{kx}] = \frac{1}{2}C_1.A^*_{kx} \tag{7.13}$$

Thus, with Constant C_1, (7.6a-b) is rewritten as

$$y = C_1\cos N(kx, kx) = \frac{C_1}{2}A^*_{kx}$$

from which, differentiating successively

$$y' = -C_1.k\,B^*_{kx}$$
$$y'' = -C_1.k^2\,C^*_{kx}$$
$$y''' = 2\,C_1.k^3\,D^*_{kx} \tag{7.14}$$

Boundary Condition–2 (shear criterion) that:

$-EI\,y''' = -P/2$, just right of the origin at the symmetry point at which the point load P acts, get

$$C_1 = \frac{P}{4EI.k^3} = \frac{P.s}{k}, \text{ where } 4k^4 = s\,/\,(EI)$$

Thus finally we arrive at the well known solution from (7.8) as

$$y = C_1 \cos N(kx, kx) = \frac{C_1}{2} A^*_{kx} = \frac{P.s}{2k} \cdot A^*_{kx} \tag{7.15}$$

Solution for Beams on Elastic Foundation

(7.15) when translated using (7.10a), can be identified as

$$y = \frac{P.s}{2k} A^*_{kx} = \frac{P.s}{2k} e^{-kx} [\cos kx + \sin kx] \tag{7.16}$$

This is the solution of the well known equation of the beams on elastic foundation.

❑❑❑

8

MATRIX REPRESENTATION OF Nbic FUNCTIONS

8.1 GENERAL

Let us represent the complex Nbic functions in the matrix notation similar in the way as any complex number, $z = a + i\,b$, is done.

First, take the case of the basic complex number, $z = a + i\,b$, where, $i = \sqrt{-1}$.

The difference between real and complex numbers lies in the fact that, the real numbers are one dimensional, that is, they can be represented on the straight line, called number line, (extending from, $-\infty$ to $+\infty$), whereas, the complex numbers are points on a plane. This plane is bounded by real axis, and imaginary axis. These two axes are at right angle to each other.

There are ways of representing the complex number, namely, classical form, ordered pair form, vector form, and, matrix form. They are :

8.1.1 Types of Representations

Classical form:

$$z = a + i\,b$$

a = real part, b = imaginary part, $i = \sqrt{-1}$ = a sort of *tag* to separate the real and imaginary parts.

$$a = a + i.0,\ i = 0 + i.1,\ 0 = 0 + i.0$$

$$(a + i\,b) + (c + i\,d) = (a + c) + i\,(b + d)$$

$$(a + i\,b).(c + i\,d) = (ac - bd) + i\,(bc + ad)$$

$$\text{Modulus} = |a+ib| = \sqrt{a^2+b^2} \tag{8.1}$$

Ordered pair form:

$$a + i\,b \Rightarrow (a, b) = (a, 0) + (0, b) = a + b\,(0, 1)$$

$$a \Rightarrow (a, 0),\ i \Rightarrow (0, 1),\ 0 \Rightarrow (0, 0),$$

$$(a + i\,b) + (c + i\,d) \Rightarrow (a, b) + (c, d) = [(a + c), (b + d)]$$

$$(a + i\,b).(c + i\,d) \Rightarrow (a, b).(c, d) = [(ac - bd), (ad + bc)] \tag{8.2}$$

Vector form:

$$a + i\,b \Rightarrow \begin{bmatrix} a \\ b \end{bmatrix}, \quad c + i\,d \Rightarrow \begin{bmatrix} c \\ d \end{bmatrix}$$

$$a \Rightarrow \begin{bmatrix} a \\ 0 \end{bmatrix}, \quad i \Rightarrow \begin{bmatrix} 0 \\ 1 \end{bmatrix}, \quad 0 \Rightarrow \begin{bmatrix} 0 \\ 0 \end{bmatrix}$$

$$(a + i\,b) + (c + i\,d) = (a + c) + i\,(b + d)$$

$$\Rightarrow \begin{bmatrix} a \\ b \end{bmatrix} + \begin{bmatrix} c \\ d \end{bmatrix} = \begin{bmatrix} a+c \\ b+d \end{bmatrix}$$

$$(a + i\,b).(c + i\,d) = (a + c) + i\,(b + d)$$

$$\Rightarrow \begin{bmatrix} a \\ b \end{bmatrix} \cdot \begin{bmatrix} c \\ d \end{bmatrix} = \begin{bmatrix} ac - bd \\ ad + bc \end{bmatrix}$$

$$\text{Modulus of } \begin{bmatrix} a \\ b \end{bmatrix} = \sqrt{a^2 + b^2} \tag{8.3}$$

Matrix form:

real part

$$1 \Rightarrow [\mathrm{I}] = \text{identity matrix (size : } 2 \times 2) = \begin{bmatrix} 1 & 0 \\ 0 & 1 \end{bmatrix}$$

$$a \Rightarrow [a] = \text{real number matrix (size : } 2 \times 2) = a\begin{bmatrix} 1 & 0 \\ 0 & 1 \end{bmatrix},$$

$$b \Rightarrow [b] = \text{real number matrix (size : } 2 \times 2) = b\begin{bmatrix} 1 & 0 \\ 0 & 1 \end{bmatrix} \tag{8.4a}$$

imaginary part

$$i \Rightarrow [i] = \text{imaginary symbol matrix (size: } 2 \times 2) = \begin{bmatrix} 0 & -1 \\ 1 & 0 \end{bmatrix},$$

$$i\,b \Rightarrow \text{imaginary number matrix (size: } 2 \times 2) = [i]\ [b]$$

$$= \begin{bmatrix} 0 & -1 \\ 1 & 0 \end{bmatrix} \cdot \begin{bmatrix} b & 0 \\ 0 & b \end{bmatrix} = \begin{bmatrix} 0 & -b \\ b & 0 \end{bmatrix} \tag{8.4b}$$

complex number

for any real number (a and b)

$$a + i\,b \Rightarrow \begin{bmatrix} a & 0 \\ 0 & a \end{bmatrix} + \begin{bmatrix} 0 & -b \\ b & 0 \end{bmatrix} = \begin{bmatrix} a & -b \\ b & a \end{bmatrix}$$

$$a - i\,b \Rightarrow \begin{bmatrix} a & 0 \\ 0 & a \end{bmatrix} - \begin{bmatrix} 0 & -b \\ b & 0 \end{bmatrix} = \begin{bmatrix} a & b \\ -b & a \end{bmatrix} \tag{8.4c}$$

Though the representative methods for the the complex number $(a + i\,b)$ may be different, the three forms mentioned, viz., the point form $\begin{bmatrix} a \\ b \end{bmatrix}$, the ordered pair form (a, b), and the matrix form $\begin{bmatrix} a & -b \\ b & a \end{bmatrix}$, are the one and the same.

8.1.2 Rotation

Consider that a point P on a plane P_1, is defined by the set of rectangular coordinates, (x, y). If the same point P, is defined in another plane P_2, oriented at an angle θ, (measured anticlockwise, + ve), from the plane P_1, which is bounded by a different set of rectangular coordinates, (u, v), the interrelation of the coordinate system, (x, y) and (u, v), is given by

$$x = u \cos\theta - v \sin\theta$$
$$y = u \sin\theta + v \cos\theta \tag{8.5}$$

which in matrix notation

$$\begin{bmatrix} x \\ y \end{bmatrix} = \begin{bmatrix} \cos\theta & -\sin\theta \\ \sin\theta & \cos\theta \end{bmatrix} \begin{bmatrix} u \\ v \end{bmatrix} \tag{8.6a}$$

or

$$[X] = [T]\,[U]$$

where, $[X]_{2\times1}$, and $[U]_{2\times1}$ = coordinate matrix,

$[T]_{2\times2}$ = transformation matrix

= rotation matrix = orientation matrix (8.6b)

That the plane P_2 is rotated by an angle θ, off the plane P_1, is synonymous with the rotation of the u-axis with respect to the x-axis, in the same sense.

At $\theta = 0$ deg, u-axis coincides with the positive x-axis, (directed from 0 to $+\infty$), and at $\theta = 90$ deg off from x-axis, the positive y-axis is reached. At further 90 deg off from the y-axis, (*i.e.*, $\theta = 180$ deg), the negative x-axis is again reached, (which extends from 0 to $-\infty$).

Then speaking vectorially about the point, $\begin{bmatrix} x \\ y \end{bmatrix}$, which is all the same as, $\begin{bmatrix} x & -y \\ y & x \end{bmatrix}$, when expressed in matrix form, representing the complex number $(x + i\,y)$, the rotation matrix (of modulus =1), with the understanding that, x-axis and y-axis are the real axis and imaginary axis, respectively, becomes

$$[T] = \begin{bmatrix} \cos\theta & -\sin\theta \\ \sin\theta & \cos\theta \end{bmatrix} \tag{8.6c}$$

For $\theta = 0$ deg:

$$[T_0] \quad = \quad \begin{bmatrix} x \\ 0 \end{bmatrix} \Rightarrow \begin{bmatrix} 1 & 0 \\ 0 & 1 \end{bmatrix} \tag{8.6d}$$

The resultant rotation matrix can be easily identified as matrix representing the real number 1, which in other words indicates that:

the value of $x = +1$, and the value of $y = 0$, so that the number is: $(+1) + i.(0)$

This shows that the positive real axis is at $\theta = 0$ deg.

For $\theta = 90$ deg:

$$[T_{90}] \quad = \begin{bmatrix} 0 \\ y \end{bmatrix} \Rightarrow \begin{bmatrix} 0 & -1 \\ 1 & 0 \end{bmatrix} \tag{8.6e}$$

The resultant rotation matrix can be easily identified as the same imaginary symbol matrix, $i = [i]$, which in other words, indicates that:

the value of $x = 0$, and the value of $y = i\ .1$, so that the number is : $(0) + i.(+1)$

This shows that positive imaginary axis is oriented at 90 deg off from the positive real axis.

For $\theta = 180$ deg:

$$[T_{180}] \quad = \quad \begin{bmatrix} x \\ 0 \end{bmatrix} \Rightarrow \begin{bmatrix} -1 & 0 \\ 0 & -1 \end{bmatrix} \tag{8.6f}$$

The resultant rotation matrix can be easily identified as the matrix, representing -1.

This means that:

the value of $x = -1$, and the value of $y = i.\ 0$, so that the number is: $(-1) + i.(0)$

It is interesting to note that, to reach the imaginary axis, a rotation of 90 deg from 0 deg line, (which represents the real number line), is required. This is indicated by the matrix for $[i]$, *i.e.*, $\begin{bmatrix} 0 & -1 \\ 1 & 0 \end{bmatrix}$.

This is indicated by $[T_{90}]$ matrix (8.6e).

Again, a further rotation of 90 deg from this imaginary axis, the negative real axis is reached. This should be confirmed from the resultant matrix $[T_{180}]$, as to be that of real number -1.

In fact, it is so, and is represented by (8.6f), that is, $[T_{180}] = \begin{bmatrix} -1 & 0 \\ 0 & -1 \end{bmatrix}$.

To verify this, let us multiply, $[i]$ matrix with $[i]$ matrix, corresponding to two 90 deg rotations to reach $\theta = 180$ deg from $\theta = 0$ deg line of real numbers.

$$[i]\ [i] = [T_{90}]\ [T_{90}] \quad \begin{bmatrix} 0 & -1 \\ 1 & 0 \end{bmatrix} \cdot \begin{bmatrix} 0 & -1 \\ 1 & 0 \end{bmatrix} = \begin{bmatrix} -1 & 0 \\ 0 & -1 \end{bmatrix}$$

$$\Rightarrow i\ .\ i = i^2 = -\ 1 \tag{8.7}$$

This resultant matrix represents the real number, -1, as given by (8.6f), which stands for $[T_{180}]$.

This shows that the negative real axis (extending from 0 to – ∞), is oriented at 90 deg off from the positive imaginary axis, and 180 deg off from the positive real axis.

8.1.3 For Functions and Complex Functions

Complex circular function

Recall (2.2)

$$e^{ix} = \cos x + i \sin x$$
$$e^{-ix} = \cos x - i \sin x$$

Referring to the matrix notation (8.4a and 8.4c), and remembering the transformation matrix (8.6c) written in terms of circular functions

$$c = \cos x = \begin{bmatrix} c & 0 \\ 0 & c \end{bmatrix}, \; s = \sin x = \begin{bmatrix} s & 0 \\ 0 & s \end{bmatrix} \tag{8.8a}$$

$$\begin{aligned} e^{ix} &= \cos x + i \sin x \\ &= \begin{bmatrix} c & 0 \\ 0 & c \end{bmatrix} + \begin{bmatrix} 0 & -1 \\ 1 & 0 \end{bmatrix} \cdot \begin{bmatrix} s & 0 \\ 0 & s \end{bmatrix} = \begin{bmatrix} c & -s \\ s & c \end{bmatrix} \\ &= \begin{bmatrix} \cos x & -\sin x \\ \sin x & \cos x \end{bmatrix} = [\mathrm{T}_{x_1}] \end{aligned} \tag{8.8b}$$

$$\begin{aligned} e^{-ix} &= \cos x - i \sin x \\ &= \begin{bmatrix} c & 0 \\ 0 & c \end{bmatrix} - \begin{bmatrix} 0 & -1 \\ 1 & 0 \end{bmatrix} \cdot \begin{bmatrix} s & 0 \\ 0 & s \end{bmatrix} = \begin{bmatrix} c & s \\ -s & c \end{bmatrix} \\ &= \begin{bmatrix} \cos x & \sin x \\ -\sin x & \cos x \end{bmatrix} = [\mathrm{T}_{x_2}] \end{aligned} \tag{8.8c}$$

where, $[\mathrm{T}_{x_1}]_{2\times2}$ and $[\mathrm{T}_{x_2}]_{2\times2}$, are the respective transformation matrices, written in terms of circular functions.

Now proceeding with the matrix multiplication, and writing the result with (8.4a) meant for [I] matrix

$$e^{ix}\, e^{-ix} = [\mathrm{T}_{x_1}]\,[\mathrm{T}_{x_2}] \tag{8.9a}$$

$$\begin{bmatrix} c & -s \\ s & c \end{bmatrix} \begin{bmatrix} c & s \\ -s & c \end{bmatrix}$$

$$= \begin{bmatrix} c^2 + s^2 & c.s - c.s \\ s.c - s.c & c^2 + s^2 \end{bmatrix} = \begin{bmatrix} 1 & 0 \\ 0 & 1 \end{bmatrix} = [\,\mathrm{I}\,] \tag{8.9b}$$

where, $c = \cos x$, $s = \sin x$, $[\,\mathrm{I}\,]_{2\times2}$ = identity matrix = 1.

Hence, $e^{ix}\, e^{-ix} = 1$ (8.9c)

Complex hyperbolic function

Recall (2.4a)

$$f^{iy} = chy + i\ shy$$
$$f^{-iy} = chy - i\ shy \tag{8.10}$$

Substituting

$$C = chy,\ S = shy \tag{8.11}$$

Rewrite (8.10), in terms of the matrix of hyperbolic functions, $[H_{y_1}]$ or, $[H_{y_2}]$, as the case may be, we have

$$f^{iy} = chy + i\ shy$$
$$= \begin{bmatrix} C & 0 \\ 0 & C \end{bmatrix} + \begin{bmatrix} 0 & -1 \\ 1 & 0 \end{bmatrix} \cdot \begin{bmatrix} S & 0 \\ 0 & S \end{bmatrix} = \begin{bmatrix} C & -S \\ S & C \end{bmatrix}$$
$$= \begin{bmatrix} chy & -shy \\ shy & chy \end{bmatrix} = [H_{y_1}] \tag{8.12a}$$
$$f^{-iy} = chy - i\ shy$$
$$= \begin{bmatrix} C & 0 \\ 0 & C \end{bmatrix} - \begin{bmatrix} 0 & -1 \\ 1 & 0 \end{bmatrix} \cdot \begin{bmatrix} S & 0 \\ 0 & S \end{bmatrix} = \begin{bmatrix} C & S \\ -S & C \end{bmatrix}$$
$$= \begin{bmatrix} chy & shy \\ -shy & chy \end{bmatrix} = [H_{y_2}] \tag{8.12b}$$

where, $[H_{y_1}]_{2\times2}$ and $[H_{y_2}]_{2\times2}$, are the respective transformation matrices, written in terms of hyperbolic functions.

Now proceeding with the matrix multiplication, and writing the result with (8.4a) meant for [I] matrix

$$f^{iy} f^{-iy} = [H_{y_1}]\ [H_{y_2}]$$
$$= \begin{bmatrix} C & -S \\ S & C \end{bmatrix} \begin{bmatrix} C & S \\ -S & C \end{bmatrix}$$
$$= \begin{bmatrix} C^2+S^2 & CS - C.S \\ S.C - S.C & C^2+S^2 \end{bmatrix} = \begin{bmatrix} ch\ 2y & 0 \\ 0 & ch\ 2y \end{bmatrix}$$
$$= ch\ 2y. \begin{bmatrix} 1 & 0 \\ 0 & 1 \end{bmatrix} = ch\ 2y.[\ I\ ] = ch\ 2y \tag{8.13a}$$

where, $C = chy,\ S = shy,\ [\ I\ ]_{2\times2}$ = identity matrix = 1.

Hence,

$$f^{iy} f^{-iy} = ch\ 2y \tag{8.13b}$$

$$f^{iy} f^{iy} = [H_{y_1}]\ [H_{y_1}]$$

$$= \begin{bmatrix} chy & -shy \\ shy & chy \end{bmatrix} \begin{bmatrix} chy & -shy \\ shy & chy \end{bmatrix}$$

$$= \begin{bmatrix} C^2 - S^2 & -2SC \\ -2SC & C^2 - S^2 \end{bmatrix} = \begin{bmatrix} 1 & -sh\ 2y \\ sh\ 2y & 1 \end{bmatrix}$$

$$= \begin{bmatrix} 1 & 0 \\ 0 & 1 \end{bmatrix} + \begin{bmatrix} 0 & -1 \\ 1 & 0 \end{bmatrix} \begin{bmatrix} sh\ 2y & 0 \\ 0 & sh\ 2y \end{bmatrix} \tag{8.13c}$$

In view of the substitution, $a = 1,\ b = sh\ 2y$ (8.13d)

we have, $a + i\,b\ =\ 1 + i\ sh\ 2y$

Hence, $f^{iy} f^{iy}\ = 1 + i\ sh\ 2y = [f^{iy}]^2 = f^{2iy}$ (8.13e)

Single Nbic function

Single Nbic functions, $N(x, y)$, $[N_1(x, y)]$ and $[N_2(x, y)]$

$N_1(x, y)$: (1.33a)

$$\cos N(x, y) = \cos x\ chy - \sin x\ shy$$
$$\sin N(x, y) = \sin x\ chy + \cos x\ shy \tag{8.14a}$$

$$\begin{bmatrix} \cos N(x, y) \\ \sin N(x, y) \end{bmatrix} = \begin{bmatrix} chy & -shy \\ shy & chy \end{bmatrix} \begin{bmatrix} \cos x \\ \sin x \end{bmatrix} \tag{8.14b}$$

$N_2(x, y)$: (1.33b)

$$\cos N(x, -y) = \cos x\ chy + \sin x\ shy$$
$$\sin N(x, -y) = \sin x\ chy - \cos x\ shy \tag{8.14c}$$

$$\begin{bmatrix} \cos N(x, -y) \\ \sin N(x, -y) \end{bmatrix} = \begin{bmatrix} chy & shy \\ -shy & chy \end{bmatrix} \begin{bmatrix} \cos x \\ \sin x \end{bmatrix} \tag{8.14d}$$

Combining (8.14b) and (8.14d) in matrix form

$$\begin{bmatrix} N_1(x, y) \\ N_2(x, y) \end{bmatrix} \Rightarrow \begin{bmatrix} \cos N(x, y) \\ \sin N(x, y) \\ \cos N(x, -y) \\ \sin N(x, -y) \end{bmatrix} = \begin{bmatrix} chy & -shy \\ shy & chy \\ chy & shy \\ -shy & chy \end{bmatrix} \begin{bmatrix} \cos x \\ \sin x \end{bmatrix} \tag{8.14e}$$

Particular case

$$y = x$$

Single Nbic functions, $N(x, x)$, $[N_1 x]$ and $[N_2 x]$

$N_1 x$: (1.35a)

$$\cos N(x, x) = \cos Nx = \cos x\ ch\ x - \sin x\ sh\ x$$
$$\sin N(x, x) = \sin Nx = \sin x\ ch\ x + \cos x\ sh\ x \tag{8.15a}$$

$$\begin{bmatrix} \cos N(x, x) \\ \sin N(x, x) \end{bmatrix} = \begin{bmatrix} chx & -shx \\ shx & chx \end{bmatrix} \begin{bmatrix} \cos x \\ \sin x \end{bmatrix} \tag{8.15b}$$

N_2x : (1.35b)

$$\cos N(x, -x) = \cos x\, chx + \sin x\, shx$$
$$\sin N(x, -x) = \sin x\, chx - \cos x\, shx \tag{8.15c}$$

$$\begin{bmatrix} \cos N(x,-x) \\ \sin N(x,-x) \end{bmatrix} = \begin{bmatrix} chx & shx \\ -shx & chx \end{bmatrix} \begin{bmatrix} \cos x \\ \sin x \end{bmatrix} \tag{8.15d}$$

Combining (8.15b) and (8.15d), or substituting $y = x$, in (8.14e), the matrix form becomes

$$\begin{bmatrix} N_1(x,x) \\ N_2(x,x) \end{bmatrix} \Rightarrow \begin{bmatrix} \cos N(x,x) \\ \sin N(x,x) \\ \cos N(x,-x) \\ \sin N(x,-x) \end{bmatrix} = \begin{bmatrix} chx & -shx \\ shx & chx \\ chx & shx \\ -shx & chx \end{bmatrix} \begin{bmatrix} \cos x \\ \sin x \end{bmatrix} \tag{8.15e}$$

Complex Single Nbic function, N/(*x*, *y*) [$N_1/(x, y)$] and [$N_2/(x, y)$]:

Recall Cosn 2 functions, $N_1/(x, y)$ and $N_2/(x, y)$, from (2.29)

$$N_1/(x, y) = \cos N(x, y) + i \sin N(x, y) = e^{iN/(x, y)}$$
$$N_2/(x, y) = \cos N(x, y) - i \sin N(x, y) = e^{-iN/(x, y)} \tag{8.16a}$$

Substitute

$$a = \cos N(x, y),\ b = \sin N(x, y) \tag{8.16b}$$

In view of the substitution (8.16b) and (8.4c), we write the complex form into matrix form as

$$\cos N(x, y) + i \sin N(x, y) = a + i\, b$$

$$\Rightarrow \begin{bmatrix} a & -b \\ b & a \end{bmatrix} = \begin{bmatrix} \cos N(x,y) & -\sin N(x,y) \\ \sin N(x,y) & \cos N(x,y) \end{bmatrix} \tag{8.17a}$$

$$\cos N(x, y) - i \sin N(x, y) = a - i\, b$$

$$\Rightarrow \begin{bmatrix} a & b \\ -b & a \end{bmatrix} = \begin{bmatrix} \cos N(x,y) & \sin N(x,y) \\ -\sin N(x,y) & \cos N(x,y) \end{bmatrix} \tag{8.17b}$$

Now (8.16a) in matrix form becomes

$$\begin{bmatrix} N_1/(x,y) \\ N_2/(x,y) \end{bmatrix} = \begin{bmatrix} 1 & i \\ 1 & -i \end{bmatrix} \begin{bmatrix} \cos N(x,y) \\ \sin N(x,y) \end{bmatrix} \tag{8.18a}$$

Using two unit matrices (2 × 2), and submatrices for $i\,(2 \times 2)$ and $-i\,(2 \times 2)$:

$$\begin{bmatrix} N_1/(x,y) \\ N_2/(x,y) \end{bmatrix} \Rightarrow \begin{bmatrix} N_1/(x,y) \\ 0 \\ N_2/(x,y) \\ 0 \end{bmatrix} = \begin{bmatrix} 1 & 0 & 0 & -1 \\ 0 & 1 & 1 & 0 \\ 1 & 0 & 0 & 1 \\ 0 & 1 & -1 & 0 \end{bmatrix} \begin{bmatrix} \cos N(x,y) \\ 0 \\ \sin N(x,y) \\ 0 \end{bmatrix} \tag{8.18b}$$

Particular case

$$y = x$$

Substituting $y = x$, the complex single Nbic function, N/(x, x), can be obtained accordingly as:

$$N_1/(x, x) = \cos Nx + i \sin Nx = e^{iN/x}$$

$$N_2/(x, x) = \cos Nx - i \sin Nx = e^{-iN/x}$$

Using two unit matrices (2 × 2), and submatrices for i (2 × 2) and $-i$ (2 × 2):

$$\begin{bmatrix} N_1/(x,x) \\ N_2/(x,x) \end{bmatrix} \Rightarrow \begin{bmatrix} N_1/(x,x) \\ 0 \\ N_2/(x,x) \\ 0 \end{bmatrix} = \begin{bmatrix} 1 & 0 & 0 & -1 \\ 0 & 1 & 1 & 0 \\ 1 & 0 & 0 & 1 \\ 0 & 1 & -1 & 0 \end{bmatrix} \begin{bmatrix} \cos N(x,x) \\ 0 \\ \sin N(x,x) \\ 0 \end{bmatrix} \quad (8.18c)$$

8.2 TRIGONOMETRIC–HYPERBOLIC INTERRELATION OF Nbic FUNCTIONS

Recollect once again (8.4a-c) representing the matrix form (2 × 2) as:

$$1 \Rightarrow [I] = \text{identity matrix} = \begin{bmatrix} 1 & 0 \\ 0 & 1 \end{bmatrix}$$

$$a \Rightarrow [a] = \text{real number matrix} = a \begin{bmatrix} 1 & 0 \\ 0 & 1 \end{bmatrix},$$

$$b \Rightarrow [b] = \text{real number matrix} = b \begin{bmatrix} 1 & 0 \\ 0 & 1 \end{bmatrix} \quad (8.19a)$$

$$i \Rightarrow [i] = \text{imaginary symbol matrix} = \begin{bmatrix} 0 & -1 \\ 1 & 0 \end{bmatrix},$$

$$i\,b \Rightarrow \text{imaginary number matrix} = [i]\ [b] \begin{bmatrix} 0 & -b \\ b & 0 \end{bmatrix} \quad (8.19b)$$

so that

$$a + i\,b \Rightarrow \begin{bmatrix} a & 0 \\ 0 & a \end{bmatrix} + \begin{bmatrix} 0 & -b \\ b & 0 \end{bmatrix} = \begin{bmatrix} a & -b \\ b & a \end{bmatrix}$$

$$a - i\,b \Rightarrow \begin{bmatrix} a & 0 \\ 0 & a \end{bmatrix} - \begin{bmatrix} 0 & -b \\ b & 0 \end{bmatrix} = \begin{bmatrix} a & b \\ -b & a \end{bmatrix} \quad (8.19c)$$

[n] and [m] matrix and usage

Similar to $i = [i]$ matrix, let us identify two more additional matrices viz., $n = [n]$ and $m = [m]$, along with $[i]$ matrix, each of size (2 × 2). Omitting, matrix symbol [] for i, n and m, they are witten without loss of generality as:

$$i \Rightarrow [i] = \begin{bmatrix} 0 & -1 \\ 1 & 0 \end{bmatrix}$$

$$i^{T} = i^{-1} = -i$$

$$i^2 = i^6 = \ldots = -1,\ i^4 = i^8 = \ldots = 1,\ i^3 = i^7 = \ldots = -i,\ i^5 = i^9 = \ldots = i,$$

$$n \Rightarrow [n] = \begin{bmatrix} 1 & 0 \\ 0 & -1 \end{bmatrix}$$

$$n^{T} = n^{-1} = n$$

$$n^k = n \text{ (for } k = \text{ odd), or, } =1 \text{ (for } k = \text{even)}$$

$$m \Rightarrow [m] = \begin{bmatrix} 0 & 1 \\ 1 & 0 \end{bmatrix}$$

$$m^{T} = m^{-1} = m$$

$$m^k = m \text{ (for } k = \text{odd), or, } = 1 \text{ (for } k = \text{even)} \tag{8.19d}$$

so that

$$i.n = m \Rightarrow \begin{bmatrix} 0 & -1 \\ 1 & 0 \end{bmatrix}\begin{bmatrix} 1 & 0 \\ 0 & -1 \end{bmatrix} = \begin{bmatrix} 0 & 1 \\ 1 & 0 \end{bmatrix},\ n.i = -m$$

$$m.i = n \Rightarrow \begin{bmatrix} 0 & 1 \\ 1 & 0 \end{bmatrix}\begin{bmatrix} 0 & -1 \\ 1 & 0 \end{bmatrix} = \begin{bmatrix} 1 & 0 \\ 0 & -1 \end{bmatrix},\ i.m = -n$$

$$m.n = i \Rightarrow \begin{bmatrix} 0 & 1 \\ 1 & 0 \end{bmatrix}\begin{bmatrix} 1 & 0 \\ 0 & -1 \end{bmatrix} = \begin{bmatrix} 0 & -1 \\ 1 & 0 \end{bmatrix},\ n.m = -i \tag{8.19e}$$

With these

$$a + i\,b \Rightarrow \begin{bmatrix} a & 0 \\ 0 & a \end{bmatrix} + \begin{bmatrix} 0 & -b \\ b & 0 \end{bmatrix} = \begin{bmatrix} a & -b \\ b & a \end{bmatrix}$$

$$a - i\,b \Rightarrow \begin{bmatrix} a & 0 \\ 0 & a \end{bmatrix} - \begin{bmatrix} 0 & -b \\ b & 0 \end{bmatrix} = \begin{bmatrix} a & b \\ -b & a \end{bmatrix}$$

$$a + n\,b \Rightarrow \begin{bmatrix} a & 0 \\ 0 & a \end{bmatrix} + \begin{bmatrix} b & 0 \\ 0 & -b \end{bmatrix} = \begin{bmatrix} a+b & 0 \\ 0 & a-b \end{bmatrix}$$

$$a - nb \Rightarrow \begin{bmatrix} a & 0 \\ 0 & a \end{bmatrix} - \begin{bmatrix} b & 0 \\ 0 & -b \end{bmatrix} = \begin{bmatrix} a-b & 0 \\ 0 & a+b \end{bmatrix}$$

$$a + mb \Rightarrow \begin{bmatrix} a & 0 \\ 0 & a \end{bmatrix} + \begin{bmatrix} 0 & b \\ b & 0 \end{bmatrix} = \begin{bmatrix} a & b \\ b & a \end{bmatrix}$$

$$a - mb \Rightarrow \begin{bmatrix} a & 0 \\ 0 & a \end{bmatrix} - \begin{bmatrix} 0 & b \\ b & 0 \end{bmatrix} = \begin{bmatrix} a & -b \\ -b & a \end{bmatrix} \quad (8.19f)$$

Check these two products

1. $(a + mb)\,(a + nb) = a^2\,[\mathrm{I}] + ab[m + n] + b^2\,[mn]$

$$\text{LHS} = \begin{bmatrix} a & b \\ b & a \end{bmatrix}\begin{bmatrix} a+b & 0 \\ 0 & a-b \end{bmatrix} = \begin{bmatrix} a(a+b) & b(a-b) \\ b(a+b) & a(a-b) \end{bmatrix}$$

$$\text{RHS} = a^2\begin{bmatrix} 1 & 0 \\ 0 & 1 \end{bmatrix} + ab\begin{bmatrix} 1 & 1 \\ 1 & -1 \end{bmatrix} + b^2\begin{bmatrix} 0 & -1 \\ 1 & 0 \end{bmatrix} = \begin{bmatrix} a(a+b) & b(a-b) \\ b(a+b) & a(a-b) \end{bmatrix} \quad (8.19g)$$

Since LHS = RHS, check is OK.

2. $[n + m]\,[n - m] = 2[i]$

Use: $m = i.n = -\,n.i$, $n^2 = n.n = 1$

$$\text{Directly: LHS} = \begin{bmatrix} 1 & 1 \\ 1 & -1 \end{bmatrix}\begin{bmatrix} 1 & -1 \\ -1 & -1 \end{bmatrix} = \begin{bmatrix} 0 & -2 \\ 2 & 0 \end{bmatrix} = 2i = \text{RHS}$$

Alternatively: LHS = $(n - n.i)(n + n.i) = [n(1 - i)]\,[n(1 + i)]$

$$= n^2.\begin{bmatrix} 1 & 1 \\ -1 & 1 \end{bmatrix}\begin{bmatrix} 1 & -1 \\ 1 & 1 \end{bmatrix} = 2i = \text{RHS}$$

The usefulness of the matrix form for $[n]$ and $[m]$, as given in (8.19d), is described with an Example, in which the trigonometric functions are fully written in terms of the hyperbolic functions, and vice versa, through the Nbic functions.

Example: Given the single Nbic functions

$\cos\mathrm{N}x = \cos\mathrm{N}(x, x) = \cos x\; chx - \sin x\; shx$

$\sin\mathrm{N}x = \sin\mathrm{N}(x, x) = \sin x\; chx + \cos x\; shx$ (8.20a)

Express the trigonometric functions ($\cos x$, $\sin x$), in terms of the hyperbolic functions (chx, shx), and vice versa.

Let us put

$\mathrm{P} = \cos\mathrm{N}(x, x) = \cos\mathrm{N}x$, $\mathrm{Q} = \sin\mathrm{N}(x, x) = \sin\mathrm{N}x$

$x = \cos x$, $y = \sin x$

$x = chx$, $y = shx$

so that: $\mathrm{X}^2 + \mathrm{Y}^2 = 1$, $\mathrm{X}^2 - \mathrm{Y}^2 = \cos 2x$, $x^2 + y^2 = ch\,2x$, $x^2 - y^2 = 1$ (8.20b)

With these, the single Nbic functions (8.20a), when rewritten become

$\mathrm{P} = \mathrm{X}x - \mathrm{Y}y$, and, $\mathrm{Q} = \mathrm{Y}x + \mathrm{X}y$

CHAPTER 8

$$\text{Thus, } = \begin{bmatrix} P \\ Q \end{bmatrix} = \begin{bmatrix} X & -Y \\ Y & X \end{bmatrix}\begin{bmatrix} x \\ y \end{bmatrix}, \text{ or } \begin{bmatrix} P \\ Q \end{bmatrix} = \begin{bmatrix} x & -y \\ y & x \end{bmatrix}\begin{bmatrix} X \\ Y \end{bmatrix} \quad (8.20c)$$

Here, $\begin{bmatrix} X & -Y \\ Y & X \end{bmatrix}$, or $\begin{bmatrix} x & -y \\ y & x \end{bmatrix}$ conform to the form: $a + ib = \begin{bmatrix} a & -b \\ b & a \end{bmatrix}$, as in (8.19f).

Now, eliminate (x, y) in (8.20a), get (X,Y) expressed in terms of (x, y) as

$$\begin{bmatrix} X \\ Y \end{bmatrix} = \frac{1}{x^2+y^2}\begin{bmatrix} P & Q \\ Q & -P \end{bmatrix}\begin{bmatrix} x \\ y \end{bmatrix} = \frac{1}{P^2+Q^2}\begin{bmatrix} P & Q \\ Q & -P \end{bmatrix}\begin{bmatrix} x \\ y \end{bmatrix} = \frac{1}{P^2+Q^2}[N]\begin{bmatrix} x \\ y \end{bmatrix}$$

$$Þ \begin{bmatrix} \cos x \\ \sin x \end{bmatrix} = \frac{1}{ch\ 2x}\begin{bmatrix} \sin Nx.\ shx + \cos Nx.\ chx \\ \sin Nx.\ chx - \cos Nx.\ shx \end{bmatrix} = \frac{1}{ch\ 2x}\begin{bmatrix} \cos N_{2f}(x, -x) \\ \sin N_{2f}(x, -x) \end{bmatrix} \quad (8.21a)$$

Or, eliminate (X, Y) in (8.20a), get (x, y) expressed in terms of (X,Y) as

$$\begin{bmatrix} x \\ y \end{bmatrix} = \begin{bmatrix} P & Q \\ Q & -P \end{bmatrix}\begin{bmatrix} X \\ Y \end{bmatrix} = [N]\begin{bmatrix} X \\ Y \end{bmatrix}$$

$$Þ \begin{bmatrix} chx \\ shx \end{bmatrix} = \begin{bmatrix} \sin Nx.\ \sin x + \cos Nx.\ \cos x \\ \sin Nx.\ \cos x + \cos Nx.\ \sin x \end{bmatrix} = \begin{bmatrix} \cos N_{2e}(x, -x) \\ \sin N_{2e}(x, -x) \end{bmatrix} \quad (8.21b)$$

For the given Nbic functions (8.20a), (8.21a) and (8.21b) are the required interrelation of the trigonometric and hyperbolic functions.

From these, in view of the relations (8.20b), we get

$$\begin{aligned} &\cos N_{2f}^2\ (x,-x) + \sin N_{2f}^2\ (x,-x) = ch^2\ 2x \\ &\cos N_{2e}^2\ (x,-x) - \sin N_{2e}^2\ (x,-x) = 1 \end{aligned} \quad (8.21c)$$

The matrix $[N] = \begin{bmatrix} P & Q \\ Q & -P \end{bmatrix}$, common in (8.21a) or (8.21b) when reviewed, shows

$$[N] = \begin{bmatrix} P & Q \\ Q & -P \end{bmatrix} = Q\begin{bmatrix} 0 & 1 \\ 1 & 0 \end{bmatrix} + P\begin{bmatrix} 1 & 0 \\ 0 & -1 \end{bmatrix}$$

$= m\text{Q} + n\text{P} = i.n\text{Q} + n\text{P} = -\ n.i.\text{Q} + n\text{P} = n(\text{P} - i\text{Q})$

$[N] = n\ (\text{P} - i\text{Q})$

$$= \begin{bmatrix} 1 & 0 \\ 0 & -1 \end{bmatrix}\begin{vmatrix} P & Q \\ -Q & P \end{vmatrix} = \begin{vmatrix} P & Q \\ Q & -P \end{vmatrix} \quad \textbf{(Checked)} \quad (8.22)$$

The arithmetic check of (8.21a) and (8.21b) can be performed easily with (8.22).

This requires to prove that: [N] [N] = $[N]^2 = [P^2 + Q^2]$

Now, $[N]^2 = [n(P - iQ)]^2 = [n(P - iQ)]\ [n(P - iQ)]$

$$= \begin{bmatrix} P & Q \\ -Q & P \end{bmatrix}\begin{bmatrix} P & Q \\ -Q & P \end{bmatrix} = \begin{bmatrix} P^2+Q^2 & 0 \\ 0 & P^2+Q^2 \end{bmatrix} = (P^2+Q^2)\begin{bmatrix} 1 & 0 \\ 0 & 1 \end{bmatrix}$$

$$\Rightarrow N^2 = P^2 + Q^2 \qquad \textbf{(Checked)} \qquad (8.23)$$

❑❑❑

CONCLUDING REMARKS

The present work on the Nbic functions and Nbic trigonometric relations is just a very small step beyond the horizon of our mathematical umbrella, and naturally, it gives rise to many questions rather than the solution as the answers. To get the full-fledged total work in hand, to be called as a self-content subject, lot much work is need to be supplemented. To identify them all, is a difficult task.

Since the unit plane circle forms the basis of the circular functions, and the relations like, the position of the points, the equation and length of a line between two points, area of a triangle, and so on and so forth, in the respective coordinate system of the (X, Y) is clearly known and forms a whole hearted subject, similar relations in the Nbic coordinate system (X, Y), is yet to be studied and supplemented, on the basis of the respective Nbic circles.

Nbic Geometry—a maiden step

Let us step into a new arena of geometry, *viz.*, Nbic Geometry, with a small beginning. The description of the various circles, namely, Hyperbolic as well as Nbic circles (discussed in Chapter 6), using the corresponding Nbic coordinates (X,Y) expressed in terms of the corresponding Nbic functions, actually are the illustrations coming under the scope of the Nbic Geometry.

First, take the known case of plane circle: $X^2 + Y^2 = 1$, where the coordinate system (X,Y) is defined through circular function *viz.*, $X = \cos x$, $Y = \sin x$. The chord distance d = AB, between two points $A(X_1, Y_1)$, and $B(X_2, Y_2)$ on the periphery, is given by:

$$d = \sqrt{(X_2 - X_1)^2 - (Y_2 - Y_1)^2} = \sqrt{2}.\sqrt{1 - (\cos x_2 \cos x_1 + \sin x_2 . \sin x_1)} = 2\sin\frac{(x_2 - x_1)}{2}$$

Thus, we arrive at the well known result, *viz.*,

$$d = \sqrt{2} \text{ for } x_1 = 0,\ x_2 = \pi/2, \text{ and } d = 2 \text{ (= diameter) for } x_1 = 0,\ x_2 = \pi.$$

Similarly, for hyperbolic circle, $X^2 + Y^2 = 1$, where $X = \frac{chx}{\sqrt{ch\ 2x}}$, $Y = \frac{shx}{\sqrt{ch\ 2x}}$, the chord distance d is written as

$$d=\sqrt{2}.\sqrt{1-\frac{chx_2.chx_1+shx_2.shx_1}{ch\ 2x_2.ch\ 2x_1}}=\sqrt{2}.\sqrt{\frac{ch(x_2+x_1)}{ch\ 2x_2 ch\ 2x_1}}$$

For all other varieties of Nbic circles, defined under their corresponding coordinate systems (X,Y), satisfying $X^2 + Y^2 = 1$, the chord distance d = AB, can be expressed as

$$d=\sqrt{2}.\sqrt{1-\frac{\cos N_K(x_2).\cos N_K(x_1)+\sin N_K(x_2).\sin N_K(x_1)}{(ch\ 2x_2.ch\ 2x_1)^m}}$$

where, the suffix K, and exponent m, are for:

N-circle: K = 1, (but omitted), m = 1,

N_{2e}-circle: K = 2e, m = 1

N_{2f}-circle: K = 2f, m = 2

N_{2e}-circle: K = 3e, m = 2

N_{2f}-circle: K = 3f, m = 3

The general pattern of the chord distance comes out to be

$$d=\sqrt{2}.\sqrt{1-\frac{C(x_1).C(x_2)+S(x_1).S(x_2)}{[c(2x_2).c(2x_1)]^m}}, \text{ (for hyperbolic and Nbic circles)}$$

$$d=\sqrt{2}.\sqrt{1-C(x_1).C(x_2)+S(x_1).S(x_2)} \text{ (for plane circles)}$$

[The characters (S, C, c) used are for (sin, cos *ch*), respectively.]

In addition to this, an interesting indication as the extension of the Nbic functions of a different kind, (*viz.*, B-function), is presented here. A full form of mathematical relations is also is needed to be carried out for further development of the present work.

B-functions and its Relation with Nbic Functions

The type of the functions, sinN(x, x), cosN(x, $-x$), etc., dealt elaborately in Chapters 1–8, is developed considering that the basic unit Nbic functions as a product of trigonometric functions (sin$x$, cos$x$, etc.), and hyperbolic functions (shx, chx, etc.) in the form, (sin$x$.chx), etc. Again, it is interesting to study the functions, which when is derived as a division of the trigonometric functions and hyperbolic functions, in the form, (sin$x$ / chx), etc. These new functions are still trigonometric-hyperbolic type, but certainly are different from the Nbic functions. Let them be identified by the name **B-functions**, and the operators are by, sin**B**(x, x), cos**B** (x, $-x$), and so on.

Depending on the arrangement of the basic units of the trigonometric-hyperbolic functions as the product form or division form, the Nbic functions and the B-functions can be categorized accordingly.

Nbic Functions of the First Kind and Second Kind

If the Nbic functions (product form of the trigonometric-hyperbolic functions) are termed as the Nbic functions of the First Kind, the B-functions (division form of the trigonometric-hyperbolic functions) then, will be better identified as Nbic functions of the Second Kind.

That the Nbic functions (First Kind) alternatively, can also be generated from the B-functions (Second Kind), and also that, the complex B-functions, and their subsequent powers, follow the similar form as those of the Coci-functions, *viz.*, $(\cos x + i \sin x)^n = (\cos nx + i \sin nx) = e^{inx}$, (for $n = 1, 2, 3\ldots$), have been shown here.

Let us define the basic B-functions in two different ways.

Definition Type 1

$$\text{sinB}(x, x) = \text{sinB}x = \frac{\sin x}{shx}$$

$$\text{cosB}(x, x) = \text{cosB}x = \frac{\cos x}{chx}$$

and

$$\text{tanB}(x, x) = \text{tanB}x = \frac{\sin \text{B}(x,x)}{\cos \text{B}(x,x)} = \frac{\tan x}{thx}$$

$$\text{cosecB}(x.\ x) = \text{cosecB}x = 1 / \text{sinB}x$$

$$\text{secB}(x, x) = \text{secB}x = 1 / \text{cosB}x$$

$$\text{cotB}(x, x) = \text{cotB}x = 1 / \text{tanB}x \qquad \text{(a.1.1)}$$

Nbic functions of the First Kind, can be generated from the B-functions [Nbic functions of the Second Kind]. This can be shown as follows:

$$\text{sinB}(x, x) + \text{cosB}(x, x) = \frac{\sin x.chx + \cos x.shx}{shx.chx} = \frac{2\text{sinN}(x,x)}{sh\ 2x}$$

$$\text{sinB}(x, x) - \text{cosB}(x, x) = \frac{\sin x.chx - \cos x.shx}{shx.chx} = \frac{2\text{sinN}(x,-x)}{sh\ 2x}$$

Thus

$$\text{sinN}(x, \pm x) = \frac{1}{2}[\text{sinB}(x, x) \pm \text{cosB}(x, x)].sh\ 2x \qquad \text{(a1.2)}$$

Define further

$$\text{sinB}_1(x, x) = \frac{\text{sinN}(x,x)}{shx}$$

$$\text{cosB}_1(x, x) = \frac{\text{cosN}(x,x)}{chx} \qquad \text{(a1.3)}$$

Then

$$\sin B_1(x, x) + \cos B_1(x, x) = \frac{\sin Nx.chx + \cos Nx.shx}{shx.chx} = \frac{2\sin N_{2f}(x,x)}{sh\ 2x}$$

$$\sin B_1(x, x) - \cos B_1(x, x) = \frac{\sin Nx.chx - \cos Nx.shx}{shx.chx} = \frac{2\sin N_{2f}(x,-x)}{sh\ 2x}$$

Thus

$$\sin N_{2f}(x, \pm x) = \frac{1}{2}\,[\sin B_1(x, x) \pm \cos B_1(x, x)].sh\ 2x \tag{a1.4}$$

Define further

$$\sin B_2(x, x) = \frac{\sin N_{2f}(x,x)}{shx}$$

$$\cos B_2(x, x) = \frac{\cos N_{2f}(x,x)}{chx} \tag{a1.5}$$

$$\sin B_2(x, x) + \cos B_2(x, x) = \frac{\sin N_{2f}x.chx + \cos N_{2f}x.shx}{shx.chx} = \frac{2\sin N_{3f}(x,x)}{sh\ 2x}$$

$$\sin B_2(x, x) - \cos B_2(x, x) = \frac{\sin N_{2f}x.chx - \cos N_{2f}x.shx}{shx.chx} = \frac{2\sin N_{3f}(x,-x)}{sh\ 2x}$$

Thus

$$\sin N_{3f}(x, \pm x) = \frac{1}{2}\,[\sin B_2(x, x) \pm \cos B_2(x, x)\,].sh\ 2x \tag{a1.6}$$

Few Observations

Observation 1

$$\cos B^2(x, x) + \sin B^2(x, x) = \frac{\sin^2 x}{sh^2 x} + \frac{\cos^2 x}{ch^2 x}$$

$$= \frac{\sin^2 x.ch^2 x + \cos^2 x.sh^2 x}{sh^2 x.ch^2 x} \tag{a.2.1}$$

$$1 + \tan B^2(x, x) = 1 + \frac{\sin^2 x.ch^2 x}{\cos^2 x.sh^2 x} = \frac{\sin^2 x.ch^2 x + \cos^2 x.sh^2 x}{\cos^2 x.sh^2 x} \tag{a.2.2}$$

From (a2.1) and (a2.2)

$$[\cos B^2(x, x) + \sin B^2(x, x)]\,(sh^2 x\ .\ ch^2 x) = [1 + \tan B^2(x, x)]\,(\cos^2 x\ .\ sh^2 x)$$

or,

$$\cos B^2(x,\ x) + \sin B^2(x,\ x) = \frac{1 + \tan B^2(x,x)}{\sec B^2(x,x)} \tag{a2.3}$$

In plane trigonometry, we have

$$1 + \tan^2 x = \sec^2 x$$

so that when (a2.3) is viewed in terms of plane trigonometry, turns up to the identity

$$\cos^2 x + \sin^2 x = 1$$

But since

$$1 + \tan B^2(x,\ x) \neq \sec B^2(x,\ x)$$

(a2.3) shows

$$\cos B^2(x,\ x) + \sin B^2(x,\ x) \neq 1 \tag{a2.4}$$

Case 1: Let us see what would happen when

$$\cos B^2(x,\ x) + \sin B^2(x,\ x) = 1$$

Then $\qquad \dfrac{\sin^2 x}{sh^2 x} = 1 - \dfrac{\cos^2 x}{ch^2 x}$

$$= 1 - \frac{1-\sin^2 x}{1+sh^2 x} = \frac{\sin^2 x + sh^2 x}{1+sh^2 x}$$

Equating both sides

$$\sin^2 x = sh^4 x$$

$$\Rightarrow \sin B(x,\ x) = \frac{\sin x}{shx} = \pm shx \tag{a2.5}$$

Case 2: Let us see what would happen when

$$1 + \tan B^2(x,\ x) = \sec B^2(x,\ x)$$

Expanding both sides with the help of (a1.1), we arrive at

$$\cos B^2(x,\ x) + \sin B^2(x,\ x) = 1$$

$$\text{LHS} = \frac{\sin^2 x}{sh^2 x} + \frac{\cos^2 x}{ch^2 x} = \frac{1-\cos^2 x}{ch^2 x - 1} + \frac{\cos^2 x}{ch^2 x}$$

$$= \frac{1 - \dfrac{\cos^2 x}{ch^2 x}}{ch^2 x - 1} = \frac{1 - \cos B^2(x,x)}{ch^2 x - 1} = 1$$

From which

$$\cos B(x,\ x) = \frac{\cos x}{chx} = \pm\sqrt{2 - ch^2 x} \tag{a2.6}$$

It is thus clear that, to have equation (a2.3) = 1 valid, (a2.5) and (a2.6) are to be satisfied. That is, if

$$\sin B(x,\ x) = \pm\ shx$$

$$\cos B(x, x) = \pm\sqrt{2-ch^2x}$$

then only

$$\cos B^2(x, x) + \sin B^2(x, x) = \frac{1+\tan B^2(x,x)}{\sec B^2(x,x)} = 1$$

Observation 2

Let the expansion of the multiple angle of the B-functions be viewed.

Expansion of sinB 2(x, x) [= sinB 2x]:

$$\sin B\ 2(x, x) = \sin B[(x, x) + (x, x)]$$

Expanding according to sin (A + B) formula of plane trigonometry

$$\begin{aligned}\sin B\ 2(x, x) &= \sin B(x, x) \cos B(x, x) + \cos B(x, x) \sin B(x, x)\\ &= 2 \sin B(x, x) \cos B(x, x)\\ &= 2\frac{\sin x}{shx}\cdot\frac{\cos x}{chx} = 2\frac{\sin\ 2x}{sh\ 2x}\end{aligned} \quad \text{(a2.7)}$$

Expansion of cosB 2(x, x) [= cosB 2x]:

$$\cos B\ 2(x, x) = \cos B[(x, x) + (x, x)]$$

Expanding according to cos (A + B) formula of plane trigonometry

$$\begin{aligned}\cos B\ 2(x, x) &= \cos B(x, x) \cos B(x, x) - \sin B(x, x) \sin B(x, x)\\ &= \cos B^2(x, x) - \sin B^2(x, x)\\ &= [\frac{\cos^2 x}{ch^2x} - \frac{\sin^2 x}{sh^2x}]\end{aligned} \quad \text{(a2.8)}$$

Expansion of tanB 2(x, x) [= tanB 2x]:

$$\tan B\ 2(x, x) = \frac{\sin B\ 2(x,x)}{\cos B\ 2(x,x)}$$

Expanding according to tan (A + B) formula of plane trigonometry

$$\begin{aligned}\tan B\ 2(x, x) &= \frac{2\sin B(x,x).\cos B(x,x)}{\cos B^2(x,x)-\sin B^2(x,x)}\\ &= \frac{2\tan B(x,x)}{1-\tan B^2(x,x)}\end{aligned} \quad \text{(a2.9)}$$

Expansion of sinB 3(x, x) [= sinB 3x]:

$$\begin{aligned}\sin B\ 3(x, x) &= \sin B[2(x, x) + (x, x)]\\ &= \sin B\ 2(x, x) \cos B(x, x) + \cos B\ 2(x, x) \sin B(x, x)\\ &= 2\frac{\sin\ 2x}{sh\ 2x}.\frac{\cos x}{chx} + [\frac{\cos^2 x}{ch^2x} - \frac{\sin^2 x}{sh^2x}].\frac{\sin x}{shx}\\ &= 3\frac{\cos^2 x}{ch^2x}.\frac{\sin x}{shx} - \frac{\sin^3 x}{sh^3x}\end{aligned}$$

$$= 3 \sin B(x, x) \cos B^2(x, x) - \sin B^3(x, x) \tag{a2.10}$$

From (a2.7), we have $\quad \sin B\ 2(x, x) = 2\dfrac{\sin 2x}{sh 2x}$

but, $\quad \sin B\ 3(x, x) \neq 3\dfrac{\sin 3x}{sh 3x}$

Expansion of $\cos B\ 3(x, x)$ [= cosB 3x]:

$$\begin{aligned}\cos B\ 3(x, x) &= \cos B[2(x, x) + (x, x)]\\ &= \cos B\ 2(x, x) \cos B(x, x) - \sin B\ 2(x, x) \sin B(x, x)\\ &= [\cos B^2(x, x) - \sin B^2(x, x)].\ \cos B(x, x)\\ &\quad - 2 \sin B^2(x, x).\cos B(x, x)\\ &= \cos B^3(x, x) - 3 \sin B^2(x, x).\cos B(x, x)\end{aligned} \tag{a2.11}$$

The expressions (a2.7) through (a2.11) show that, the **B-functions follow the plane trigonometric rule of expansion.**

Observation 3

Now let us see whether DeMoivre's form, *viz.*, $(\cos x + i \sin x) = e^{ix}$, can be extended to the similar forms of the B-functions, or not.

Let

$$e^{iB/(x,x)} = \cos B(x, x) + i \sin B(x, x) \tag{a3.1}$$

$$\begin{aligned}[e^{iB/(x,x)}]^2 &= [\cos B\ (x, x) + i \sin B\ (x, x)]^2\\ &= [\cos B^2(x, x) - \sin B^2(x, x)] + 2\, i \sin B(x, x) \cos B(x, x)\\ &= \cos B\ 2(x, x) + i \sin B\ 2(x, x) = e^{i2B/(x,x)}\end{aligned} \tag{a3.2}$$

[where (a2.7) and (a2.8) have been used]

$$\begin{aligned}[e^{iB/(x,x)}]^3 &= [\cos B\ (x, x) + i \sin B\ (x, x)]^3\\ &= [\cos B^3(x, x) - 3 \sin B^2(x, x).\cos B(x, x)]\\ &\quad + i\, [3 \sin B(x, x) \cos B^2(x, x) - \sin B^3(x, x)]\\ &= \cos B\ 3(x, x) + i \sin B\ 3(x, x) = e^{i3B/(x,x)}\end{aligned} \tag{a3.3}$$

[where (a2.10) and (a2.11) have been used] and so on.

Thus, **similar to the complex circular functions (Coci-functions), the B-functions too follow similar pattern of expression.**

For multiple angles as in sinB(nx, nx); n = 2, 3, 4, ..., if the expansion is performed according to the B-function format, [instead of expansion according plane trigonometric way, as applied in (a2.7 – a2.11)], we get

$$\sin B\ (2x, 2x) = \sin B\ 2x = \frac{\sin 2x}{sh\ 2x} = \frac{2 \sin 2x.ch\ 2x}{sh\ 4x}$$

$$\cos B\ (2x, 2x) = \cos B\ 2x = \frac{\cos\ 2x}{ch\ 2x} = \frac{2 \cos\ 2x.sh\ 2x}{sh\ 4x}$$

$$\tan B\ (2x, 2x) = \tan B\ 2x = \frac{\tan\ 2x}{th\ 2x} \tag{a4.1}$$

Generalising this way, for $n = 1, 2, 3, \ldots$

$$\text{sinB } nx = \frac{\sin nx}{sh\ nx} = \frac{2\sin(nx).ch(nx)}{sh\ 2nx}$$

$$\text{cosB } 2x = \frac{\cos nx}{ch\ nx} = \frac{2\cos(nx).sh(nx)}{sh\ 2nx}$$

$$\text{tanB } nx = \frac{\tan 2x}{th\ nx} \tag{a4.2}$$

so that

$$[\text{sinB } nx \pm \text{cosB } nx] = \frac{2}{sh\ 2nx} \cdot \text{sinN}(nx, \pm nx) \tag{a4.3}$$

Definition Type 2

Ignoring completely the earlier definition, when the unit B-functions are now expanded according to B-function format:

$$\text{sinB}x = \frac{\sin x}{shx} + \frac{\cos x}{chx} = \frac{2}{shx}\ \text{sinN}x$$

$$\text{cosB}x = \frac{\cos x}{shx} - \frac{\sin x}{chx} = \frac{2}{shx}\ \text{cosN}x$$

$$\text{tanB}x = \frac{\text{sinB}(x)}{\text{cosB}(x)} = \text{tanN}x \tag{a5.1}$$

Generalising this way, for $n = 1, 2, 3, \ldots$

$$(\text{sinB } nx, \text{cosB } nx) = \frac{2}{sh(nx)}\text{sinN } nx, \text{cosN } nx)$$

$$\text{tanB } nx = \text{tanN } nx \tag{a5.2}$$

So

$$\frac{\text{sinB}(nx) + \text{cosB}(nx)}{\text{sinN}(nx) + \text{cosN}(nx)} = \frac{2}{sh(2nx)} \tag{a5.3}$$

Thus it is clearly seen that, the **B-functions can be written in terms of the N-functions** (Nbic functions), and *vice versa*.

Nbic-Algebraic Polynomial

The Nbic functions discussed in this book is a form of trigonometric-hyperbolic type of function. Side by side, as a matter of interest, let us note the form of trigonometric or hyperbolic fuctions mixed with algebraic functions resulting in trigonometric-algebraic or hyperbolic-algebraic functions, respectively, as polynomial equations. An extension of this class of function with Nbic functions will be of another maiden step towards a new direction.

Trigonometric-Algebraic Form

Find the roots (x_1, x_2) of the quadratic equation: $4x^2 - 4x \cos\theta + 1 = 0$

$$(x_1, x_2) = \frac{\cos\theta \pm \sqrt{\cos^2\theta - 1}}{2} = \frac{\cos\theta \pm i.\sin\theta}{2}$$

The two roots are: $x_1 = \frac{1}{2}e^{i\theta}, \quad x_2 = \frac{1}{2}e^{-i\theta}$

Hyperbolic-Algebraic Form

Find the roots (x_1, x_2) of the quadratic equation: $4x^2 - 4x\ ch\ \theta + 1 = 0$

$$(x_1, x_2) = \frac{ch\theta \pm \sqrt{ch^2\theta - 1}}{2} = \frac{ch\theta \pm sh\theta}{2}$$

The two roots are: $x_1 = \frac{1}{2}e^{\theta}, x_2 = \frac{1}{2}e^{-\theta}$

Rewrite the first as well as the second equation as: $(\cos\theta, ch\theta) = x + \frac{1}{4x}$.

Observe that: each of the roots (x_1 or x_2) of these two form of polynomial equations $4x^2 - 4x. (\cos\theta, ch\ \theta) + 1 = 0$, when added to one-fourth of its reciprocal, follows cosine or cosine hyperbolic variations, (given by, $\cos\theta$ or $ch\theta$), respectively.

In general, the given polynomial equation can be : $4x^2 - 4nx .(\cos\theta, ch\theta) + n^2 = 0$, and the roots are : $(x_1, x_2) = \frac{n}{2}\ (e^{i\theta}, e^{-i\theta})$, and $\frac{n}{2}\ (e^{\theta}, e^{-\theta})$, respectively.

Further, let us frame the hyperbolic-algebraic polynomial with the well known numerical values of the roots, namely, $(x_1, x_2) = (e^{\theta}, e^{-\theta}) = (1.618, 0.618)$.

[**Note:** These two numbers represent actually the ratio of any two consecutive numbers of the famous Fibonacci series, *viz.*, 1, 1, 2, 3, 5, 8, 13, 21, …, in which every third number is the sum of the previous second and first numbers. That is, $t_n = t_{n-1} + t_{n-2}$.

As we go for larger terms t_n, we get the ratio $[t_{n+1}/t_n]$ = **1.618**, or alternatively, $[t_n/t_{n+1}]$ = 1/1.618 = **0.618**. These two values of the ratios, are the answer of the following quarry: What is the number which differs from its reciprocal by unity? Mathematically, it is represented by: $x - \frac{1}{x} = 1$, or, $x^2 - x - 1 = 0$. The number is 1.618, or 0.618.

Along the similar line, the question may be extended in the field of matrix algebra as: What is the structure of the square matrix A, if the difference of A and its inverse A^{-1}, is an identity matrix I? Mathematically, it is represented by: $A - A^{-1} = I$.]

Thus, $\theta = 0.4812118$, $sh\theta = (e^{\theta} - e^{-\theta})/2 = 0.5$, and $ch\theta = (e^{\theta} + e^{-\theta})/2 = 1.1180339$, (remember, $ch\theta > 1$)

Hence, the required hyperbolic-algebraic polynomial is: $x^2 - 2\ x\ ch\theta + 1 = 0$

In this interesting direction of trigonometric-algebraic and hyperbolic-algebraic polynomials, in-depth study is needed to be carried out, with a view to extend the same concept for the **Nbic-algebraic functions** like, $ax^2 + bx$ [cosN, sinN] $\theta + c = 0$, as a future research on **Nbic functions**.

❑❑❑

Appendix A

PYTHAGOREAN TRIPLETS WITH A PAIR OF CONSECUTIVE NUMBERS

A-1 INTRODUCTION

Three numbers are called Pythagorean triplets if they represent the sides of a right triangle, x, y, z, where, x and y are the two adjacent sides meeting at right angle, and z being the hypotenuse satisfying the criterion, $x^2 + y^2 = z^2$.

The sets of Pythagorean triplets (x, y, z) can be very easily obtained by employing two variables (u, v) replacing three *viz.*, (x, y, z). This is possible if formula of simple algebra if followed. That is

$$(u^2 - v^2)^2 + (2uv)^2 = (u^2 + v^2)^2 \tag{A.1}$$

Substituting

$$u^2 - v^2 = x,\ 2uv = y, \text{ and } u^2 + v^2 = z \tag{A.2}$$

an infinite sets of Pythagorean triplets can be readily obtained satisfying, $x^2 + y^2 = z^2$, just adopting any two values to the new variables, u and v.

For example, take numerical values of u an v, and represent symbolically the triplets (x, y, z) along with the values of (u, v) as $(u, v : x, y, z)$:

With $u = 7,\ v = 3,$

$x = u^2 - v^2 = 40,\ y = 2uv = 42,\ z = u^2 + v^2 = 58,$

so that, $40^2 + 42^2 = 58^2$

The triplet is (7, 3 : 40, 42, 58)

With $u = 11,\ v = 8,$

$x = u^2 - v^2 = 57,\ y = 2uv = 176,\ z = u^2 + v^2 = 185,$

so that, $57^2 + 176^2 = 185^2$

The triplet is (11, 8 : 57, 176, 185)

With $u = 5,\ v = 2,$

$x = u^2 - v^2 = 21,\ y = 2uv = 20,\ z = u^2 + v^2 = 29,$

so that, $20^2 + 21^2 = 29^2$

The triplet is $(5,\ 2 : \overline{20,21},\ 29)$.

In these three examples, the third one shows that, $x = 20$ and $y = 21$, so that they are consecutive numbers, whereas the first two examples do not show this property. Whether the consecutiveness of the two adjacent sides, specially identified as $(\overline{x, y},\ z)$, is to met with by accident, or can be found out at will, is the quarry which is answered in the following paragraphs.

A-1.1 Separate Class of Pythagorean Triplets

There exists an unending number of Pythagorean triplets (x, y, z) with a pair of consecutive numbers, $(\overline{x, y})$ shown under bar sign.

Typical examples are: $(\overline{3,4},\ 5)$, $(\overline{119,120},\ 169)$, $(\overline{803760,803761},\ 1136689)$, and so on.

About the consecutiveness of two sides of a right triangle, reference is made to the approach of Sierpinski[1] (1962), and Dye and Nickalls[2] (1998).

Sierpinski[1] considers two cases:

(a) difference of, z and larger of x and y, is 1, and...

(b) difference of x and y is 1.

So far as item (b) is concerned, the triplets are described as : $(3x + 2z + 1,\ 3x + 2z + 2,\ 4x + 3z + 2)$, and the generation of successive triplets requires some basic triplet with known numerical value of x, y, and z, to start with. The basic triplet is $(\overline{3,4},\ 5)$.

Thus using formula

with $x = 3$, $y = 4$, $z = 5$, the first set is $(\overline{3,4},\ 5)$,

with $x = 3$, $z = 5$, the second set is $(\overline{20,21},\ 29)$,

with $x = 20$, $z = 29$, the third set is $(\overline{119,120},\ 169)$,

with $x = 119$, $z = 169$, the third set is $(\overline{696,697},\ 985)$,

and so on.

The severe drawback that Sierpinski's approach suffers, is that until and unless the previous values of the $(n - 1)$-th triplets is known, the next n-th one cannot be calculated. So, all the sets prior to the required n-th set of triplets must be available beforehand. That is to say, this approach fails to provide just the n-th set directly.

Dye and Nickalls[2] also adopt the iteration technique to generate the sets of triplets in which, similar to Sierpinski's approach, the value of the previous set of triplet is to be fed.

The suggested iterative formulas of Dye and Nickalls, are capable of generating all the relevant Pythagorean triplets. However, all the generated triplets do not satisfy the criterion that, the two adjacent sides x and y should differ by unity. The unwanted ones are to be screened out manually. This method thus, not only suffers the same difficulties of the previous method cited, but also additionally requires sieving out manually the unwanted sets of triplets.

That is to say that, none of the two methods can provide the sets of the special category of the Pythagorean triplets, if wanted to calculate the n-th set directly (*i.e.*, without using the values of the previous $(n - 1)$-th set, or any other stipulated condition).

In this direction, the approach of Bairagi (the author of this book) to get this special class of Pythagorean triplets directly, on eliminating altogether the drawbacks as encountered in the two previous methods, is presented as follows.

A-1.2 Proposed Method (Bairagi)

The present method describes the numbers both u and v, as well as x, y, z in the form of infinite series. Typical triplet set is represented by: $(u, v : \overline{x, y}, z)$. Typical interesting properties of each of these numbers have been probed into and described.

Selection of u and v

The requirement is that, in the number pair $(\overline{x, y})$ in the triplet $(\overline{x, y}, z)$, the difference of x and y is 1. It may so happen that, $x > y$, or $y > x$, *i.e.*, $x - y = 1$, or $y - x = 1$.

We still use the general formula (A.2) for any Pythagorean triplet

$$x = u^2 - v^2,\ y = 2uv, \text{ and } z = u^2 + v^2 \tag{A.3}$$

For Case 1, that is,

$$\begin{aligned} x - y &= 1, \\ u^2 - v^2 - 2uv &= 1 \\ u^2 - 2uv - (v^2 + 1) &= 0 \end{aligned}$$

For Case 2, that is,

$$\begin{aligned} y - x &= 1, \\ u^2 - v^2 - 2uv &= 1 \\ u^2 - 2uv - (v^2 - 1) &= 0 \end{aligned} \tag{A.4a}$$

Solving the quadratic equation in u for both the cases,

$$u = v + \text{A}, \text{ where A} = \sqrt{2v^2 + 1}, \text{ (for Case 1)}$$

$$u = v + \text{B}, \text{ where B} = \sqrt{2v^2 - 1}, \text{ (for Case 2)} \tag{A.4b}$$

from which x can be determined as

$$\begin{aligned} x &= u^2 - v^2 \\ &= 2\,v\text{A} + \text{A}^2, \text{ (for Case 1)} \\ &= 2\,v\text{B} + \text{B}^2, \text{ (for Case 2)} \end{aligned} \tag{A.4c}$$

where, A and B are given by (A.4b)

If the value of u as in (A.4b), has to be a positive integer, it is necessary that, the expressions of A or B, must yield a positive integer value. That is, $(2v^2 + 1)$, or $(2v^2 - 1)$, as the case may be, must come out to be a perfect square. Once the proper values of v are available in hand, the corresponding u-values can be readily obtained using (A.4b).

Let us see for what values of v, A or B are perfect squares.

Substituting 1, 2, 3, 4, ..., for v, the effective ones are:

For v = 1, A = no, B = 1, so that, u = 2
For v = 2, A = 3, B = no, so that, u = 5
For v = 5, A = no, B = 7, so that, u = 12
For v = 12, A = 17, B = no, so that, u = 29
For v = 29, A = no, B = 41, so that, u = 70

and so on.

[where, 'no' stands for 'no value']

This way, knowing the trends of the effective values of v, and the corresponding values of u, the corresponding sets of Pythagorean triplets $(\overline{x, y}, z)$ can be directly evaluated. It is interesting to note that, not only both the effective values of u and v can be described through infinite sequence, all the three quantities of the triplets, *viz.*, x, y and z follow their respective trends in infinite sequence.

Thus, the appropriate infinite sequence for v and u, as well as x, y and z can be worked out. In the following **Table A.1**, their corresponding numerical figures up to first ten terms, are demonstrated.

Table A.1 Pythagorean triplets $(u, v : \overline{x, y}, z)$

n	v_n	u_n	Pythagorean triplets $(\overline{x, y}, z)$			Number ratio z_{n+1}/z_n
			Consecutive pairs $(\overline{x, y})$		Hypotenuse z_n	
			x_n	y_n		
1	1	2	3	4	5	5.8000000
2	2	5	21	20	29	5.8275862
3	5	12	119	120	169	5.8284024
4	12	29	697	696	985	5.8284264
5	29	70	4059	4060	5741	5.8284264
6	70	169	23661	23660	33461	5.8284271
7	169	408	137903	137904	195025	5.8284271
8	408	985	803761	803760	1136689	5.8284271
9	985	2378	4684659	4684660	6625109	5.8284271
10	2378	5741	27304197	27304196	38613965	5.8284271

The ratio, (z_{n+1}/z_n) converges to: $(3+2\sqrt{2}) = 5.828\ 427\ 124\ 746$ (up to 12 decimal places).

A-2 PROPERTIES OF THESE NUMBER CHAINS

A-2.1 *v*-sequence

(a) The number chain of v, namely, 1, 2, 5, 12, 29, 70, 169, 408, 985, 2378, ….., for different values of n (= 1, 2, 3, ….), is seen to follow a recurring sequence, having the scale of relation: (1, –2, –1). That is

$$(1).v_{n+2} - (2).v_{n+1} - (1).v_n = 0 \tag{A.5a}$$

For example, (**Table A.1**) for $n = 5$, collect $v_5 = 29$, $v_6 = 70$, $v_7 = 169$, so that, (A.7) gives: 1.(169) – 2.(70) – 1.(29) = 0

(b) Any n-th term of the v-sequence is given by

$$v_n = \left[\frac{(\sqrt{2}+1)^n}{2\sqrt{2}} + (-1)^{n-1}\frac{(\sqrt{2}-1)^n}{2\sqrt{2}}\right], \quad (n = 1, 2, 3, \ldots, \infty) \tag{A.5b}$$

Thus, $v_1 = 1$, $v_2 = 2$, $v_4 = 12$, $v_6 = 70$, $v_{10} = 2378$, and so on.

For example, v_3 and v_6 are calculated using scientific calculator. Call (A.5b)

$$v_3 = \left[\frac{(\sqrt{2}+1)^3}{2\sqrt{2}} + (-1)^2 \frac{(\sqrt{2}-1)^3}{2\sqrt{2}}\right]$$

with, $\sqrt{2} = 1.414213562$, $2\sqrt{2} = 2.828427125$,

$\sqrt{2} + 1 = 2.414213562$, $\sqrt{2} - 1 = 0.414213562$,

$$v_3 = \left[\frac{(2.414213562)^3}{2.828427125} + (-1)^2 \frac{(0.414213562)^3}{2.828427125}\right]$$

$$= 4.974873731 + 0.025126265 = \mathbf{4.999999996} = 5$$

$$v_6 = \left[\frac{(2.414213562)^6}{2.828427125} + (-1)^5 \frac{(0.414213562)^6}{2.828427125}\right]$$

$$= 70.0017856 - 1.785668724 \times 10^{-3} = \mathbf{69.999999993} = 70$$

Note that, error in decimal places is inevitable, at times, in calculator calculation, and the result may not be the whole number. However, to eliminate this, try the formula algebraically, on expanding the powers binomially, or otherwise. This is done for v_6.

$$v_6 = \left[\frac{(\sqrt{2}+1)^6}{2\sqrt{2}} - \frac{(\sqrt{2}-1)^6}{2\sqrt{2}}\right]$$

$$= \frac{1}{2\sqrt{2}} [(\sqrt{2} + 1)^3 + (\sqrt{2} - 1)^3]\,[(\sqrt{2} + 1)^3 - (\sqrt{2} - 1)^3]$$

$$= \frac{1}{2\sqrt{2}} [(\sqrt{2} + 1 + \sqrt{2} - 1)^3 - 3\,(2 - 1)\,(\sqrt{2} + 1 + \sqrt{2} - 1)]$$

$$\times [(\sqrt{2} + 1 - \sqrt{2} - 1)^3 - 3\,(2 - 1)\,(\sqrt{2} + 1 - \sqrt{2} + 1)]$$

$$= \frac{1}{2\sqrt{2}} [(2\sqrt{2})^3 - 3\,(2\sqrt{2})][8 + 3 \times 2]$$

$$= \frac{14 \times 10\sqrt{2}}{2\sqrt{2}} = \mathbf{70.}$$

[This procedure yields the correct result in whole number.]

Incidentally, it is worth mentioning that, the well known **Fibonacci sequence** *viz.*, 1, 1, 2, 3, 5, 8, 13, 21, …, is similar to this v-sequence, but the scale of relation is (1, –1, –1).

(c) Every third number of the v-sequence is the sum of the first, and twice of the second. Thus,

$$v_{n+2} = 2.\, v_{n+1} + v_n\,,\ (n = 1, 2, 3, \ldots, \infty) \qquad \text{(A.5c)}$$

For example, for $n = 6$, $v_n = 70$, $v_{n+1} = 169$, $v_{n+2} = 408$,
so that (A.5c) gives : 408 = 2.(169) + (70), and so on.
In other words, the average of the first and third number is equal to the sum of the first and second number.

Thus

$$(v_n + v_{n+2}) / 2 = v_n + v_{n+1} \tag{A.5d}$$

That is,

For $n = 3$, $(12 + 70)/2 = 12 + 29$

For $n = 7$, $(169 + 985)/2 = 169 + 408$

and so on.

A-2.2 *u*-sequence

Since, $u_n = v_{n+1}$, as observed from **Table A.1**, the number chain of the *u*-sequence, follows the same number chain of the *v*-sequence, but with the starting value as 2. In other words, $u_1 = v_2 = 2$, $u_2 = v_3 = 5$, $u_7 = v_8 = 408$, and so on.

Thus the number chain of *u* is : 2, 5, 12, 29, 70, 169, 408, 985, 2378, ….., for different values of n (= 1, 2, 3, …, ∞).

A-2.3 *x*-sequence

The number chain for n = 1, 2, 3, …, ∞, namely, 3, 21, 119, 697, 4059, 23661, ….., ∞, corresponds to that for *x*-sequence (**Table A.1**). Each value of x can be calculated using (A.4c), that is

$$x = u^2 - v^2 = (u + v)(u - v) \tag{A.6a}$$

The starting value of this sequence is 3. So the generation of the *x*-sequence is expressed as:

$$x_{n+2} = u^2_{n+2} - v^2_{n+2} = (u_{n+2} + v_{n+2})(u_{n+2} - v_{n+2}) \tag{A.6b}$$

Noting that, $u_{n+2} = v_{n+3} = 2\,v_{n+2} + v_{n+1}$, (A.6b) gives

$$x_{n+2} = (v_{n+2} + v_{n+1})(3\,v_{n+2} + v_{n+1})$$

$$= v^2_{n+1} + 4\,v_{n+1}\,v_{n+2} + 3\,v^2_{n+2}, \quad (n = 1, 2, 3, \ldots, \infty) \tag{A.6c}$$

For example, for $n = 4$,

$$x_6 = (v_6 + v_5)(3\,v_6 + v_5) = (70 + 29)(3 \times 70 + 29) = 23661$$

$$= v_5^2 + 4\,v_5\,v_6 + 3\,v_6^2 = 29^2 + 4 \times 29 \times 70 + 3 \times 70^2$$

$$= 841 + 8120 + 14700 = 23661$$

All numbers in the *x*-sequence are odd, and on the strength of (A.6a), are factorable.

A-2.4 *y*-sequence

The x and y of the Pythagorean triplets $(\overline{x, y})$, follow a definite pattern.

Notice that:

In the first set $(\overline{3,4}, 5)$, $x = 3$, $y = x + 1 = 4$. That is, $x < y$

In the second set $(\overline{21,20}, 29)$, $x = 21$, $y = x - 1 = 20$. That is, $x > y$

In the third set $(\overline{119,120}, 169)$, $x = 119$, $y = x + 1 = 120$. That is, $x < y$

This way, x is alternately less than, and greater than y, as n starts from 1, and proceeds on, in steps of 1.

Thus whether 1 is to be added to, or, subtracted from x, for getting y, depends on whether the n-th term is even or odd. If n is even, $x > y$, otherwise $x < y$. This switch is controlled by, $x_n - y_n = (-1)^n$.

A-2.5 z-sequence

(a) All numbers in the z-sequence representing the hypotenuse of the Pythagorean triangle whose adjacent sides are the consecutive number pairs $(\overline{x, y})$, are odd numbers. The number chain for z when arranged for different values of n (= 1, 2, 3, ..., ∞), appears as, 5, 29, 169, 985, 5741, 33461, 195025, 1136689,...∞, and follows a recurring sequence. These numbers are not at all new, but are positioned at regular interval in the v-sequence itself, shown already.

That is

$$(1).z_{n+2} - (6).z_{n+1} + (1).z_n = 0 \text{ , } (n \geq 3) \quad \text{(A.7a)}$$

[This relation has been arrived at both by Mills[3] (1996), as well as Dye and Nickalls (1998)].

Every third number of the z-sequence is obtained by subtracting the first number from six times the second number, suggesting that the scale of relation is: (1, – 6, 1).

For example:

Verify (A.7a) with $n = 4$, $z_4 = 985$, $z_5 = 5741$ and $z_6 = 33461$

$$(1)\ z_6 - (6)\ z_5 + (1)\ z_4 = 33461 - 6 \times 5741 + 985 = 0$$

This verifies the scale of relation also.

Calculate z_9. We require $z_7 = 195025$ and $z_8 = 1136689$.

Thus

$$z_9 = 6\ (1136689) - 1\ (195025) = 6625109 \quad \text{[This is recorded in } \textbf{Table A.1}\text{]}$$

(b) Any n-th term of the z-sequence is given by

$$z_n = \left[\frac{5\sqrt{2}-7}{2\sqrt{2}}(3-2\sqrt{2})^{n-1} + (-1)^{n-1}\frac{(5\sqrt{2}+7)}{2\sqrt{2}}(-3-2\sqrt{2})^{n-1}\right],$$

$$(n = 1, 2, 3, \ldots, \infty) \quad \text{(A.7b)}$$

Thus, $z_1 = 5$, $z_5 = 5471$, $z_6 = 33461$, $z_{10} = 38613965$, and so on.

For example, z_5 is calculated from (A.7b), using scientific calculator, and also by algebraic expansion of the formula.

z_5 (using scientific calculator):

$$\frac{5\sqrt{2}-7}{2\sqrt{2}} = \frac{7.071067812-7}{2.828427125} = 0.025126265$$

$$\frac{5\sqrt{2}+7}{2\sqrt{2}} = \frac{7.071067812+7}{2.828427125} = 4.974873733$$

$$3-2\sqrt{2} = 0.171572875$$

$$-3-2\sqrt{2} = -5.828427125$$

For $n = 5$, using calculator

$$\begin{aligned} z_5 &= (0.025126265)\,(0.171572875)^4 \\ &\quad + (-1)^4\,(4.974873733)\,(-5.828427125) \\ &= 2.1773320946 \times 10^{-5} + 5740.99978 \\ &= \mathbf{5740.999802} = 5741 \end{aligned}$$

[This procedure yields inherent round off error, however negligible, it may be.]

For $n = 5$, using algebraic expansion

$$z_5 = \left[\frac{5\sqrt{2}-7}{2\sqrt{2}}(3-2\sqrt{2})^4 + \frac{(5\sqrt{2}+7)}{2\sqrt{2}}(-3-2\sqrt{2})^4\right]$$

Now

$$(3-2\sqrt{2})^4 = [(3-2\sqrt{2})^2]^2 = [17-12\sqrt{2}]^2 = 577-408\sqrt{2}$$

$$(-3-2\sqrt{2})^4 = [(-3-2\sqrt{2})^2]^2 = [17+12\sqrt{2}]^2 = 577+408\sqrt{2}$$

$$z_5 = \frac{5\sqrt{2}-7}{2\sqrt{2}}(577-408\sqrt{2}) + \frac{5\sqrt{2}+7}{2\sqrt{2}}(577+408\sqrt{2})$$

$$= \frac{1}{2\sqrt{2}}[5741\sqrt{2}-8119]+[5741\sqrt{2}+8119]$$

$$= \frac{2\times 5741\sqrt{2}}{2\sqrt{2}} = \mathbf{5741}$$

[This procedure yields correct whole number.]

The calculation procedure for obtaining the numerical value of z_n, for larger and larger values of n, becomes too involved to handle manually, and so, an alternative path is sought for in order to avoid the cumbersome, lengthy and tedious arithmetic. Fortunately, help comes from the fact that, the z_n-sequence and the v_n-sequence, have simple link, which is as follows.

(c) It is interesting to note that the numbers of the z-sequence can be taken from those of the v-sequence on dropping the alternate terms. The starting number is $z_1 = 5$. The subsequent numbers are: $z_2 = v_5 = 29$, $z_3 = v_7 = 169$, and so on. That is

$$z_n = v_{2n+1}\,, \quad (n = 1, 2, 3, \ldots, \infty) \tag{A.7c}$$

(d) **Table A.1** for z_n-sequence reveals that, the ratio (z_{n+1}/z_n) tends to converge very rapidly to the number, 5.828427124746... $(= 3+2\sqrt{2})$, after few terms onwards.

Remember that, Fibonacci number which is the ratio of (F_{n+1}/F_n), for the sequence, 1, 1, 2, 3, 5, 8, 13, 21, 34, ..., converges to $1.618 = \left(\frac{\sqrt{5}+1}{2}\right)$.

Fibonacci number can be defined as:

From which number when its reciprocal is subtracted, yields 1? Obviously, we know the number is, $x = 1.618...$, and its reciprocal is, $\frac{1}{x} = 0.618...$To find the answer, the connecting quadratic equation can be very easily framed as:

$$x - \frac{1}{x} = 1 \Rightarrow x^2 - x - 1 = 0 \Rightarrow x = \frac{\sqrt{5}+1}{2}, \text{ (taking the positive root only)}$$

The same quadratic equation also can be arrived at in two ways:

First, let the solution of the required quadratic equation. is $x = \frac{\sqrt{5}+1}{2}$.

Then, $2x - 1 = \sqrt{5} \Rightarrow 4x^2 - 4x + 1 = 5 \Rightarrow 4x^2 - 4x - 4 = 0 \Rightarrow x^2 - x - 1 = 0$

Alternatively,

Let the quadratic equation is : $ax^2 + bx + c = 0$, (take $a = 1$),

whose solution is, $x = \frac{\sqrt{5}+1}{2}$. Then

$$-\frac{b}{2} = \frac{1}{2} \Rightarrow b = -1, \text{ and}$$

$$\frac{\sqrt{b^2 - 4c}}{2} = \frac{\sqrt{5}}{2} \Rightarrow = \quad 1 - 4c = 5 \Rightarrow c = -1$$

So the quadratic equation is : $x^2 - x - 1 = 0$

In the same way, taken from the number sequence for z, representing the hypotenuse for this special Pythagorean triplets, we screen the number as the ratio of (z_{n+1}/z_n) which is obtained as **5.828427124746...** $= (3+2\sqrt{2})$. This number, henceforth is proposed to be termed as **Bairagi Number**.

Now let, $ax^2 + bx + c = 0$, has the solution $x = 3+2\sqrt{2}$

Let, $a = 1$, then

$$-\frac{b}{2} = 3 \Rightarrow b = -6, \Rightarrow b^2 = 36, \text{ and}$$

$$\frac{\sqrt{b^2 - 4c}}{2} = 2\sqrt{2} \Rightarrow \sqrt{b^2 - 4c} = 4\sqrt{2} \Rightarrow b^2 - 4c = 32 \Rightarrow c = 1$$

So the quadratic equation is : $x^2 - 6x + 1 = 0$. (A.7d)

Observe that:

If the coefficients of this equation are arranged as: $(a, b, c) = (1, -6, 1)$, the form gives the **scale of relations** for z-sequence. Refer to (A.7a).

Further dividing both sides of (A.7d) by x,

we get, $x + \frac{1}{x} = 6$ (A.7e)

With the solution, $x = 5.828427124746... = (3+2\sqrt{2})$,

and its reciprocal, $\frac{1}{x} = 0.171572875... = (3-2\sqrt{2})$,

the addition of the number $(3+2\sqrt{2})$ and its reciprocal $(3-2\sqrt{2})$, [or alternatively, the number $(3-2\sqrt{2})$ and its reciprocal $(3+2\sqrt{2})$], yields = 6, justifying (A.7e).

Definition of Bairagi Number:

This observation (A.7e) immediately suggests a **definition for Bairagi number**, (similar to the one done for Fibonacci number), in the form of an enquiry as:

"Which number when added to its reciprocal, yields 6?"

A-3 CONCLUDING REMARKS

A very simple and straight forward direct procedure is proposed by Bairagi (the author of this book) for obtaining sets of infinite number of the special category of Pythagorean triplets $(\overline{x, y}, z)$, in which a pair $(\overline{x, y})$, being the consecutive numbers.

The triplets $(\overline{x_n, y_n}, z_n)$, for any particular value of n, $(n = 1, 2, 3, ..., \infty)$, can be very quickly and readily written down, as described in the following steps :

Step 1

Primarily, the number sequence for v_n is the most sought out sequence. Once, v_n is determined, all other number chains of u_n, and x_n, y_n and z_n, all of which are linked with v_n, will then be available. Out of the choice (A.5a), or alternatively (A.5b), preference goes to the later expression.

First get v_n using (A.5c). Note that, evaluation of v_n starts from v_3 onwards. There is no need of calculating v_1 and v_2, since $v_1 = 1$, $v_2 = 2$.

So $$v_{n+2} = 2.\ v_{n+1} + v_n,\ (n = 1, 2, 3, ..., \infty)$$

Step 2

Since u_n is greater than v_n, get

$$u_n = v_{n+1},\ (n = 1, 2, 3, ..., \infty)$$

Step 3

Use (A.3) to obtain x_n, y_n, and z_n

$$x_n = u_n^2 - v_n^2 = (u_n + v_n)(u_n - v_n),\ (n = 1, 2, 3, ..., \infty)$$

There is no need of calculating x_1, since $x_1 = 3$. Start evaluating x_n from x_2 onwards, using

$$x_n = (v_n + v_{n-1})(3v_n + v_{n-1}),\ (n = 2, 3, ..., \infty)$$

Step 4

$$y_n = 2\, u_n v_n, \ (n = 1, 2, 3, \ldots, \infty)$$

Alternatively,

$$y_n = x_n - (-1)^n, \ (n = 1, 2, 3, \ldots, \infty)$$

Step 5

$$z_n = u_n^{\,2} + v_n^{\,2}, \ (n = 1, 2, 3, \ldots, \infty)$$

Alternatively, use (A.7c)

$$z_n = v_{2n+1}, \quad (n = 1, 2, 3, \ldots, \infty)$$

Thus, the **10-th triplet is** calculated as follows :

$$v_{10} = 2378 \ , \ u_{10} = v_{11} = 2378$$

$$v_{10}^2 = 5654884, \ u_{10}^2 = 32959081, \ u_{10}\, v_{10} = 13652098$$

$$x_{10} = u_{10}^2 - v_{10}^2 = 27304197$$

$$y_{10} = x_{10} - 1 = 27304196$$

$$z_{10} = v_{21} \ (= u_{10}^2 + v_{10}^2) = 38613965$$

The 10-th triplet is then $(\overline{27304197, 27304196}, 38613965)$

The suggested present method confirms that the number of this special category of Pythagorean triplets is infinite, and any *n*-th set of such triplets can be directly written down, using the suggested simple sequence of *v* or *z*.

Similar to the Fibonacci Number expressed as the ratio of the *n*-th to the (*n*–1)-th number of the Fibonacci sequence, *viz*., 1, 1, 2, 3, 5, 8, 13, 21, 34, …, a number as the ratio of (z_{n+1}/z_n) of the number sequence for *z*, representing the hypotenuse for this special Pythagorean triplets, is obtained as **5.828427124746** = $(3+2\sqrt{2})$, which henceforth is proposed to be termed as **Bairagi Number. This number and its reciprocal when added, gives the number 6.**

References:

1. Sierpinski, Waclaw. Pythagorean Triangles. *Graduate School of Science*, Yeshira Univ., New York, 1962.
2. Dye, R and Nickalls, R.W.D. A New Algorithm for Generating Pythagorean Triples, *Math. Gazette*, 82, pp. 86–91, March, 1998.
3. Mills, J.T.S., Another Family Tree for Pythagorean Triples, *Math. Gazette*, 8, pp. 545–548, November, 1996.

□□□

Appendix B

ROOTS OF POLYNOMIAL EQUATIONS—APPLICATION TO THE RESULTANTS OF FORCE SYSTEMS

B-1 INTRODUCTION

Polynomial root analogy of like parallel force systems each of unit magnitude

Any equation of the type

$$P_n = P_n(x) = a_n x^n + a_{n-1} x^{n-1} + a_{n-2} x^{n-2} + \ldots + a_2 x^2 + a_1 x + a_0 = 0 \quad (1)$$

where, the maximum power of x is n (known as degree), which decreases gradually to zero, is known as polynomial equation, and is represented by P_n. Since n is the degree of the polynomial equation, the total number of roots (*viz.*, $x_1, x_2, x_3, \ldots x_n$), is also equal to n.

Yet until probed, other interesting property of the roots of this type of polynomial equation of constant coefficients, as observed by Bairagi (the present author), has been demonstrated in this article.

Considering that the given polynomial P_n as the basic polynomial, which is not differentiated at all, (that is, at 0-th stage of differentiation), be renamed as P_n^0. If the polynomial P_n^0, is differentiated with respect to x, once, twice, thrice, and so on, successively, a family of polynomials will result following a gradually reduced order, *viz.*, $(n-1)$, $(n-2)$, $(n-3)$, etc. Designate these polynomials obtained at any k-th stage of differentiation as, $P_n^1, P_n^2, P_n^3, \ldots, P_n^k$ complying with the 1st, 2nd, 3rd …, k-th successive stages of differentiation of the polynomial P_n, with respect to x.

Obviously, each of the polynomial equation $P_n^k = 0$, obtained by successive differentiation of $P_n = P_n^0 = 0$, has exactly $(n - k)$ number of roots, (where $k = 1, 2, 3\ldots, n$). Thus, the basic polynomial when differentiated successively up to $(k-1)$-th stage of differentiation, yields a linear equation in x, which provide a single value of the root. This way, it is possible to differentiate P_n^0 up to $(k-1)$ -th stage, when finally differentiation is not possible for getting root since the polynomial equation ceases to exist. It is interesting to observe that, the average of the roots of each of the polynomials $P_n^k = 0$, for $k = 0, 1, 2, 3, \ldots(n-1)$, equals to

$$x_{av} = -\frac{1}{n}.\frac{a_{n-1}}{a_n}$$

where, a_{n-1} and a_n are the coefficients of x^{n-1} and x^n, respectively.
Thus remembering that

$$R = \text{total number of roots} = \text{degree of the polynomial} = n$$

$$S = \text{sum of the roots} = \sum_{i=1}^{n} x_i,$$

$$x_{av} = \frac{S}{R} = -\frac{a_{n-1}}{n.a_n}, \text{ we have}$$

0-th Stage of Differentiation (k = 0):

$$P_n^0 = a_n\, x^n + a_{n-1}\, x^{n-1} + a_{n-2}\, x^{n-2} + \ldots + a_2\, x^2 + a_1\, x + a_0 = 0$$

$$R = n,\ S = -\frac{a_{n-1}}{a_n},\ x_{av} = -\frac{a_{n-1}}{n.a_n}$$

1-st Stage of Differentiation (k = 1):

$$P_n^1 = n\, a_n\, x^{n-1} + (n-1)\, a_{n-1}\, x^{n-2} + (n-2)\, a_{n-2}\, x^{n-3} + \ldots + 2\, a_2\, x + a_1 = 0$$

$$R = n-1,\ S = -\frac{a_{n-1}}{a_n}.\frac{n-1}{n},\ x_{av} = -\frac{a_{n-1}}{n.a_n}$$

2-nd Stage of Differentiation (k = 2):

$$P_n^2 = n\,(n-1)\, a_n\, x^{n-2} + (n-1)\,(n-2)\, a_{n-1}\, x^{n-3} + (n-2)\,(n-3)\, a_{n-2}\, x^{n-4} + \ldots$$
$$+\, 6\, a_3 x + 2a_2 = 0$$

$$R = n-2,\ S = -\frac{a_{n-1}}{a_n}.\frac{(n-1)(n-2)}{n(n-1)} = -\frac{a_{n-1}}{a_n}.\frac{n-2}{n},\ x_{av} = -\frac{a_{n-1}}{n.a_n}$$

3-rd Stage of Differentiation (k = 3):

$$P_n^3 = n\,(n-1)\,(n-2)\, a_n\, x^{n-3} + (n-1)\,(n-2)\,(n-3)\, a_{n-1}\, x^{n-4}$$
$$+\, (n-2)\,(n-3)\,(n-4)\, a_{n-2}\, x^{n-5} + \ldots + 6\, a_3 = 0$$

$$R = n-3,\ S = -\frac{a_{n-1}}{a_n}.\frac{(n-1)(n-2)(n-3)}{n(n-1)(n-2)} = -\frac{a_{n-1}}{a_n}.\frac{n-3}{n},\ x_{av} = -\frac{a_{n-1}}{n.a_n}$$

(n–1)th Stage of Differentiation (k = n–1):

$$P_n^{n-1} = n\,(n-1)\,(n-2) \ldots 4.3.2.a_n\, x + (n-1)\,(n-2)\,(n-3) \ldots 3.2.1 = 0$$

$$R = 1,\ S = -\frac{a_{n-1}}{a_n}.\frac{(n-1)(n-2)\ldots.3.2.1}{n(n-1)(n-2)\ldots.4.3.2} = -\frac{a_{n-1}}{a_n}\frac{1}{n},\ x_{av} = -\frac{a_{n-1}}{n.a_n}.$$

It can be easily verified that, whatever be the stage of differentiation, the average of the roots (x_{av}) of that stage of polynomial equation is unique, and is equal to, $-\frac{a_{n-1}}{n.a_n}$, and to arrive at this value, it is not necessary to evaluate all the roots of that polynomial equation. Instead, the same can be found readily by solving the linear equation obtained by successive $(n-1)$-th differentiation of the polynomial equation $P_n^0 = 0$.

In **Table B.1**, this property of roots has been demonstrated with the help of a 5-th degree polynomial equation, and those obtained by its successive differentiation. In other words, the value, $-\frac{a_{n-1}}{a_n}\cdot\frac{1}{n}$, is unique for the average of all the roots x_{av} for a particular polynomial equation P_n^k at any k-th stage of differntiation. The best way to get this value is to differentiate the given polynomial equation upto (n – 1)-th stage, and obtain the value of x, of the very linear equation on setting it to zero.

Table B.1. Average roots of polynomial equations $P_n^k = 0$, $k = 1, 2, 3 \ldots (n-1)$

Stage (k)	Polynomial equation ($P_n^k = 0$)	No. of roots (n)	Sum of Roots (S)	Av. of roots (S/n)
0	$a_5 x^5 + a_4 x^4 + a_3 x^3 + a_2 x^2 + a_1 x + a_0 = 0$	5	$-\frac{a_4}{a_5}$	$-\frac{1}{5}\cdot\frac{a_4}{a_5}$.
1	$5\,a_5 x^4 + 4\,a_4 x^3 + 3\,a_3 x^2 + 2a_2 x + a_1 = 0$	4	$-\frac{4}{5}\cdot\frac{a_4}{a_5}$	$-\frac{1}{5}\cdot\frac{a_4}{a_5}$.
2	$5.4\,a_5 x^3 + 4.3\,a_4 x^2 + 3.2a_3 x + 2\,a_1 = 0$	3	$-\frac{4}{5}\cdot\frac{3}{4}\cdot\frac{a_4}{a_5}$	$-\frac{1}{5}\cdot\frac{a_4}{a_5}$.
3	$5.4.3\,a_5 x^2 + 4.3.2\,a_4 x + 3.2.1\,a_3 x = 0$	2	$-\frac{4}{5}\cdot\frac{3}{4}\cdot\frac{2}{3}\cdot\frac{a_4}{a_5}$	$-\frac{1}{5}\cdot\frac{a_4}{a_5}$.
4	$5.4.3.2\,a_5 x^2 + 4.3.2.1\,a_4 = 0$	1	$-\frac{4}{5}\cdot\frac{3}{4}\cdot\frac{2}{3}\cdot\frac{1}{2}\cdot\frac{a_4}{a_5}$	$-\frac{1}{5}\cdot\frac{a_4}{a_5}$.

B-1.1 Analogous Representation of Like Parallel Force System

This particular property of the roots of the polynomial equations of constant coefficients can be gainfully applied to find the location of the resultant of parallel force analogously. This is illustrated with a particular example:

Example B.1

Let us take a fourth degree polynomial, $P_4 = P_4^0$, with constant coefficients.

$$P_4^0 = x^4 - 5x^3 + 0 + 20x - 16 = 0,$$

[basic polynomial, (*i.e.*, 0th stage of differentiation)]

$$(x - 1)(x - 2)(x + 2)(x - 4) = 0$$

Four roots are: $x_1 = 1$, $x_2 = 2$, $x_3 = -2$, $x_4 = 4$: $x_{av} = 1.25$

$$P_4^1 = 4x^3 - 15x^2 + 0 + 20 = 0, \text{ [1st stage of differentiation]}$$

$$= (x - 1.148)\ (x - 3.291)\ (x + 1.021) = 0$$

Three roots are: $x_1 = 1.148$, $x_2 = 3.291$, $x_3 = -1.021$: $x_{av} = 1.25$

$$P_4^2 = 2x^2 - 5x = 0, \quad \text{[2nd stage of differentiation]}$$
$$= 2x.(x - 2.5) = 0$$

Two roots are: $x_1 = 0$, $x_2 = 2.5$: $x_{av} = 1.25$

$$P_4^3 = 4x - 5 = 0, \quad \text{[3-rd stage of differentiation]}$$

One root is: $x_1 = 1.25$: $x_{av} = 1.25$

It is interesting to observe that the average of the roots, x_{av}, for all the polynomials, P_4^0, and its successive stages of derivatives, namely, $P_4^1, \ldots, P_4^3$, is of the same value 1.25. This feature can be used analogously to represent the location of the resultant of **like parallel force systems** simply by considering that, if through each of the root points, x_i, corresponding to any given polynomial equation, $P_n^k = 0$, (where, k represents the basic polynomial, or that at any stage of differentiation), a force of unit magnitude, parallel to one another acts, the resultant R of magnitude = n (which is also the degree of the polynomial, so also is the total number of roots), of which then acts through the point given by x_{av}, which is the average of the roots. This x_{av} actually represents the **centroid** of these parallel force system.

Thus, on a horizontal line representing x-axis (**Figure B.1**), on which four vertically downward forces ($a_1, a_2, \ldots, a_4$), each of magnitude 1, are considered to act at the four root points, $x = 1, 2, -2$ and 4, representing, $P_4^0 = 0$. The line of action (downwards) of the resultant $R = (a_1 + a_2 + \ldots + a_4)$, of magnitude 4, then acts through $x = 1.25$, which is the average of the four roots, and at the same time, is identified as the centroid of these forces.

Similarly, in $P_4^2 = 0$, two downward forces (a_1 and a_2), each of magnitude 1, acts at $x = 0$ and 2.5, and the resultant $R = (a_1 + a_2)$, of magnitude 2, then acts through $x = 1.25$, which is the average of two roots.

The linear equation represented by $P_4^3 = 0$, itself has one root, namely 1.25, and at this value of x, the line of action of the single force of magnitude 1, passes through.

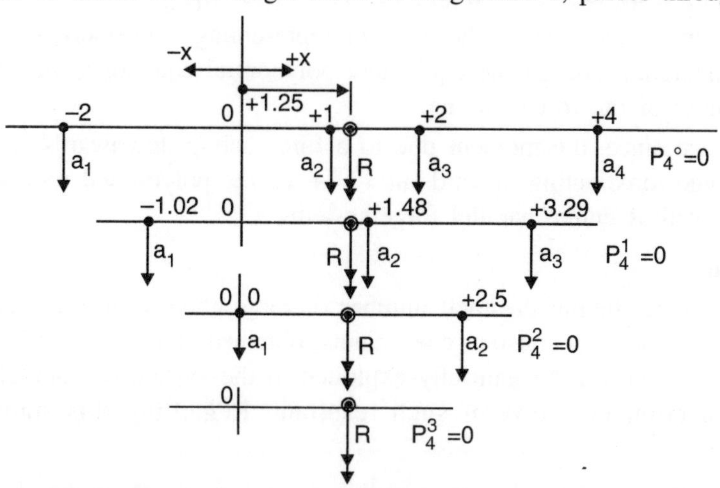

Fig. B.1: Resultant of like parallel force systems analogous to roots of successively differentiated polynomial equations $P_n^k = 0$, $k = 0,1,\ldots(n - 1)$.

Note: Here the roots of the polynomial equation selected, are real numbers. But it is possible that some roots turn out to be complex, and in that case, they will come in conjugate pairs so that while averaging, the imaginary part is cancelled out automatically, yielding only the remaining real part.

For example, say the in a 5-th degree polynomial,

$$P_5^0 = x^5 - 3.5x^4 - 6x^3 + 21x^2 - 40x - 62.5 = 0, \text{ [0th stage of differentiation]}$$
$$= (x - 1)(x + 1)(x + 2.5)(x - 3 - 4i)(x - 3 + 4i) = 0$$

Five roots are: 1, – 1, – 2.5, 3 + 4i, 3 – 4i, [i = sqrt(– 1)]:

The sum of roots S = 3.5

Here, the imaginary parts, +4i and – 4i, of the roots (3 + 4i), (3 – 4i) cancels out leaving only the sum of the real parts, *i.e.*, 3 + 3 = 6.

B-1.2 Case 1: Polynomial Root Analogy of Like Parallel Force System

The polynomial equation, $P_n = P_n^0 = 0$, of degree n, can be obtained from its n–number of roots, x_i, (i = 1 to n), by writing down the chain product of all the n-number of $(x - x_i)$ functions, which when simplified and arranged suitably in terms of ascending or descending order of the powers of x, and set to zero, needless to mention, speaks for the same polynomial function. That is :

$$P_n^0 = (x - x_1)\,(x - x_2)\,(x - x_3)\,.....(x - x_n)$$
$$= a_n\,x^n + a_{n-1}\,x^{n-1} + a_{n-2}\,x^{n-2} + ... + a_2x^2 + a_1x + a_0 = 0$$

For Example, P_4^1 of Example B.1 is taken.

Three roots are: 1.48, 3.291, – 1.021.

In terms of the roots, the polynomial equation when constructed, becomes:

$$P_4^1 = (x - 1.148)\,(x - 3.291)\,(x + 1.021) = 0$$

which when is simplified, gives:

$$P_4^1 = 4x^3 - 15x^2 + 0 + 20 = 0$$

This observation immediately paves the way for representing conversely, parallel force system (each of unit magnitude) through the equivalent polynomial equations, the degree of which is same as the number of the forces acting.

Remember that since, the moment due to a force acting downwards at $x = -x_1$ is same as that for the same force acting upwards at $x = +x_1$, the polynomial root analogy is equally valid for, like as well as unlike parallel force systems.

Rules Proposed

This interesting feature linking the total number of roots, their values and average values for any polynomial equation with constant coefficients, obtained at any stage of differentiation *viz.*, k = 0, 1, 2, 3, ...(n –1), can be gainfully exploited to the system of parallel forces regarding the resultant and point of action of such resultant. Regarding this statement, two rules are proposed:

Rule 1:

The value of x derived from the $(n-1)$-th derivative of a polynomial equation is equal to the average of the n-number of roots of the given polynomial equation. This average of the roots is equal to those of each of the polynomial equation derived at any k-th successive stage of differentiation, [where, k is less than or equal to $(n-1)$], including the basic given polynomial equation, (for which $k = 0$)].

Rule 2:

If n number of like parallel forces, each of unit magnitude, acting normal to a certain straight line, is expressed by a polynomial equation of degree n, which is formed by the position of its individual roots, x_i, ($i = 1$ to n), then the centroid, that is, the location through which the line of action the resultant R of magnitude n, of this force system passes, can be obtained analogously, either: (1) by averaging the roots, or, (2) by solving the $(n-1)$-th derivative, [this always yields linear equation], set to zero.

[**Note**: It is equally true that, each polynomial equation of higher degree obtained by successive integration of the given polynomial equation, will yield the same, x_{av} value, and hence is also applicable to the description of the parallel force system.]

B-1.3 Case 2: Polynomial Root Analogy of Unequal Parallel Force Systems

If instead, parallel forces of magnitudes $a_1, a_2, a_3, \ldots, a_n$, act at root points $x_1, x_2, x_3, \ldots, x_n$, it is possible to convert the system into the parallel force system each of unit magnitude (*viz.*, $a_1 = a_2 = a_3 = \ldots = a_n = 1$), acting at the modified root points given by,

$$\frac{a_1x_1}{\text{C}}, \frac{a_2x_2}{\text{C}}, \frac{a_3x_3}{\text{C}}, \ldots, \frac{a_nx_n}{\text{C}}; \text{where}, \frac{1}{\text{C}} = \frac{n}{\sum_{i=1}^{n} a_i}$$

Then the polynomial equation P_n^0 becomes

$$\left(x - \frac{a_1x_1}{\text{C}}\right)\left(x - \frac{a_2x_2}{\text{C}}\right)\left(x - \frac{a_3x_3}{\text{C}}\right)\ldots\left(x - \frac{a_nx_n}{\text{C}}\right) = 0$$

Application to the Centroid Problem of Regular Lamina

This modified form of polynomial equation can be adopted as the analogous mathematical model of the parallel force systems. Typical examples, in which it is required to find the magnitude, direction and point of action of the resultant, are described with the following numerical example.

Example B.2

Locate the centroid of the lamina ABCDEA (drawn with solid lines) as shown in **Figure B.2** for the cases when the area is composed of:

(1) two laminas, (2) three laminas, (3) four laminas, *viz.*,

(1) Area composed of 2 laminas [ABFEA – CFD]
(2) Area composed of 3 laminas [BCIH + AHDE + ICD]
(3) Area composed of 4 laminas [BCIH + HIJA + JIDE + DIC]

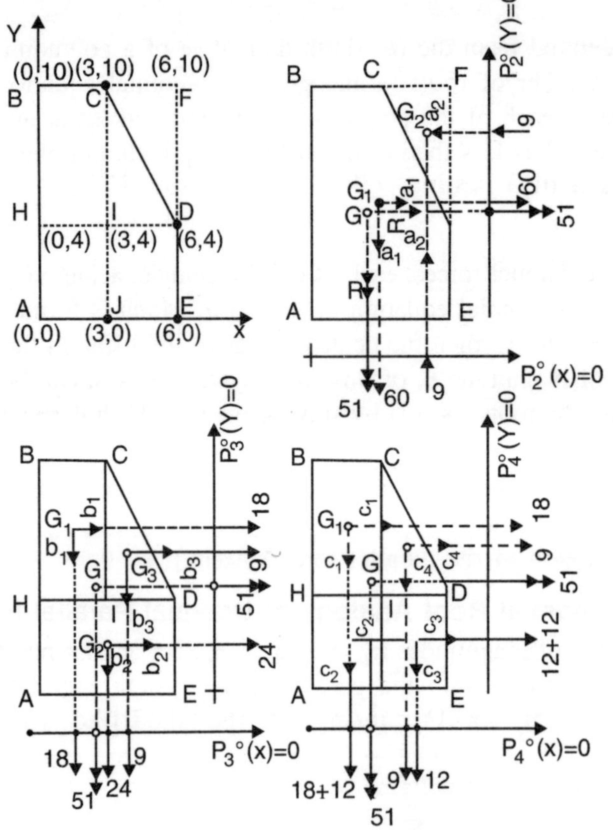

Fig. B.2: Centroid of lamina using polynomial root analogy equation

(1) Area composed of 2 laminas [ABFEA – CFD]

$$n = 2$$
$$a_1 = \text{area ABFEA} = 6 \times 10 = 60,\ x_1 = 3,\ y_1 = 5$$
$$a_2 = \text{area CFD} = -6 \times 3 / 2 = -9,\ x_2 = 5,\ y_2 = 8$$

$$a_1 + a_2 = 51,\ \frac{1}{C} = \frac{2}{51}$$

$$\frac{a_1x_1}{C} = 7.0533,\ \frac{a_2x_2}{C} = -1.7647,\ \frac{a_1y_1}{C} = 11.7641,\ \frac{a_2y_2}{C} = -2.8235$$

P_2^0 along x:
$$(x - \frac{a_1x_1}{C})(x - \frac{a_2x_2}{C}) = 0$$
$$(x - 7.0533)(x + 1.7647) = 0$$
$$x^2 - 5.2941\,x + \ldots = 0$$

Differentiating once with respect to x,

$$2\,x - 5.2941 = 0,\ \text{yields: } \bar{x} = 2.6471$$

P_2^0 along y:
$$(x - \frac{a_1y_1}{C})(x - \frac{a_2y_2}{C}) = 0$$
$$(y - 11.7641)(y + 2.8235) = 0$$

$$y^2 - 8.9412\, y + \ldots = 0$$

Differentiating once with respect to y

$$2y - 8.9412 = 0, \text{ yields : } \bar{y} = 4.4706$$

Hence the centroid is at (2.6471, 4.4706)

(2) Area composed of 3 laminas [BCIH + AHDE + ICD]

$$n = 3$$
$$b_1 = \text{area BCIH} = 6 \times 3 = 18,\ x_1 = 1.5,\ y_1 = 7$$
$$b_2 = \text{area AHDE} = 6 \times 4 = 24,\ x_2 = 3,\ y_2 = 2$$
$$b_3 = \text{area ICD} = 6 \times 3/2 = 9,\ x_3 = 4,\ y_3 = 6$$

$$b_1 + b_2 + b_3 = 51,\ \frac{1}{C} = \frac{3}{51}$$

$$\frac{b_1x_1}{C} = 1.5882,\ \frac{b_2x_2}{C} = 4.2353,\ \frac{b_3x_3}{C} = 2.1176$$

$$\frac{b_1y_1}{C} = 7.4118,\ \frac{b_2y_2}{C} = 2.8235,\ \frac{b_3y_3}{C} = 3.1765$$

P_3^0 along x:

$$(x - \frac{b_1x_1}{C})(x - \frac{b_2x_2}{C})(x - \frac{b_3x_3}{C}) = 0$$
$$(x - 1.5882)(x - 4.2353)(x - 2.1176) = 0$$
$$x^3 - 7.9411\, x^2 + \ldots = 0$$

Differentiating twice with respect to x,

$$2[3x - 7.9411] = 0, \text{ yields : } \bar{x} = 2.6471$$

P_3^0 along y:

$$(y - \frac{b_1y_1}{C})(y - \frac{b_2y_2}{C})(y - \frac{b_3y_3}{C}) = 0$$
$$(y - 7.4118)(y - 2.8235)(y - 3.1765) = 0$$
$$y^3 - 13.4118\, y^2 + \ldots = 0$$

Differentiating twice with respect to y

$$2[3y - 13.4118] = 0, \text{ yields : } \bar{y} = 4.4706$$

Hence the centroid is at (2.6471, 4.4706)

(3) Area composed of 4 laminas [BCIH + HIJA + JIDE + DIC]

$$n = 4$$
$$c_1 = \text{area BCIH} = 6 \times 3 = 18,\ x_1 = 1.5,\ y_1 = 7$$
$$c_2 = \text{area HIJA} = 3 \times 4 = 12,\ x_2 = 1.5,\ y_2 = 2$$
$$c_3 = \text{area JIDE} = 4 \times 3 = 12,\ x_3 = 4.5,\ y_3 = 2$$
$$c_4 = \text{area DIC} = 6 \times 3/2 = 9,\ x_4 = 4,\ y_3 = 6$$

$$c_1 + c_2 + c_3 + c_4 = 51,\ \frac{1}{C} = \frac{4}{51}$$

$$\frac{c_1x_1}{C}=2.1176,\ \frac{c_2x_2}{C}=1.4118,\ \frac{c_3x_3}{C}=4.2353,\ \frac{c_4x_4}{C}=2.8235$$

$$\frac{c_1y_1}{C}=9.8828,\ \frac{c_2y_2}{C}=1.8824,\ \frac{c_3y_3}{C}=1.8824,\ \frac{c_4y_4}{C}=4.2353$$

P_4^0 along x : $(x-\frac{c_1x_1}{C})(x-\frac{c_2x_2}{C})(x-\frac{c_3x_3}{C})(x-\frac{c_4x_4}{C})=0$

$$(x - 2.1176)(x - 1.4118)\ (x - 4.2353)(x - 2.8235) = 0$$

$$x^4 - 10.5882\ x^3 + \ldots = 0$$

Differentiating thrice with respect to x,

$24x - 3 \times 10.5882 = 0$, yields : $\bar{x} = 2.6471$

$P_4{}^0$ along y: $(y-\frac{c_1y_1}{C})(y-\frac{c_2y_2}{C})(y-\frac{c_3y_3}{C})(y-\frac{c_4y_4}{C})=0$

$$(y - 9.8828)(y - 1.8824)(y - 1.8824)(y - 4.2353) = 0$$

$$y^4 - 17.8825\ y^3 + \ldots = 0$$

Differentiating thrice with respect to y

$24y - 3 \times 17.8825 = 0$, yields : $\bar{y} = 4.4706$

Hence the centroid G $(\bar{x}, \bar{y})$ of the lamina ABCDEA is at (2.6471, 4.4706).

B-2 SUMMARY OF POLYNOMIAL-ROOT ANALOGY

The polynomial root analogy links two different fields of mathematics. On one hand stands the parallel force system, its resultant and its location. On the other side, is the equivalent polynomial equation which could be suitably differentiated successively up to the desired stage.

Once the suitable polynomial equation is framed based on the locations through which parallel force system acts, it becomes a simple routine matter of differentiation of the said polynomial equation to get the centriod through which the resultant would pass.

The centroid of a given lamina is a unique point, and is independent of the subdivision of the said lamina into component parts. This fact is established by the average of the roots of the analogous polynomial equation of certain degree, which, depending on the number of the equivalent parallel forces, may vary.

Certainly, the number of subdivision of the given lamina gives rise to the same number of equivalent forces, which again, call for the same degree of the analogous polynomial equation. The successive differentiation to arrive at the linear form from it, though is a simple and straight forward matter, invites increasing number of steps to perform.

The analogous equivalence between the system of parallel forces, and the polynomial equation of constant coefficients are:

System of parallel forces ⇔ Polynomial equation with constant coefficients

Number of parallel forces (n) ⇔ Degree of polynomial (n)

Magnitude of the resultant (R) ⇔ Degree of polynomial (n)

Location of the line of action of the resultant ($\bar{x}$) ⇔ Average of the n-number of roots (x_{av}).

❑❑❑

Appendix C

BAIRAGI'S THEOREM ON TRISECTION OF AN ANGLE

C-1 INTRODUCTION

The geometric construction of bisecting any plane angle using only an unmarked ruler and a compass is an easy, straight forward and direct process.

By **geometric construction,** it is understood that, the only instruments that are permitted and restriction imposed on their use for construction, are:

- a **ruler**, which does not have any marking or calibration of length units on it, and is permitted to draw only straight lines with its straight edge, and
- a **compass** which only draws circular arcs around a fixed point (pivot), which is not allowed to slide during the drawing of the arc.

However, using the same unmarked ruler and compass, the geometric construction for trisection of any plane angle, to the knowledge of the author, still remains an unsolved problem, even to day. Great mathematicians and geometers throughout the world, since time immemorial, are engaged in the task of finding a general process of construction procedure for trisection of a general angle.

Yes, few construction procedures, no doubt, are available on the trisection of an angle, but they do not conform to be geometric, simply because, either they are approximate, or are not permissible when judged from the angle of the criteria and restrictions imposed for the geometric constructions.

For certain specific angles, it is possible to get the geometric construction for trisection. These angles are: 180°, 90°, 72°, 18°(as central angle of a 20-sided regular polygon), 15°(as central angle of a 24-sided regular polygon), and 3° (as by superposing 18° and 15° cases).

C-1.1 Archimedean Model for Trisection of an Angle

The earliest construction model for trisection of an angle is due to **Archimedes** who suggested the name as **Neusis construction**. The construction model is described as follows:

In **Figure C.1a**, a circle with centre C is shown, in which angle ACB = c, is required to be trisected. Taking A as pivot, a straight line AED through A is so drawn up to D (the meeting point of the line BC when extended) as to ensure that, the portion DE is equal to the radius R = AC (= BC). At this position angle ADC = a, thus obtained, is the trisection of the angle c.

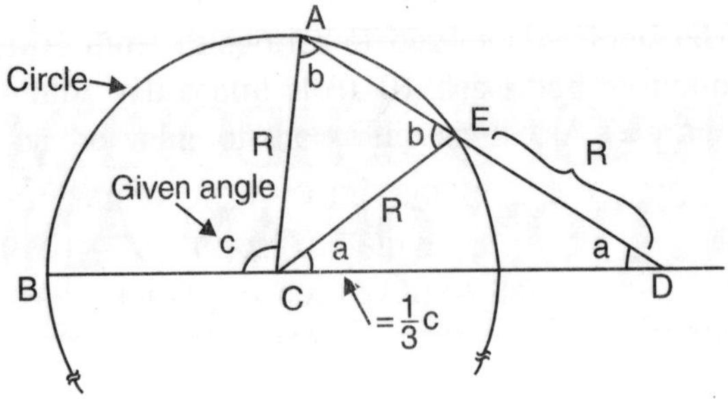

Fig. C.1a: Archimedean model

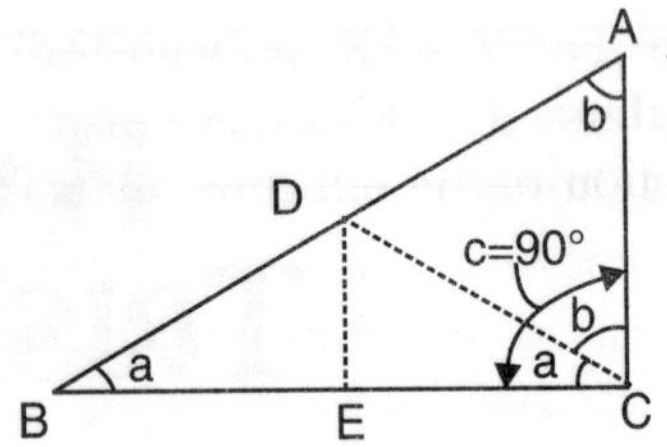

Fig. C.1b: Right angled triangle (Hypotenuse =2×perpendicular)

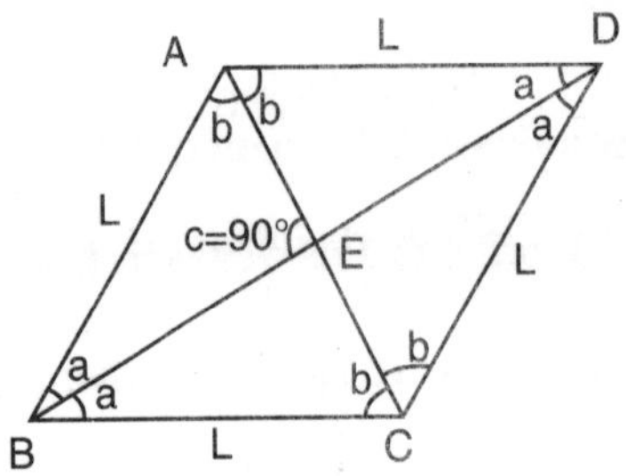

Fig. C.1c: General rhombus

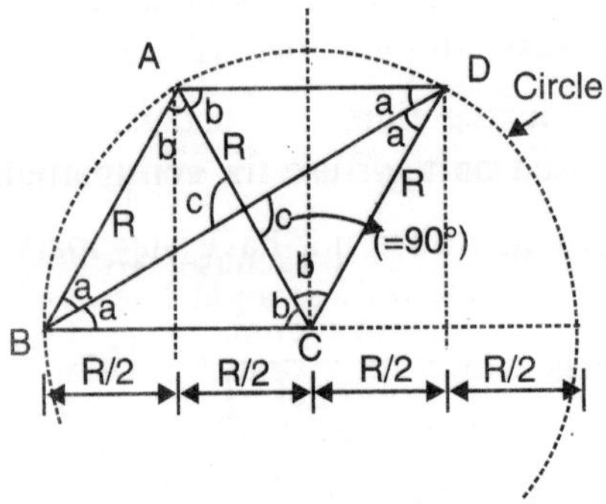

Fig. C.1d: Quasi-cylic rhombus

Proof:

Join CE.

Agreed that, ED = CE = R

⇒ angle ECD = angle EDC = a

Since, CE = CA = R

⇒ angle CAE = angle CEA (external angle of the triangle CED) = $b = 2a$

and, angle ACE = $d = 180 - 2b = 180 - 4a$

So, angle ACB = $c = 180 - (a + d) = 3a$

$\Rightarrow a = c/3$.

Observation 1

Thus, the angle EDC = a, produced by the straight line AED (as the extension of the chord AE), and the diametrical line BCD (as the extension of the radius BC), is the trisection of the external angle ACB = c, of the triangle ADC.

Here, the problem arises, how to get the point D on BD, so that portion ED of the line AED is equal to R, with an unmarked ruler.

Archimedean construction, as can be seen, provides a criterion. It is invalid so far the geometric construction method is concerned. It means that, if an angle ACB = c is trisected, then the criterion that, AC = CE = ED = R, is satisfied. But, to get an angle c trisected using the reverse procedure, in compliance with the Neusis construction, taking help of the said criterion (AC = CE = ED = R), is not possible. Wantzel (1936) proved mathematically that this Neusis construction of Archimedes, obtained this way, is not possible.

In this Appendix, a new model proposed as **Bairagi's model**, for trisection of angle, is presented. The suggested model of Bairagi (author of this book), is on the geometric property of a specific type of trapezium, drawn inside a circle, three vertices of which lie on the circle, while the fourth one forms the centre of the same circle.

C-2 BAIRAGI'S MODEL FOR TRISECTION OF ANGLE

Few preliminaries are required to be elaborated before discussing the proposed model.

C-2.1 Right Angled Triangle, where the Hypotenuse is Twice the Length of the Perpendicular

Consider a right angled triangle ABC, which is right angled at C where BC = base, AC = perpendicular, AB = hypotenuse, so that, angle ABC = a, and angle BAC = $b = 90 - a$, providing, $a + b = 90$.

Further consider that, the hypotenuse AB = 2 AC (**Figure C.1b**). Join the mid point D of AB with C. Then AD = DB = AC.

Drop a perpendicular DE on BC.

In ΔDBC, since D bisects AB, E also bisects BC, so that DE is the line of symmetry of the area DBC. Then, angle DBE = angle DCE = a, and DB = DC.

Hence angle BAC = $90 - a$.

In ΔADC, AD = AC, so that angle ADC = angle ACD = $90 - a$.

Hence angle DAC = $180 - 2(90 - a) = 2a$.

Now in right angled ΔABC, angle ABC = a, and, angle BAC = $2a$. Hence, $3a = 90$,

or $$a = \frac{1}{3}(90) = 30.$$

Trigonometrically, it can be shown that

$$\sin a = AC/AB = 1/2 \Rightarrow a = 30.$$

Observation 2

Thus, **in a right angled triangle, where the hypotenuse is twice the perpendicular, the angle opposite to the perpendicular trisects the angle situated at the foot of the perpendicular**. [This angle a is 30° always.]

C-2.2 Rhombus, in which Three Vertices are Equidistant from the Fourth Vertex

This statement is the same as that for a rhombus, where the shorter diagonal is equal to the length of the sides.

First, take the case of a general rhombus ABCD (**Figure C.1c**). It is a parallelogram (opposite sides are equal and parallel), with specific property that, all sides are of same length. That is: AB = BC = CD = DA, and AD is parallel to BC, and AB is parallel to DC.

The intersecting diagonals BD (longer diagonal) and AC (shorter diagonal) meet at E at right angles, and divide each other in equal parts, *i.e.*, for BD, BE = ED, and for AC, AE = EC.

The diagonals bisect the vertex angles.

Considering this, and additionally with the parallelism of opposite sides, and equality of the lengths of the sides, we get:

angle EDA = angle EDC = angle EBA = angle EBC = a, and

angle ECB = angle ECD = angle EAD = angle EAB = b,

angle AEB = c = 90 (so that at this junction at E, all angles = 90).

With respect to the base BC, the rhombus ABCD (as shown tilted towards right), the angle AEB = c = 90. So, in ΔEBC, $a + b = c = 90$.

For different tilt (towards right) of the rhombus, BD remains larger than AC.

Specific Cases

If the diagonals BD and AC are equal, the rhombus ABCD turns out to be a square, in which case, angle a = angle b = 45 = angle $c/2$.

If the semi shorter diagonal EC is half of the side BC, then as per the property of the right angled triangle as given in **Figure C.1b**, angle a = angle $b/2$ = 30 = angle $c/3$.

The later condition is synonymous with the fact that, the shorter diagonal (AC) is of equal length with the length of the side of the rhombus.

That is

$$\text{EC [= AE]} = \frac{1}{2}\,[\text{AC = AB = BC = CD = DA}].$$

So, in a rhombus, if the shorter diagonal is equal to the sides of the rhombus, the problem of trisection of right angles reduces to a particular problem, and is automatically performed, since the whole area is equally divided into four equal right angled triangles, each one of which follows the criterion same as that described through **Figure C.1b**.

That is to say, $a = \frac{1}{2}b = \frac{1}{3}c = 30$.

Quasi-Cyclic Rhombus

Now visualise the same problem of the rhombus ABCD [**Figure C.1d**], in which the shorter diagonal AC is equal to the sides AB [= BC = CD = DA], as:

With centre at C, draw a circle with radius R = CB (= CA = CD). This ensures that:

1. Three vertex points B, A, and D of the rhombus ABCD lie on the circumference of the circle, while the fourth vertex C, is the centre, so that the rhombus ABCD, in this position cannot be termed as cyclic, but being within the circle, let it be called **hypo-cyclic**, or **quasi-cyclic**.

2. The two triangles ABC and CAD having all the sides = R, are equilateral triangles, and are equal in all respects.

3. Additionally, a perpendicular dropped from the vertex A of the triangle ABC, on BC, divides BC into two equal parts, and so also, the perpendicular dropped from the vertex C of the triangle CAD, on AD divides BC into two equal parts.

Observation 3

Here, in this particular case of rhombus, it can be concluded that:

The trisection of the angle included between the two intersecting diagonals in a rhombus, having the shorter diagonal equal to the side, and the three vertices are equidistant from the fourth vertex, is given by the angle between the longer diagonal and one of the boundary line passing the fourth vertex.

C-3 TRAPEZIUM, IN WHICH THREE VERTICES ARE EQUIDISTANT FROM THE FOURTH VERTEX

The generalization of the observation done on the rhombus, in which three vertices are equidistant from the fourth vertex, is discussed now, by considering trapezium instead of rhombus, through the proposed model, henceforth to be known as **Bairagi's model**, which is now described as follows.

Bairagi's Model

In this model which aims at for the trisection of the angle AEB = *c*, let us take the geometry of the trapezium ABCD, as shown in **Figure C.2**.

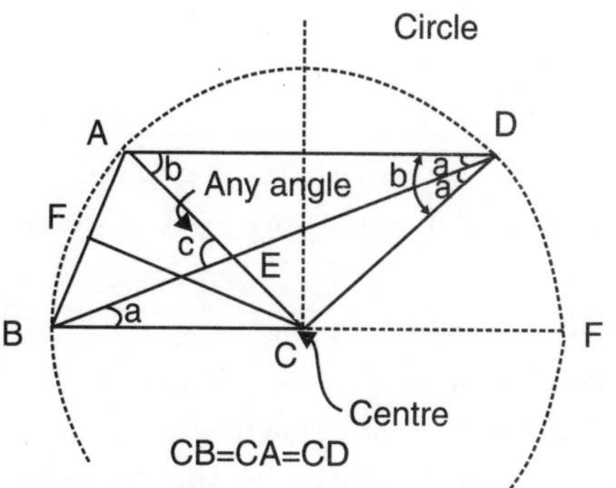

Fig. C.2: Quasi-cyclic trapezium in Bairagi's model

As a ready reckoner, the different parts of the trapezium are identified for reference as:

Side **BC** as the **base** and also as the **bottom boundary line** (length = R)

Side **AB** as the **left boundary line.**

Side **AD** as the **top boundary line**
Side **CD** as the **right boundary line** (length = R)
Diagonal **AC** as **shorter diagonal** (length = R)
Diagonal **BD** as the **longer diagonal**
Angle **AEB** (= c) as the angle, just opposite to the left boundary line, produced by the two intersecting diagonals meeting at E.

Angle **DBC** = angle **BDC** (= a) as the angle produced by the longer diagonal (BD), with the adjacent boundary line (BC or CD), passing through the fourth vertex (C), equidistant (distance = R), from the other three vertices (B, A, D).

Notice that, unlike the rhombus case [**Figure C.1d**], [in which the two triangles ABC and CAD, having all the sides = R, are equilateral triangles, and are identical in all respects], the trapezium ABCD (**Figure C.2**), in which three vertices (B, A, D) are equidistant from the fourth vertex (C), the two triangles ABC and CAD, turns out to be isosceles triangles, but with different apex angles BCA and ACD, respectively.

Now

In triangle ABC, CA = CB, so, angle CAB = angle ABC
In triangle CAD, CA = CD, so, angle CAD = angle CDA = b
In triangle BCD, CB = CD, so, angle CBD = angle CDB = a
But, AD is parallel to BC, so angle DAC = angle ACB = b
Also, angle DBC = angle BDA = a
Thus, the longer diagonal BD bisects the vertex angle CDA (= b)

where, angle CDB = angle BDA = $a = \frac{1}{2}\,b$

In triangle EBC

angle EBC (a) + angle ECB ($b = 2a$) = external angle BEA (c)

$$\text{This gives, } 3a = c \Rightarrow a = \frac{1}{3}c.$$

Further notice that, since AD and BC (**Figure C.2**) are parallel, the triangles EBC and EDA in the trapezium ABCD of Bairagi's model, satisfying A-A-A condition, are similar in all respect, in which:

angle BEC = angle DEA = 180° – c
angle EBC = angle EDA = a = c/3
angle ECB = angle EAD = b = $2a$ = $2c/3$

This shows that, **angle DBC** (or, **angle BDC**) **is the trisection of the angle BEA.**

Because of the fact that, the three vertices (B, A, and D) of the trapezium ABCD in the suggested Bairagi's model are equidistant from C (the fourth vertex), these three vertices lie on the circumference of the circle drawn with centre at C and radius = CB (or CA or CD) = R. Thus, Bairagi's model for trisection of an arbitrary angle AEB (= c), as the angle CBD (or angle CDB = angle BDA = a), using the quasi-cyclic trapezium ABCD, fits absolutely well for the purpose.

Bairagi's Theorem

In a trapezium, where the two adjacent sides, one of which is the parallel side, and one of the diagonals meeting at the common corner (vertex) are equal, the angle subtended by one of these adjacent sides and the other diagonal, is the trisection of the angle included in between the intersecting diagonals, which is just opposite to the other non parallel side.

To clarify the Theorem, refer to the trapezium ABCD with the two diagonals AC and BD intersecting at E as in **Figure C.2**. The two parallel sides are BC and AD, and the two non-parallel sides are CD and AB.

The two other non parallel sides CB and CD, and one of the diagonal (shorter) CA meeting at a common corner C are equal (= R).

The **angle** (angle EBC = ***a***) subtended by one of these adjacent sides (BC) and the other diagonal (BD) = $\frac{1}{3}$ of the **angle** (angle AEB = ***c***) which is included in between intersecting diagonals AC and BD, and is opposite to the other non parallel side AB.

That is, according to the statement of the Theorem, $a = \frac{1}{3}c$.

Since CB = CA = CD, points B, A and D will lie on the circumference of the circle drawn with the centre at C, and radius R. Then this trapezium ABCD (**Figure C.2**) turns up to be **quasi-cyclic**, and thus satisfies **Bairagi's model**.

Quasi-cyclic trapezium (B-trapezium) formulas

The quasi-cyclic trapezium (**Figure C.2**), which satisfies the criterion for Bairagi's Theorem based on Bairagi's model, will henceforth be termed as **B-trapezium**.

In the B-trapezium ABCD, we have: BC = AC = CD = R, AD is parallel to BC and BE when extended meets at D on the circumference of the circle of radius R.

From similar triangles BCE and DEA, the sides (BE: EC: BC) = k.(ED : AE : AD), where k = constant. The angle AEB = c, is formed by the legs AE = l_1 and BE = l_2, so that, $r = l_2 / l_1$. Angle EBC = angle CDE = angle EDA = $a = c/3$, angle ECB = angle EAD = $2a$.

Then,

AE = l_1, EC = $k\, l_1$, AC = $(1 + k)\, l_1$, BE = $l_2 = rl_1$, ED = l_2/k, BD = $(1 + k)\, l_2 /k$,
BC = $l_2 \cos(a) + k\, l_1 \cos(2a)$

Also, $l_2 \sin(a) = k\, l_1 \sin(2a)$ gives:

$$r = 2k \cos (a) \tag{C.1}$$

Further, from the property of the two intersecting chords in a circle
AE.(2R – AE) = BE.ED gives:

$$2k^2 + k = r^2 \tag{C.2}$$

Solving for k,

$$k = \left[-1+\sqrt{1+8r^2}\right]/4 \quad \text{C.3}$$

From (C.1) and (C.3), eliminating r, we get

$$k = \frac{1}{2\cos(2a)} \quad \text{C.4}$$

Combining (C.1) and (C.4), we get **cosine rule** (for triangle EBC) as:

$$r = \frac{l_2}{l_1} = \frac{\cos(a)}{\cos(2a)}, \text{ or } \frac{l_2}{\cos(a)} = \frac{l_1}{\cos(2a)} \quad \text{C.5}$$

$$\text{EC : BE : BC} = 1 : \frac{\sqrt{2k^2+k}}{k} : \frac{1+k}{k} \quad \text{C.6}$$

Verification of the B-trapezium formulas

Checks of the formulas (C.1–C.6), have been shown with the help of three Examples using B-trapezium representation for trisecting an angle c (lengths are in cm, and angles in deg).

Example 1

Given c = 61.2 deg, and the lengths: AB = 4.9, BC = CD = 7.0, AD = 10.55. Get the other lengths and compare.

k = BC/AD = 7.0 / 10.55 = 0.664

$r^2 = 2k^2 + k$ = 1.546, r = 1.243 **(1.25)**

$r = 2\,k\,\cos(a)$, gives

$\cos(a)$ = 1.243 / 1.328 = 0.936, a = 2061**(20.4)** deg

[alternatively, $r = \cos(a) / \cos(2a)$ = 0.936 / 0.753 = 1.243]

BC = $l_3 = l_2 \cos(a) + k\, l_1 \cos(2a)$ gives

$l_2 \cos(a) + k \cos(2a)/r]$ = 7.0, from which, l_2 = 5.217 **(5.25)**

AE = $l_1 = l_2/r$ = 5.217 / 1.243 = 4.197**(4.2)**

[alternatively, $k\, l_1 \sin(2a) = l_2 \sin(a)$ gives, l_1 = 3.944]

EC = $k\, l_1$ = (0.664) (4.197) = 2.786 **(2.8)**

AC = AE + EC = 4.197 + 2.786 = 6.973**(7.0)**

ED = l_2/k = 5.217/0.664 = 7.857

BD = $(1 + k)\, l_2 /k$ = 13.07**(13.1)**

Check for : AE.(2R – AE) = BE.ED

LHS = (4.197)(14.0 – 4.197) = 41.143

RSH = (5.217) (7.857) = 41.0. Hence, LHS = RHS.

Example 2

The measurements of the sides l_1, l_2 and l_3 in the B-trapezium representation are: 2.4, 6.65 and 6.5, respectively. Find c, and other dimensions of the trapezium, and compare.

$l_1 = 2.4$, $l_2 = 6.65$, from which, $r = l_1/l_2 = 2.771$.

Using r-value in (C.3), get $k = 1.721$.

[alternatively, $R = l_3 = (1 + k)\, l_1$, yields $k = 1.708$]

AC = BC = CD = R = 6.5,

AD = l_3/k = 3.78(**3.7**), BD = $(1 + k)\, l_2/k$ = 10.4 (**10.51**)

$r = 2k \cos(a)$ gives: $\cos(a) = 0.8051$,

from which, $a = 36.38$ deg, $c = 3a = 109.15$ (**110**) deg

AB = 2R $\sin(a)$ = 7.71 (**7.8**)

Example 3

For a 60 deg B-trapezium construction model, the dimension noted for AB = 4.5, and the ratio of the diagonals BD : AC = 1.874. Calculate all the sides and verify.

BD/AC = $[(1 + k)l_2/k] / [(1 + k)l_1] = r/k = 1.874$

Using (C.2) with this r/k value, get $k = 0.661$ (**0.664**)

So, $r = 1.874\, k = 1.239$ (**1.2436**)

Now, $r = 2k \cos(a)$ gives : $\cos(a) = 0.937$, $a = 20.44$ (**20**) deg

AB = 4.5 = $2l_3 \sin(a)$ gives : AB = l_3 = 6.44 (**6.5**)

AD = $[(1 + k)\, l_1/k$ =9.74 (**9.9**)

$(1 + k)l_1 = l_3$ gives: AE = l_1 = 3.877 (**3.9**)

BE = $l_2 = rl_1$ = (1.239) (3.877) = 4.8 (**4.85**)

BD = $(1 + k)\, l_2/k$ = 12.06 (**12.25**)

C.4. TRISECTION OF ARBITRARY ANGLE—USE OF BAIRAGI'S MODEL

For trisecting an arbitrary angle AEB = c, using only unmarked ruler and compass, refer to **Bairagi's model** as given in **Figure C.2**.

Given angle AEB = c.

Mark off point B and A from E such that, EB is slightly larger than EA.

Extend both the lines AE and BE.

Join AB, and find its mid point F, from which draw a perpendicular line FC, which meets the extended line AE at C.

With C as centre, and CA or, CB as radius, draw a circle.

Locate the point D on this circular periphery, where the extended line BE meets, simultaneously satisfying that the line drawn from A parallel to the line BC passes also through D.

[**Note:** In case, the point D is not thus uniquely located, repeat the drawing procedure by adjusting legs EA and EB.]

This completes the trapezium ABCD with diagonals AC and BD, representing **Bairagi's model** for trisecting arbitrary angle AEB = c, where the angle EBC (or angle EDC) = $a = \frac{1}{3}c$, represents the intended angle.

[**Note**: If EB = EA, the point E will have no relevance, since E actually will fall on the centre C, and as a result, what would have been the portion of the diagonal (BE), now becomes the base line BC. In this situation, the trisection of the angle AEB can never be executed.]

Observation 4

The angle AEB = c, which is intended to be trisected using Bairagi's model, can have any value between 0° to135° (the two extreme values excluded). It depends on the **gap** (perpendicular distance) between the two parallel lines, *viz.*, top line AD, and the bottom base line BC, of the trapezium ABCD, which depends on the inclination of the diagonal BD with respect to the base line BC (anticlockwise), representing the angle ***a***, or alternatively, on the inclination of the diagonal CA with respect to the base line CB (clockwise), representing the angle ***b***. In the trapezium ABCD, BC = CA = CD = radius = R.

In **Figure C.3**, the angle AEB = c, between the two intersecting diagonals BD and AC, is shown to have acute angle, right angle and obtuse angle, progressively.

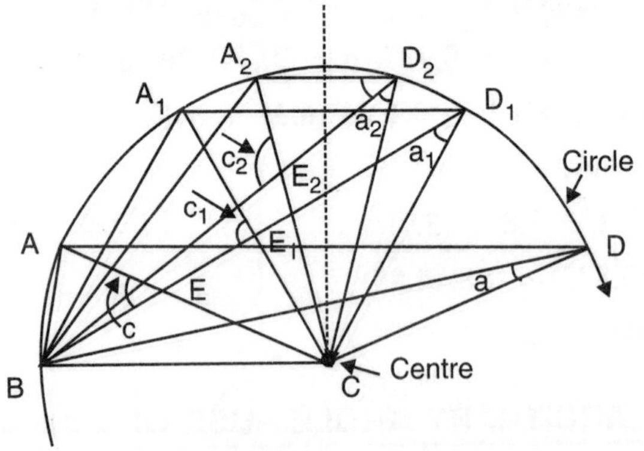

$$\begin{cases} [c, c_1, c_2] \longrightarrow [<,=,>]\dfrac{\pi}{2} \\ \text{BD, BD}_1, \text{BD}_2 \longrightarrow \text{Longer Diagonal} \\ \text{AC, AC}_1, \text{AC}_2 \longrightarrow \text{Shorter Diagonal} \end{cases}$$

Fig. C.3: Relative situations of the angle c in Bairagi's model

Thus a screening of **Figure C.3**, reveals:

1. **For $c \to 0°$**

   $a \to 0°$, $b \to 0°$, $h \to 0$

   (where, h = distance between the two parallel lines AD and BC)

   In the trapezium ABCD **[Figure C.3]:**

   AD (now diameter = 2R) will lie on BC (radius = R). AB (= CD) → 0, (AD / BC) →2.

   The trapezium ABCD reduces to overlapping diameter on radius along the base line BC.

   AB = CD = 0, (AD/BC) = 2.

2. **For c = acute angle [$0 < c < 90°$]**

   $0 < a < 30°,\ 0 < b < 60°,\ 0 < h < \frac{\sqrt{3}}{2}R$

   (where, h = distance between the two parallel lines AD and BC)

   In the trapezium ABCD [In **Figure C.3**, it is shown as ABCD]:

   AD > BC, AB < CD, BD > AC, 2 < (AD/BC) < 1

3. **For c = right angle = 90°**

   $a = 30°,\ b = 60°,\ h = \frac{\sqrt{3}}{2}R$

   (where, h = distance between the two parallel lines AD and BC)

   In trapezium ABCD [In **Figure C.3**, it is shown as A_1BCD_1]:

   AD = BC = R, the trapezium ABCD reduces to a rhombus **(Figure C.1b)**.

   AD = BC = R, AB = CD = R, BD > AC, (AD/BC) = 1

4. **For c = obtuse angle [$90° < c < 135°$]**

   $90° < a < 45°,\ 60° < b < 90°,\ \frac{\sqrt{3}}{2}R < h < R$

   (where, h = distance between the two parallel lines AD and BC)

   In the trapezium ABCD [**In Figure C.3**, it is shown as A_2BCD_2]:

   AD < BC, AB > CD, BD > AC, (AD/BC) = 0

5. **For c = obtuse angle [→135°]**

   $a \to 45°$, $b \to 90°$, $h = R$

   (where, h = distance between the two parallel lines AD and BC)

   In the traperium ABCE: AD reduces to point AB > CD (=R), BD > AC.

6. For obtuse angles > 135°

The trisection of angle using Bairagi's model, for any obtuse angle, x (preferably greater than 135°), can be performed as follows:

Get the acute supplementary angle, $y = 180° - x$, and perform the drawing operation for getting $\frac{y}{3}$.

Since, drawing an equilateral triangle for getting angle 60° is a straight forward and valid method of geometrical construction, the final trisection of the angle x, will then be the angle obtained after superposing this $\frac{y}{3}$, from 60°.

Alternatively, the proposed **Theorem** can be stated in a different fashion, which finally, as will be seen, will speak for **Bairagi's model**. The statement is as follows:

On the base AB of an isosceles triangle ABC (with apex at C) of equal sides AC = BC = R, if another triangle ABD (with apex at D), maintaining a distance between the two apexes CD = R, stands in such a way that, the outermost sides of the two triangles AD and BD are parallel, then the apex angle ADB is one-third of the angle AEB at E, formed by the intersecting inner sides AC and BD of the two triangles, opposite to the base of the isosceles triangle.

Here since, for the isosceles triangle ABC, AB = BC = R, and also CD = R (where C and D are the apexes of the triangles ABC and ABD respectively), a circle can be drawn with centre C and radius = R, on the periphery of which will lie the points A, B and D (all are the three vertices of the non-isosceles triangle ABD).

Moreover since, the outer side AD of the triangle ABD and that of BC of the triangle ABC, is parallel; the figure ABCD becomes a trapezium, in which the inner sides of the triangles, namely, BD and AC become the two intersecting diagonals at E.

In effect, the **alternative form of the proposed Theorem thus, satisfies fully the proposed Bairagi's Model.**

Then if the apex angle ADB of the triangle ADB = a, and the angle AEB at the intersection of the two diagonals AC and BD at E = c, we arrive at, $a = \frac{1}{3}c$.

K-Model

Also observe that, **Bairagi's Model** can be simply be represented by the English letter **K**, where the two equal hands CA and CD of length = R, meet at the mid height of the vertical line BF at C, such that CB = CF = R. Hence, a nick name **K-Model** for **Bairagi's Model** seems to be appropriate.

The principle based on which the trisection of an angle is proposed using Bairagi's Model, (*viz.*, quasi-cyclic trapezium), a mechanical apparatus, to be called as **Angle Trisector** is shown in **Figure C.4**.

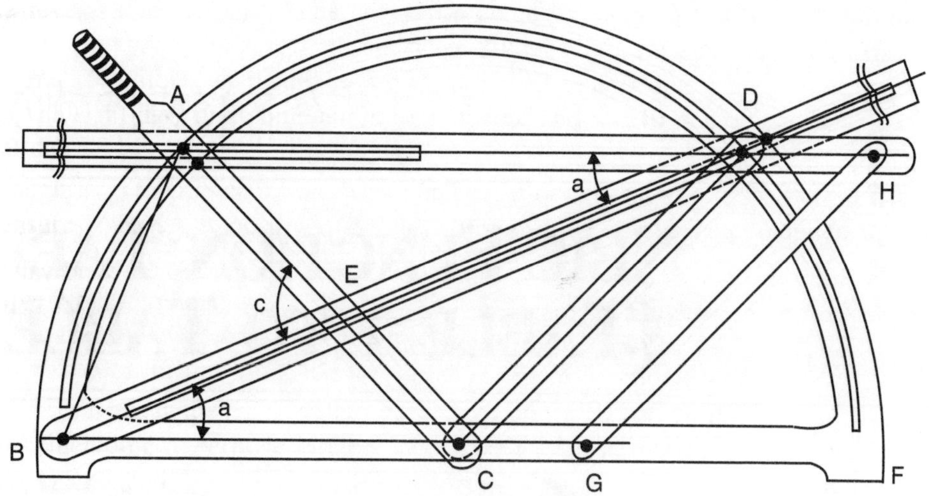

Fig. C.4: Angle trisector

The semicircular inverted **D**, (BADF), with a circular slot along the curved periphery, and a horizontal arm along the diameter BF, monolithic with the curved part, is shown in **Figure C.4**. Pinned at the centre C, about which the end A of a lever arm CA (= CB = radius) can only travel along the slotted curved path of **D**. Also an arm CD (= radius) is pinned with its one end at the centre C, the other hand D of which is allowed to move along the slotted curved path DAB. The ends D of CD, and A of CA, again pass through a straight arm ADH (shown partly slotted towards A). To keep AD parallel to BC, a link device GH is provided connecting ADH and BCG through pins H and G, such that DH = CG.

Any angle *c* to be trisected at E, between the arms AC and BD, is made by moving the lever arm AC about C. The parallel linkage system then keeps AD parallel to BC. The pins A, B, C, and D make the figure ABCD as the **B-trapezium**, which justifies the angle DBC = ADB = *a*, as one-third of the angle *c*.

A simpler alternative form for the angle trisector adopting B-trapezium linkage is shown in **Figure C.5.**

PINS at C and D are fixed.

PIN at A fixed on CA, moves only in the slot of DA.

PIN at B fixed on CB, moves only in the slot of DB.

CD = CA = CB, and DA || CB, $\therefore \angle a = \frac{1}{3}\angle c$

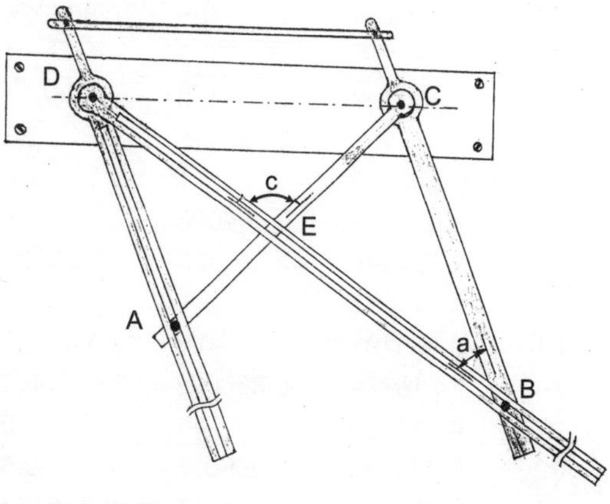

Fig. C.5: B-Trapezium linkage as angle trisector

❑❑❑

Appendix D

NBIC FUNCTION TABLES [TABLE 1 TO TABLE 7]

TABLE D.1 TRIGONOMETRIC AND HYPERBOLIC FUNCTIONS OF (x)

x deg	x rad	sin x	cos x	tan x	sh x	ch x	th x
–360	–6.28319	0	1	0	–267.745	267.7468	–0.99999
–330	–5.75959	0.5	0.866025	0.5773503	–158.607	158.6101	–0.99998
–300	–5.23599	0.866025	0.5	1.7320508	–93.9547	93.95998	–0.99994
–270	–4.71239	1	–1.8E–16	–5.44E+15	–55.6544	55.66338	–0.99984
–240	–4.18879	0.866025	–0.5	–1.732051	–32.9639	32.97906	–0.99954
–210	–3.66519	0.5	–0.86603	–0.57735	–19.519	19.54461	–0.99869
–180	–3.14159	0	–1	0	–11.5487	11.59195	–0.99627
–150	–2.61799	–0.5	–0.86603	0.5773503	–6.81762	6.890572	–0.98941
–120	–2.0944	–0.866025	–0.5	1.7320508	–3.99869	4.121836	–0.97012
–90	–1.5708	–1	0	**U**	–2.3013	2.509178	–0.91715
–60	–1.0472	–0.866025	0.5	–1.732051	–1.24937	1.600287	–0.78071
–30	–0.5236	–0.5	0.866025	–0.57735	–0.54785	1.140238	–0.48047
0	0	0	1	0	0	1	0
30	0.523599	0.5	0.866025	0.5773503	0.547853	1.140238	0.480473
60	1.047198	0.866025	0.5	1.7320508	1.249367	1.600287	0.780714
90	1.570796	1	0	**U**	2.301299	2.509178	0.917152
120	2.094395	0.866025	–0.5	–1.732051	3.998691	4.121836	0.970124
150	2.617994	0.5	–0.86603	–0.57735	6.817623	6.890572	0.989413
180	3.141593	0	–1	0	11.54874	11.59195	0.996272
210	3.665191	–0.5	–0.86603	0.5773503	19.51901	19.54461	0.99869
240	4.18879	–0.866025	–0.5	1.7320508	32.9639	32.97906	0.99954
270	4.712389	–1	0	**U**	55.6544	55.66338	0.999839
300	5.235988	–0.866025	0.5	–1.732051	93.95465	93.95998	0.999943
330	5.759587	–0.5	0.866025	–0.57735	158.607	158.6101	0.99998
360	6.283185	0	1	0	267.7449	267.7468	0.999993

U = undefined

TABLE D.2 TRIGONOMETRIC AND HYPERBOLIC FUNCTIONS OF (2*x*)

x deg	x rad	sin 2x	cos 2x	tan 2x	sh 2x	ch 2x	th 2x
–360	–6.28319	0	1	0	–143376	143375.7	–1
–330	–5.75959	0.8660254	0.5	1.7320508	–50313.4	50313.36	–1
–300	–5.23599	0.8660254	–0.5	–1.732051	–17656	17655.95	–1
–270	–4.71239	–3.68E–16	–1	3.675E–16	–6195.82	6195.824	–1
–240	–4.18879	–0.866025	–0.5	1.7320508	–2174.24	2174.237	–1
–210	–3.66519	–0.866025	0.5	–1.732051	–762.983	762.9833	–1
–180	–3.14159	0	1	0	–267.745	267.7468	–1
–150	–2.61799	0.8660254	0.5	1.7320508	–93.9547	93.95998	–0.9999
–120	–2.0944	0.8660254	–0.5	–1.732051	–32.9639	32.97906	–0.9995
–90	–1.5708	–1.23E–16	–1	1.225E–16	–11.5487	11.59195	–0.9963
–60	–1.0472	–0.866025	–0.5	1.7320508	–3.99869	4.121836	–0.9701
–30	–0.5236	–0.866025	0.5	–1.732051	–1.24937	1.600287	–0.7807
0	0	0	1	0	0	1	0
30	0.523599	0.8660254	0.5	1.7320508	1.249367	1.600287	0.7807
60	1.047198	0.8660254	–0.5	–1.732051	3.998691	4.121836	0.9701
90	1.570796	0	–1	0	11.54874	11.59195	0.9963
120	2.094395	–0.866025	–0.5	1.7320508	32.9639	32.97906	0.9995
150	2.617994	–0.866025	0.5	–1.732051	93.95465	93.95998	0.9999
180	3.141593	0	1	0	267.7449	267.7468	1
210	3.665191	0.8660254	0.5	1.7320508	762.9826	762.9833	1
240	4.18879	0.8660254	–0.5	–1.732051	2174.237	2174.237	1
270	4.712389	0	–1	0	6195.824	6195.824	1
300	5.235988	–0.866025	–0.5	1.7320508	17655.95	17655.95	1
330	5.759587	–0.866025	0.5	–1.732051	50313.36	50313.36	1
360	6.283185	0	1	0	143375.7	143375.7	1

TABLE D.3 SINGLE Nbic FUNCTIONS [N] OF (x, x) and (X, –X)

x deg	x rad	sinN (x,x)	cosN (x,x)	tanN (x,x)	sinN (x,–x)	cosN (x,–x)	tanN (x,–x)
–360	–6.28319	267.74489	267.74676	0.999993	267.7449	267.7468	0.999993
–330	–5.75959	–58.052613	216.66391	–0.26794	216.6628	58.05692	3.731902
–300	–5.23599	34.394399	128.3471	0.26798	128.3491	–34.3871	–3.73247
–270	–4.71239	55.663381	55.654398	1.000161	55.66338	–55.6544	–1.00016
–240	–4.18879	45.042658	12.058043	3.735487	12.07876	–45.0371	–0.2682
–210	–3.66519	26.676259	–7.166622	–3.72229	–7.13165	–26.6856	0.267247
–180	–3.14159	11.548739	–11.59195	–0.99627	–11.5487	–11.592	0.996272
–150	–2.61799	2.4589488	–9.376222	–0.26225	–9.34952	–2.5586	3.654156
–120	–2.0944	–1.5702691	–5.523886	0.284269	–5.56896	1.40205	–3.97201
–90	–1.5708	–2.5091785	–2.301299	1.090331	–2.50918	2.301299	–1.09033
–60	–1.0472	–2.0105726	–0.28184	7.133733	–0.76121	1.882127	–0.40444
–30	–0.5236	–1.0445742	0.7135486	–1.46391	–0.09566	1.261402	–0.07584
0	0	0	1	0	0	1	0
30	0.523599	1.0445742	0.7135486	1.463915	0.095664	1.261402	0.07584
60	1.047198	2.0105726	–0.28184	–7.13373	0.761206	1.882127	0.404439
90	1.570796	2.5091785	–2.301299	–1.09033	2.509178	2.301299	1.090331
120	2.094395	1.5702691	–5.523886	–0.28427	5.56896	1.40205	3.972012
150	2.617994	–2.4589488	–9.376222	0.262254	9.349521	–2.5586	–3.65416
180	3.141593	–11.548739	–11.59195	0.996272	11.54874	–11.592	–0.99627
210	3.665191	–26.676259	–7.166622	3.722292	7.131653	–26.6856	–0.26725
240	4.18879	–45.042658	12.058043	–3.73549	–12.0788	–45.0371	0.268196
270	4.712389	–55.663381	55.654398	–1.00016	–55.6634	–55.6544	1.000161
300	5.235988	–34.394399	128.3471	–0.26798	–128.349	–34.3871	3.732474
330	5.759587	58.052613	216.66391	0.267939	–216.663	58.05692	–3.7319
360	6.283185	267.74489	267.74676	0.999993	–267.745	267.7468	–0.99999

TABLE D.4 DOUBLE Nbic FUNCTIONS [N_{2e}] OF (x, x)

x deg	X rad	$\sin N_{2e}(x,x)$	$\cos N_{2e}(x,x)$	$\tan N_{2e}(x,x)$
0	0	0	1	0
30	0.523599	1.261402	0.095664	–13.185754
60	1.047198	0.76121	–1.882126	–0.4046372
90	1.570796	–2.301299	–2.509179	0.917152
120	2.094395	–5.56900	1.402051	–3.972038
150	2.617994	–2.55859	9.349564	–0.273659
180	3.141593	11.548739	11.59195	0.996272
210	3.665191	26.68575	–7.13162	–3.741892
240	4.18879	12.07876	–45.03709	–0.2681958
270	4.712389	–55.654398	–55.663381	0.9996515
300	5.235988	–128.34900	34.36728	–3.7346278
330	5.759587	–58.02339	216.78778	–0.267650
360	6.283185	267.74676	267.74676	1

TABLE D.5 DOUBLE Nbic FUNCTIONS [N_{2f}] OF (*x*, *x*) and (*x*,–*x*)

x deg	x rad	sinN2f (x,x)	cosN2f (x,x)	tanN2f (x,x)	sinN2f (x,–x)	cosN2f (x,–x)	tanN2f (x,–x)
–360	–6.28319	0	143375.7	0	143375.7	1	143375.7
–330	–5.75959	–43572.1	25157.54	–1.73197	25156.68	43572.65	0.57735
–300	–5.23599	–8827.11	15291	–0.57727	15290.5	8827.977	1.732051
–270	–4.71239	1	6195.824	0.000161	6195.824	0	**U**
–240	–4.18879	1087.985	1882.445	0.577964	1882.945	–1087.12	–1.73205
–210	–3.66519	661.2623	380.6253	1.737305	381.4916	–660.763	–0.57735
–180	–3.14159	267.7449	–1	–267.745	0	–267.747	0
–150	–2.61799	80.86712	–47.8434	–1.69025	–46.98	–81.3717	0.57735
–120	–2.0944	15.61592	–29.0476	–0.5376	–28.5607	–16.4895	1.732051
–90	–1.5708	–1	–11.5487	0.08659	–11.592	0	**U**
–60	–1.0472	–2.86537	–2.96297	0.967061	–3.56961	2.060918	–1.73205
–30	–0.5236	–1.58198	0.241342	–6.55495	–0.80014	1.385889	–0.57735
0	0	0	1	0	0	1	0
30	0.523599	1.581984	0.241342	6.554949	0.800143	1.385889	0.57735
60	1.047198	2.865371	–2.96297	–0.96706	3.569615	2.060918	1.732051
90	1.570796	1	–11.5487	–0.08659	11.59195	0	**U**
120	2.094395	–15.6159	–29.0476	0.537598	28.56071	–16.4895	–1.73205
150	2.617994	–80.8671	–47.8434	1.690248	46.97999	–81.3717	–0.57735
180	3.141593	–267.745	–1	267.7449	0	–267.747	0
210	3.665191	–661.262	380.6253	–1.73731	–381.492	–660.763	0.57735
240	4.18879	–1087.98	1882.445	–0.57796	–1882.94	–1087.12	1.732051
270	4.712389	–1	6195.824	–0.00016	–6195.82	0	**U**
300	5.235988	8827.111	15291	0.577275	–15290.5	8827.977	–1.73205
330	5.759587	43572.15	25157.54	1.731971	–25156.7	43572.65	–0.57735
360	6.283185	143375.7	1	143375.7	0	143375.7	0

U = undefined

TABLE D.6 TRIPLE Nbic FUNCTIONS [N_{3e}] OF (*x*, *x*)

x deg	x rad	sinN3e (x,x)	cosN3e (x,x)	tanN2e (x,x)
–360	–6.28319	143375.7	1	143375.7
–330	–5.75959	–25155.8	43573.15	–0.57732
–300	–5.23599	8828.843	15290	0.577426
–270	–4.71239	6195.824	–1	–6195.82
–240	–4.18879	1086.253	–1883.44	–0.57674
–210	–3.66519	–382.357	–660.262	0.579099
–180	–3.14159	–267.745	1	–267.745
–150	–2.61799	–46.1113	81.86712	–0.56325
–120	–2.0944	17.34798	28.04758	0.61852
–90	–1.5708	11.54874	–1	–11.5487
–60	–1.0472	1.13332	–3.96297	–0.28598
–30	–0.5236	–1.49071	–0.58198	2.561428
0	0	0	1	0
30	0.523599	1.490709	–0.58198	–2.56143
60	1.047198	–1.13332	–3.96297	0.285978
90	1.570796	–11.5487	–1	11.54874
120	2.094395	–17.348	28.04758	–0.61852
150	2.617994	46.1113	81.86712	0.563246
180	3.141593	267.7449	1	267.7449
210	3.665191	382.3573	–660.262	–0.5791
240	4.18879	–1086.25	–1883.44	0.576737
270	4.712389	–6195.82	–1	6195.824
300	5.235988	–8828.84	15290	–0.57743
330	5.759587	25155.81	43573.15	0.577324
360	6.283185	143375.7	1	143375.7

TABLE D.7 TRIPLE Nbic FUNCTIONS $[N_{3f}]$ OF (x, x)

x deg	x rad	$\sin N_{3f}$ (x,x)	$\cos N_{3f}$ (x,x)	$\tan N_{3f}$ (x,x)
0	0	0	1	0
30	0.523599	0.095636	–0.591598	–09526734
60	1.047198	0.883580	–8.321499	–0.1061804
90	1.570796	–24.067914	–31.27915	0.7889965
120	2.094395	183.6591	–57.28608	–0.20638662
150	2.617994	878.4807	221.6535	3.9633062
180	3.141593	3092.137	3080.524	0.1272956
210	3.665191	5441.332	20346.35	0.2674353
240	4.18879	–26262.09	97945.47	–0.2681297
270	4.712389	–344880.5	344936.2	–0.9998385
300	5.235988	–2266121.0	607394.1	–3.7308907
330	5.759587	–10901031.0	–2920605.0	–37.3239895
360	6.283185	–38388111.0	38387843.0	1

❑❑❑

INDEX

D

E

F

G